Gustave Heinrich Theodor Eimer

Organic Evolution as the Result of the Inheritance of Acquired Characters

According to the Laws of Organic Growth

Gustave Heinrich Theodor Eimer

Organic Evolution as the Result of the Inheritance of Acquired Characters
According to the Laws of Organic Growth

ISBN/EAN: 9783337232146

Printed in Europe, USA, Canada, Australia, Japan

Cover: Foto ©berggeist007 / pixelio.de

More available books at **www.hansebooks.com**

ORGANIC EVOLUTION

ORGANIC EVOLUTION

AS THE RESULT OF THE

INHERITANCE OF ACQUIRED CHARACTERS

ACCORDING TO THE

LAWS OF ORGANIC GROWTH

BY

Dr. G. H. THEODOR EIMER

PROFESSOR OF ZOOLOGY AND COMPARATIVE ANATOMY IN TÜBINGEN

TRANSLATED BY

J. T. CUNNINGHAM, M.A., F.R.S.E.

LATE FELLOW OF UNIVERSITY COLLEGE, OXFORD

London
MACMILLAN AND CO.
AND NEW YORK
1890

TRANSLATOR'S PREFACE

BEFORE I became acquainted with this work of Professor Eimer, I had been for some time growing more and more dissatisfied with the uncritical acceptance accorded to Professor Weismann's theories of heredity and variation by many English evolutionists. Even before I studied Weismann's writings, I had become convinced that selection, whether natural or artificial, could not be the essential cause of the evolution of organisms. I was inclined to attach more importance to the problem of the causes of variation than to any of the other problems considered by Darwin, and among these causes it seemed to me that the most powerful were functional activity and external conditions. I was thus led to believe that a deeper insight into the phenomena of evolution would ultimately be obtained by pursuing the line of inquiry suggested by Lamarck, than by continually searching for new instances of adaptation to be explained by the Darwinian formula. When I saw that many of the ablest British biologists accepted Weismann's dogma that acquired characters are not inherited, it seemed to me that they were abandoning the richest vein of knowledge under a mistaken guide, and I cherished the hope of finding time and opportunity to add by my own researches to the evidence that the effects of the conditions of life extend beyond a single generation. I was therefore delighted to find that Weismann had to contend with a formidable opponent in his own country, and concluded that I could not for the present oppose the progress of his views in England more effectively than by publishing a translation of Professor Eimer's arguments.

I have endeavoured as far as lay in my power to express my author's ideas and reasoning in language which should be English in words, construction, and style. I am well aware that my object has not been perfectly attained, that the book contains in almost every page internal evidence of its German origin. But I hope nevertheless that the translation is sufficiently English to be readable, and that I have preserved the full force and the exact significance of Professor Eimer's exposition. I have not in any way presumed to intrude myself between my author and the reader in the text of the work, but I will take the present opportunity of expressing some of the reasons of my uncompromising opposition to Weismann's theories and views.

One of the fundamental assumptions of Weismann's theories is that acquired characters are not inherited. Only this year he has published an essay on this subject under the title *Ueber die Hypothese einer Vererbung von Verletzungen*. In this paper he criticises cases which have been adduced as evidence that scars or artificially-produced malformations are sometimes inherited. One case to which he draws particular attention is that of a peculiar cleft in the lobe of a certain man's ear, which was interpreted by Dr. Emil Schmidt as evidently due to the inheritance of a similar cleft in the ear of the man's mother. The cleft in the ear of the mother was known to have been caused by an injury.

When the mother was between six and ten years old, her ear-ring was accidentally torn out, and the lobe of the ear was split from the hole through which the ring passed down to its edge. After the two divided surfaces had healed together, a cleft was left round the external edge of the wound. Weismann gives a woodcut of each ear, taken from a photograph, indicating by reference letters the corresponding parts in the two figures, and then proceeds to argue that the cleft in the son's ear has really nothing to do with the lobe at all, but is situated between the true lobe and the termination of the helix. I do not require to appeal to experienced human anatomists to support my contention that Weismann has made an erroneous identification of the parts of the son's ear in order to suit his own argument. His own figures are sufficient to prove this to any fair-minded man who knows something of anatomy. In the first place, the line of attachment in the figure of the mother's ear is inclined downwards to the left, in the son's to the right, in consequence of which, of course, the apparent difference between the two clefts is increased. Secondly, the lower end of the helix, marked *Sp. H* in the mother's ear, is not marked at all in the son's, the same letters in the latter pointing to the right half of the lobe of the ear, separated from the left half by a cleft closely corresponding to that in the mother's ear. The only difference between the shape of the lobe in the son's ear and that in the mother's is that in the former the outer half of the lobe is shorter and smaller than the inner, while in the mother's ear they are about equal: a difference which I present to Professor Weismann to make what use of he pleases.

I mention this merely to show to what straits Professor

Weismann is reduced to find a reply to the evidence against him. For my own part, I do not think it is of very great importance whether artificial malformations are inherited or not. I think it probable that in the higher animals such inheritance does sometimes take place. Professor Weismann mentions the feet of Chinese ladies, which he says are still when uncompressed as large as if the practice of artificially compressing them had not been practised for centuries. But he does not tell us whether he ever saw a Chinese young lady, or if he has made any observations on the feet of Chinese women. The fact that artificial malformations are not usually inherited is no argument against the inheritance of acquired characters. In all animals, from the lowest up to reptiles, recrescence of lost parts takes place, and the reappearance of lost parts in the next generation in mammals and birds seems to me to be simply recrescence slightly postponed.

Weismann says that the faculty of speech and skill in pianoforte-playing are not inherited. It is true that babies are not born in the act of playing pianofortes, and no one would expect that they should be. But that particular kinds of musical skill run in particular families is admitted. It is true also that children have to be taught to speak, as they have to be taught to walk upright. But no amount of teaching would cause a young monkey to speak. The capacity for learning to speak and for speaking is inherited by children, and it is making too great a demand on our faith to ask us to believe that the peculiarities of vocal organs and nervous system on which this capacity depends are due to sexual mixture and selection.

Does Professor Weismann believe that birds inherit the

faculty of flight? Many of them are certainly unable to fly when they are first hatched, and have to practise for some little time, even when fully fledged, before they can sustain themselves in the air for any length of time. If such birds were never allowed to use their wings at all for several generations, would they not gradually lose the faculty of flight altogether? The answer is given by our domestic ducks and fowls which, although not altogether prevented from using their wings, have almost lost the power of flight.

But it may be urged that the wings of birds are certainly inherited. Quite true; and is Professor Weismann in a position to assert that no structural modifications are produced by constant pianoforte-playing during several generations, or that such structural modifications are produced but not inherited? I think it may be safely assumed that neither he nor any one else knows what changes in the muscles and nerves of the hands are produced by the practice of pianoforte-playing: although it is certain that extraordinary skill in the art is connected with peculiarities in the condition of those muscles and nerves.

Moreover, it can be shown by actual facts that all modifications, all progressive development of a structural character, are not due to "sexual mixture," that is, to the combination of the hereditary characters of the parents. I possess a female cat with six toes on every foot. One of the kittens of this cat has six toes on each hind foot, five on the left fore foot, and seven on the right fore foot. But the left foot is not normal, the inner toe or pollex is as large as two of the others. It is practically certain that the father of this kitten was an ordinary cat with the usual number of toes, and the abnormal number of toes on the right fore foot has therefore been increased in the kitten, although the action of "sexual combination" is entirely excluded.

Again, another of the assumptions on which Weismann builds is that "everything depends on adaptation." He confesses himself that this is a conviction of his own, not a demonstrated truth. In fact, it is obviously an assumption necessary to the main proposition of his theory. For Weismann means to say that every part of every animal, every structural relation, is either an advantage to its possessor under the present conditions of its life, or was once an advantage to its ancestors. No feature or character which was not at one time or other of advantage to its possessor in the struggle for existence could have been selected. Therefore, if some things were not adaptations, congenital variation and natural selection could not be the necessary and sufficient explanation of organic evolution, as Weismann and his disciples maintain them to be.

It is evident, on Weismann's own admission, that it has not yet been proved that everything is adapted, that is, that every character or structural feature in every animal has its use in the struggle for existence. It is a mere supposition, therefore, that everything has been selected. In direct opposition to Weismann, I am prepared to maintain, first, that many structural features can be pointed out which are not useful to their possessors; and secondly, that all adaptations are due to the inheritance of acquired characters.

What is the use of the coiling of the shell and the torsion of the organs in the greater number of Gasteropods? Professor Lankester has admitted recently in the pages of *Nature* that he has been teaching for many years that this torsion was due to a mechanical cause, namely, the weight of the shell leaning over to one side, without realising that the explanation was Lamarckian. Now that it has been pointed out to him that such an explanation is inconsistent with the theory of natural selection, he admits that he can at present find no explanation of the phenomenon which would be consistent with that theory.

What is the use of the scrotum in Mammalia? In what way is it better for the male mammal that its testes should descend from their original position at the back of the abdominal cavity into a membranous sac, where they are constantly liable to injury?

What is the use of phosphorescence to pelagic animals?

What are the different conditions of life to which the different structural characters of the classes of Echinodermata are adapted? The star-fishes have bodies consisting of five, eleven, or thirteen, or some other number of radiating arms connected by a central disc. In the Ophiuroidea the arms are long, slender, and so brittle as to break at the least resistance; while in the Echinoidea the body has the form of a hollow sphere. Are these differences adaptations?

It is generally assumed that the mechanism by which flat fishes change their colour so as to assimilate it to that of the ground on which they happen to lie is a beautiful instance of adaptation. And yet the specific marking of each species is perfectly distinct. A large number of different species belonging to different genera live within the same area on ground of uniform colour; to this colour they all have to assimilate their general tint, and yet each species has its own permanent characteristic marking. As these markings are all different, they cannot all aid equally in the protective resemblance; some of them at least must either be indifferent or disadvantageous to the attainment of the useful result. As far as we can judge, all the specific markings are indifferent. And the change of colour is far less important than is generally supposed. The soles in aquaria are nearly always buried in the sand on which they live: they come out most at night, when to our eyes at least no colours can be distinguished; and when they move about to seek food in the day-time, they are almost always covered by a thin layer of sand sprinkled over their upper sides, by which their colour is concealed.

Again, the lower sides of flat fishes are, except in rare abnormal specimens, of a uniform opaque white, and it is impossible to conceive any advantage which they could derive from this, since the lower sides are usually concealed. It is true that plaice and flounders often rise a short distance from the bottom to seize food; and in this condition their white lower sides are very conspicuous.

The peculiar structural condition of the flat fishes as compared with symmetrical fishes is admitted on all sides to be one of the most striking cases of adaptation that exist. Yet so obvious is the conclusion that the evolution of the asymmetry is due to the habits of the fishes, that Mr. Wallace, in his new book on *Darwinism*, attributes the distortion of the orbits to the efforts of the ancestral flat fishes during innumerable generations, at the very moment when he is repudiating with Weismann the inheritance of acquired characters. When it is pointed out that he is thus using the very principle which he denies, he justifies himself by the sophistry that what was selected in successive generations of flat fishes was the power in the skull of retaining the distortion which the efforts of the muscles produced.

In the same way we might argue that the structural conditions connected with the upright position in man are not themselves inherited, but that what is inherited is the power in the body of retaining the changes caused by the assumption of the upright position in the child.

According to Mr. Wallace's argument, the distortion of the skull in flat fishes and the peculiar asymmetry of the eyes are not inherited at all, but are due in every generation entirely to the efforts of each individual young fish acting on a peculiarly sensitive skull. But in the young turbot and brill the metamorphosis is very nearly or quite completed long before the little fish have abandoned their pelagic mode of life and retired to the sea-bottom. In these species the young have a large air-bladder, although in the adults that organ is completely wanting. In consequence of this the young fish swim at the surface of the water until they have reached a considerable size, long after the right eye has migrated to the left side. These young fishes swim at the surface in a horizontal position, as do adult flat fishes when they swim up a short distance from the sea-bottom. Mr. Wallace evidently means by the "efforts of the young fish," the efforts which it makes to use its lower eye after it has assumed the habit of lying on its side on the bottom. Now these young turbot and brill do not lie on the bottom, and do not therefore need to make efforts to bring their right eye up to the edge of the body, yet the eye migrates all the same. It may be urged that they assume the horizontal position, and therefore have to twist their right eye round : but why do they assume the horizontal position ? The reason cannot lie in the great depth of the body, as Mr. Wallace suggests with regard to flat fishes in general, for in the John Dorey

(Zeus faber), which swims in a vertical position, the vertical depth of the body is as great in proportion to the length as in the young turbot, and greater than in the young brill.

It may be thought that these facts are as much in opposition to the view that the asymmetry of flat fishes was *evolved* in consequence of the efforts of the ancestral fish to use its lower eye after it had assumed the habit of lying flat on the bottom. But it must be remembered that those who believe in the inheritance of acquired characters also believe that the modification caused by the habits of the adult is inherited at an earlier and earlier age, until it may appear long before the habits of the adult are assumed.

To some extent the controversy concerning the inheritance of acquired characters is due to an ambiguity of language. The Neo-Lamarckians do not assert that a change produced in an individual by functional activity or external conditions is inherited at once and completely by that individual's offspring; in this sense acquired characters are not usually inherited. But what they do maintain is, that when a certain functional activity produces a certain change in one generation it will produce it more easily in the next, and so on; so that if a certain constant exercise of functional activity is continued for a great number of generations, ultimately structural modifications will appear in the young even before the function which has produced them has commenced to be exercised; and this process may go on indefinitely, so that at last the structural character in question will be inherited for many generations after the exercise of the particular function has altogether ceased.

Nothing can test better the claims of the two theories, the Neo-Lamarckian and the Neo-Darwinian, to be accepted in explanation of the origin of adaptations than the case of the woodpecker. This bird lives entirely on insects, and only catches insects in one way—a very peculiar way. It probes the holes in the bark of trees made by insects, or makes holes itself, and then inserts its long pointed tongue, whose tip is provided with recurved papillæ, like a narrow bottle-brush, and with this extracts the maggots. The rapid protrusion of the tongue to a considerable distance, and its sudden retraction, are rendered possible by the elongation of the processes of the hyoid bone to which the tongue is attached. These processes are bent upwards and forwards over the back of the skull and inserted near the orbits. Now, in all birds the tongue is attached to the hyoid bone and moved by the muscles connected with that bone: in nearly all birds the tongue is muscular and mobile. It is admitted even by Weismann's adherents that the size and shape of bones, the size and shape of muscles, are in the individual modified by the use which is made of them. It will also probably be admitted that the modification is such as to facilitate the operations in which they are used. Therefore in every generation of woodpeckers—which birds, in the struggle for existence, had to be content to pick up a living on tree-trunks or starve—the constant use of the tongue in extracting insects from holes in trees must have elongated the tongue and hyoid bone, and increased the power of protrusion of the organ in each individual. The Neo-Lamarckians believe that these individual modifications were inherited in some degree, so that a greater modification in the same direction was produced in the offspring, and in this way it is easy to understand how the degree of specialisation we now see was produced.

The Neo-Darwinians say that it is quite true that such modifications were produced by the functional activity in the individuals, but these modifications were never inherited: other modifications of the same kind arose by congenital variation in some of the same individuals, and the individuals that had these survived; and then the favoured individuals pairing together, some of their offspring, inheriting from both parents, had the modification in a greater degree, and so on. Which is very much like the argument that the *Iliad* and the *Odyssey* were not written by Homer, but by another man of the same name who lived at the same time.

Professor Weismann's view of heredity is not, as he seems to think, in any sense an explanation or an approach to it. We are unable to comprehend how the individual impresses upon the germ the power of growing into another individual like itself, supposing that we regard heredity in this way. But even if it were perfectly certain that heredity had nothing to do with the influence of the "soma," but consisted entirely in a fixed tendency of the germ-plasm to develop into a certain type of adult organism, this fixed tendency would be as absolutely unexplained as the influence of the body on the germ-cells on the other hypothesis. Moreover, we have no evidence in support of the assumption of such a fixed tendency. It is a known fact that the development of an individual depends to a very great degree on the conditions under which that development takes place. The very occurrence of the process of development from the ovum to the adult is dependent upon certain conditions; outside a certain minimum and maximum of temperature, food, light, moisture, etc., the egg or embryo dies. Within those limits the result of the development varies to a very great degree with the variations of the conditions. In the cultivation of domesticated animals and plants, no degree of skill or patience in selection will produce improvement in a race, or even maintain its valuable qualities, unless favourable conditions are provided. Everybody knows that in a given district, even on a given farm, certain varieties of animals and plants cannot be produced in perfection. Individuals can be procured, the most perfect in existence, but in a particular district or on a particular farm they do not "thrive"; that is to say, the qualities for which they are valued disappear in the individuals, or, as is more often the case, in a few generations, in spite of all care and selection. Thus English races of dogs have been shown by Darwin (*Domestication*, vol. i. p. 37) to degenerate in India in a few generations, losing entirely the peculiarities of form and mental character which distinguish their race, and this in spite of the greatest care in selection and the prevention of crossing. According to Weismann's assumptions, these degenerated dogs if removed back to England would in the next generation recover entirely their lost qualities; but I do not think any sportsman would purchase one of them on Weismann's guarantee. What I wish to emphasize here is, that it is a fact that characters are only constant under approximately the same conditions; and that, when under new conditions new characters appear constantly in successive generations, we have exactly the same grounds for asserting that these new characters are inherited as for saying that the old characters under the old conditions were inherited. The very fact that characters are acquired proves that heredity is not a fixed and constant tendency independent of conditions; the fact that "acquired" characters disappear when the conditions to which they are due are removed is no more evidence against the inheritance of such characters than the fact that the original characters disappear under the new conditions is evidence against the heredity of those original characters. We know nothing more about heredity than that it is a name which we give to the generalisation that organisms remain constant in their characters within narrow limits of variation so long as the conditions of life remain unchanged, and retain their characters *for a time* after those conditions are changed.

Whether it can be proved that adaptations are due to the inheritance of acquired characters or not, it is at least certain that the Lamarckian view explains the evolution of adaptations as perfectly as Weismann's theory. Nay, more perfectly, for the theory of selection can never get over the difficulty of the origin of entirely new characters. It may be said that the necks of the giraffe's ancestors were of different lengths, and the selection of the longest produced the striking length of neck we now see. But how can it be said that the horns of ruminants arose? No other mammals have ever been stated to possess two little symmetrical excrescences on their frontal bones as an occasional variation; what then caused such excrescences to appear in the ancestors of horned ruminants? Butting with the forehead would produce them, and no other cause can be suggested which would.

There is evidence that physiological change precedes morphological. There is a climbing kangaroo in Papua which shows so little adaptation of structure to the climbing habit, that no naturalist would believe from the mere study of its body that it lived in trees. But as a matter of fact it does live entirely in trees.

Variation and survival are the explanation of adaptations, but the struggle for existence does not merely give rise to natural selection—it is the cause of the variations which survive.

Selection, whether natural or artificial, is perfectly analogous to the process of denudation in geology. It explains the extinction of innumerable forms and the consequent gaps and intervals which separate species, families, orders, etc., just as denudation explains the want of continuity in the stratified rocks. But geologists have never been blind enough to suppose that the evolution of the structure of a given rock was due to denudation; they have always believed that the structure of each rock was due to the effects of the forces which have acted upon it since its formation, and they have devoted their energies to tracing by observation and experiment the effects of the various forces. It has been the distinction of biologists to maintain the paradox that matter in a certain condition, namely, in the form of living organisms, is not subject to the action of physical forces.

But the following argument, which I recently attempted to publish in a letter to the Editor of *Nature*, seems to me to be stronger than any I have yet used. The letter containing it never appeared in the journal, which professes to represent natural science in this country. Its suppression was the more surprising because the controversial correspondence to which it belonged arose originally from a letter of mine, which was published in the same journal, on the explanation of the eyes of flat-fishes given in Wallace's *Darwinism*.

Nature has embraced the principles of Weismann's Neo-Darwinism, and while willing to devote plenty of space to favourable reviews of Weismann's essays written by undergraduates, suppresses without a word of explanation or apology contributions which argue against the fashionable creed. The argument rejected is as follows: According to the followers of Weismann, evolution is due to the selection and "sexual mixture" of congenital variations, that is to say, of variations themselves due to sexual mixture, and having nothing to do with the physiological effects of conditions or stimuli or functional activity. These congenital variations, in animals which are hatched or born, must either be present at hatching or birth, or must appear at a later period of life. If they are supposed to be present at hatching or birth, then they can have nothing to do with any of the most important and most interesting modifications which we know to have taken place in the evolution of animals. For such modifications do not begin in individual development until long after birth or hatching. For instance, flat-fishes when first hatched are perfectly symmetrical, and their metamorphosis does not begin till they are some weeks old. On the other hand, if the disciples of Weismann say that congenital variations appear at later periods of life after birth or hatching, how are such variations to be distinguished from acquired characters due to stimulus and functional activity? Can it be proved, for example, that the first slight asymmetry of the eyes of the ancestral flat-fish which appeared long after hatching, in a fish which crouched on the ground on one side, was *not* due to the use that particular fish made of its eyes? Is there any criterion by which we can distinguish among variations appearing *de novo* at a post-embryonic period of life, those which are acquired, and those which are due to variations of the germ-plasm? There is none; and therefore it would be as true to assert that there is no evidence of the inheritance of any individual variations, as to say there is no evidence of the inheritance of acquired variations.

J. T. CUNNINGHAM.

PLYMOUTH, 31*st January* 1890.

CONTENTS

	PAGE
INTRODUCTION	1

SECTION I

THE NEWEST THEORIES CONCERNING EVOLUTION	8

SECTION II

THE ORGANIC GROWTH OF THE LIVING WORLD	20
Fundamental Causes of the Manifold Variety of Organic Forms	22
Reproduction as Organic Growth	24
Individual Development as Organic Growth . . .	25
Separation of the World of Organic Forms into Species—Genepistasis	27
The Particular Causes which determine the Difference of the Directions of Evolution, and which also contribute to bring about the Division into Species . .	32
Sexual Combination—One-sided Inheritance . . .	35
Intermittent Evolution—Kölliker's Hypothesis . .	45
Constitutional Impregnation	51
Elaboration and Simplification in the Evolution of Species .	52

SECTION III

INFLUENCE OF ADAPTATION IN THE FORMATION OF SPECIES

	PAGE
Influence of Adaptation in the Formation of Species	63
Is everything adapted?	63
Death as an Adaptation	67
Further Considerations on the Adaptation and Direction of Evolution of the Markings of Caterpillars. Absence of Sexual Combination in this Evolution	72
Characters which are inessential (indifferent) to the Life of the Organism	74

SECTION IV

ACQUIRED CHARACTERS 78

Methods of Investigation—The Period of Time to be claimed for Evolution	78
Acquired Characters due to Direct External Action	87
Experiments on the Influence of Temperature on Lepidoptera	116
Further Remarks on the Causes which Change the General Colouring of Animals	135
Influences of Locality on the Variation of Animals, and thereby on the Formation of Species	139
Importance of the Stimulation of the Nervous System in relation to Adaptation and the Origin of Species	142
Particular Facts which prove the Influence of Nutrition and other external Conditions on the Variation and Formation of Species in Lepidoptera	149
Characters acquired by Use	153
Inheritance of Injuries and Diseases	173

SECTION V

DISUSE OF ORGANS—DEGENERATION—PAMMIXIS 205

Pammixis	217

SECTION VI

	PAGE
MENTAL FACULTIES AS ACQUIRED AND INHERITED CHARACTERS	221

The Function of the Brain	221
Reflex Action	222
Intelligence, Reason, Habit (Automatic Actions), Instinct	223
Particular Instances of Intelligence and Reason in Animals	231
Experiments and Observations on Instinct in newly-hatched Chickens	245
The Instinct of the Cuckoo in Laying her Eggs in other Birds' Nests	256
Reasoning Instincts	266
Further Remarks on Reasoning Instincts and Intelligent Instincts in Animals	278
Reflex Action and Instinct	294
Impulse and Instinct	301
Concluding Remarks on Instinct	304
Irritability and Sensation—Will	306

SECTION VII

ORGANIC GROWTH: THE MORPHOLOGICAL AND PHYSIOLOGICAL EVOLUTION OF THE LIVING WORLD AS THE RESULT OF FUNCTION	315

The Origin of Organisation in Unicellular Animals—The Fundamental Biological Law	315
The Evolution of Organisation in Multicellular Animals	323
Origin of Muscles	326
Origin of Striation in Muscles	328
Evolution of the various Sense-Cells from a Common Cell Layer	331
The Origin of the Central Nervous System	339
Vicarious Nerve-Centres	343
The Cell-Nucleus as a Central Nervous Organ	349

	PAGE
Origin of Nerve-Fibres and their Vicarious Action	353
The Acquisition and Inheritance of Peculiarities of Voice and Speech, and the Speech of Animals	362
Concluding Remarks	377

SECTION VIII

THE IDEA OF ORGANIC GROWTH—THE LAW OF ORGANIC FORM—RECRESCENCE   379

The Idea of Organic Growth	379
Crossing and Selection as Indirect Causes of Growth	383
The Law of Organic Form: Its Application to the Form and Structure of Plants	384
The Recrescence of Lost Parts as an Example of Organic Growth	389
CONCLUSION	409

APPENDIX

ON THE IDEA OF THE INDIVIDUAL IN THE ANIMAL KINGDOM	413

INTRODUCTION

It seemed to me long ago of the greatest importance to undertake an investigation of the question whether the modification (variation) of the species of animals is not governed by definite laws.

It had previously been assumed that variation occurred quite irregularly, in the most diverse directions, that it was abandoned completely to chance; in fact, the origin of species according to the Darwinian explanation is left entirely to chance. It was justly objected to that explanation that it asserted the predominance of chance. If, as I acknowledge, the principles of Darwinism are true because they can be shown to follow from natural laws, then it was to be expected that obedience to laws would also be discovered in that province which Darwin had surrendered to chance. But if variation were shown to follow certain laws, the same demonstration would apply to the origin of species. For species have, as every evolutionist will acknowledge, of necessity been produced from varieties. They differ, if we can draw a dividing line between the two at all, from varieties only in this, that they are separated from related forms both upwards and downwards by the impossibility of unlimited fertile sexual intercourse, or else, as in the case of forms multiplying asexually, they are defined by conspicuous morphological characteristics.

But the investigation of the laws of variation included the question of the <u>causes</u> of this variation. In this respect also Darwin had left a great gap in the explanation of the origin of species. It was because he left variation essentially to chance that he was unable to say much on the question why it occurred at all. Indeed, it was the most zealous adherents of Darwin who made, and still make, the great mistake of treating the selection depending on utility as the power which by its own action brings forth those variations of the characters of the organism which afford the possibility of that selection; or at least, the mistake of not perceiving clearly how far selection, how far Darwinism as a whole, is from being able to <u>explain these variations</u>.

The Darwinian principle of utility, the selection of the useful in the struggle for existence, does <u>not explain</u> the <u>first origin of new characters</u>. It explains only—and that in my opinion only partially—the progress and the gradually effected pre-eminence of these characters.

Darwin's assertion that every character occurring in an organism must either be now useful to it, or must once have been useful,—an assertion which Darwin himself did not adhere to, but from which latterly he withdrew more and more,—has certainly more than anything else led to the false conception we have mentioned, for he set up utility as the only ruler in the organic kingdom. By the unconditional assumption of this supremacy it was quite overlooked that utility is a purely relative conception, and that therefore it cannot possibly be the fundamental principle of the forms of the organic world.

Before anything can be useful it must first be. Why, by what means, is it brought into existence? That is the wider question which I have put to myself.

But if there are agencies independent of utility which

influence the forms of the organic world, which indeed primarily condition them, many characters of this organic world must exist which have nothing whatever to do with utility.

It was *a priori* to be expected that the result of my investigation would relate principally to such characters, hitherto but little considered.

If we could know all the natural laws which have operated in the evolution, and which operate in the existence of a single animal or a single plant, we should understand the laws of the organic world altogether.

On this ground alone the biologist may look for a rich reward in giving himself unreservedly for once to the study of a single living being, in order to penetrate into its nature as deeply as possible. The single chosen object soon tells him more than all other animals or plants together which he has observed more superficially. For the more an investigator devotes himself to a fruitful object, the richer it appears to his eyes, the more it shows new properties, the more it acquires living interest and importance, and the more do all its characters and the relations of its life show themselves to be governed by law.

The unreserved study of a single species of animal has led me to the discovery of a whole series of laws, which the extension of the investigation to other species showed to hold good generally.

As far back as the beginning of the eighth decade of the present century, I had turned my attention to the wall-lizard, whose remarkable variability is well known, as the starting-point of my investigation. The circumstance that in the spring of 1872 I became acquainted with the dark blue wall-lizard, described by me as Lacerta muralis cœrulea, on one of the Faraglione rocks at Capri determined my final decision.

In this form of the wall-lizard I had found an animal

which might be with equal justice described as species or variety, so much does it differ from the original form. Thus, at the conclusion of the paper I published on this lizard,[1] I was able to say : " The remark of some opponents of Darwinism that no one has yet observed the conversion of a variety into a species is not, I grant, contradicted by the existence of the blue wall-lizard, for it makes on the adherents of the theory a demand similar to this: we want to hear the grass grow. Besides, every naturalist will always be at liberty, within certain limits, to understand by a 'species' what he will. But an instance is afforded in this animal of undoubted natural race-production, which has evidently occurred in a relatively short period of time, and this ought to prevent any opposition, which is not expressly one of principle, from attaching any importance to the argument above quoted."

The problem of finding the causes of the modifications which this remarkable variety had undergone led me immediately into the midst of the inquiries which I had already proposed to myself. The result of my researches, which were extended to various classes of animals, was the recognition of the dominion of laws in the process of variation, not only in the lizard, but also in the most diverse tribes of the animal kingdom : these laws holding firstly in the variations of marking, previously regarded as quite indifferent, unimportant, and fortuitous, but also applying to other characters. I was able to demonstrate that variation everywhere takes place in quite definite directions which are few in number, and I was able on the basis of my observations to put forward the view that the causes which lead to the formation of new characters in organisms, and in the last result to their evolution, consist essentially in the chemico-physiological interaction between the material composition of the body and external influences.

---

[1] *Zoologische Studien auf Capri*, ii. *Lacerta muralis cœrulea*, Leipzig, Engelmann, 1874.

Finally, I succeeded, through the facts I established, in referring the separation into species, of which neither Darwin nor any of his successors had given a satisfactory explanation, in connection with the rest of my views, to natural causes.

I have expounded these views at full length in a second paper[1] on the variation of the wall-lizard.

But it seemed that very few investigators in the province of the doctrine of evolution troubled themselves about the wall-lizard, or about facts obtained from such a common animal, or about the conclusions to be drawn from them. It is possible, indeed, that the title of my papers was not very inviting. I ought to have put Darwinism first, and the wall-lizard second. Possibly the latter might then have been honoured too—possibly, for the tendency of the "scientific" zoology of to-day is to neglect the study of entire animals.[2] Anything that is not teased with the needle, or cut with the microtome, or examined with the microscope, is scarcely noticed at the present day, except by those who are exclusively systematists,—even in questions connected with the evolution theory. For, strange to say, even the doctrine of evolution is left entirely in Germany to the decision of anatomy and embryology, that is, of the microscope, or else is given up to mere speculation, although Darwin himself, the reviver of this doctrine, used neither the former nor the latter, but external

---

[1] "Researches on the Variation of the Wall-Lizard: a Contribution to the Theory of Evolution from Constitutional Causes, and also to Darwinism." *Arch. f. Naturgeschichte,* Berlin, 1881.

[2] In the zoological *Jahresbericht* of the Zoological Station in Naples for 1881, published 1883, p. 219, I read: "Th. Eimer has published a very important paper, well worth reading, on the variation of the wall-lizard. This comprehensive memoir is not of a kind to be briefly summarised, and therefore the reader is referred to the original." That is all the favourable acknowledgment of the memoirs in question which I have read up to quite recently, excepting reports in the *Naturforscher* and in the *Revue der Naturwissenschaften.* But K. Düsing has just published a report showing complete comprehension of my meaning in vol. ii. of *Kosmos* of 1886.

form, the life and the distribution of plants and animals, for his theory.

No matter. In any case, since the publication of my papers various theories concerning evolution have appeared, whose authors have not troubled themselves in the least about the facts established by me, have not even mentioned them, although they contradicted their theories. But I will not go so far as to say with Nägeli that this was "because they were unable to use these facts." In an address to the second general meeting of the Congress of German Naturalists and Physicians at Freiburg i.B., which is printed as an appendix to this book, I had developed a theory based on facts. Subsequently, in various articles in the *Zoologischer Anzeiger*, and in the journal *Humboldt*, as well as in a lecture on the markings of birds and mammals[1] delivered as early as 1882, which was afterwards published, and which is reproduced in the present book, I largely added to the number of these facts, and brought forward further irrefragable evidence for my views. But notwithstanding all this, not one of those who have set up theories or hypotheses of their own in our subject took any notice of my facts.

As we shall see, I have no special reason to complain in this matter. What has happened to me has happened to others. The botanist Nägeli makes, as I mentioned, the charge against such writers that they have not referred to the facts brought forward by him because they could not make use of them.

I might express myself in milder terms thus: The founders of theories of evolution set up since Darwin's time have as a rule deviated from Darwin's path, inasmuch as they have sought to adapt the facts to their ideas and opinions, instead

---

[1] *Ueber die Zeichnung der Vögel und Säugethiere.* Lecture delivered at the Meeting of the "Verein für Vaterländische Naturkunde in Württemberg" at Nagold on 24th June 1882. *Württembergische naturwissenschaftliche Jahreshefte,* 1883.

of adapting their ideas and opinions to the facts. Such argument leads naturally, even without any bad intention, away from the consideration of the opposing facts back to its starting-point—revolving in a circle and seeing nothing which lies without the circle. This method is most surprising in those who, like A. Wagner, with his *Law of Migration*, oppose the whole of Darwinism in order to put in its place an explanation of evolution which can only find its place in Darwinism itself as a subordinate of that theory. Isolation in space, regarded by Wagner as alone regulating the origin of species, favours and promotes the separation of the gradually produced diverse forms into species; but apart from the truth that it does not alone explain the separation, its influence cannot possibly be put in the place of that of the principle of utility.

The effective influence is not necessarily single—many can be effective together.

On this standpoint I have placed myself in my inquiries from the first, and their course and their results have fully justified me.

On account of the relations which they bear to others, I have repeated some of the facts already published by me which serve as the foundation of my theory. The last sections of this book are specially devoted to such facts.

It happens that my views on the question before us do not agree with those of some of my valued friends and honoured teachers, and if I am to uphold my own convictions I must endeavour to refute theirs.

As this opposition is little pleasing to myself, I am convinced that my frank advocacy of scientific opinions will neither by those who are aware of this personal relationship, nor by those who are immediately affected, be stigmatised as unfriendly and ungrateful.

SECTION I

THE NEWEST THEORIES CONCERNING EVOLUTION

BEFORE I proceed to the exposition of my own views I must briefly discuss some of those attempts to explain the evolution of organisms which have been made by others in recent years, namely, those of Weismann and Nägeli.

The theory of heredity introduced by Weismann under the name " Continuity of the Germ-plasm,"[1] has recently been widely discussed, partly in a favourable, partly in an adverse spirit.[2]

This theory seeks to explain the fact that the characters of the parents are inherited to so high a degree by the children, and the other fact that characters can be inherited from more remote ancestors, *e.g.* from the grandparents,

---

[1] A. Weismann, *Die Kontinuität des Keimplasma als Grundlage einer Theorie der Vererbung.* Jena, G. Fischer, 1885. The same: *Zur Frage nach der Vererbung erworbener Eigenschaften, Biolog. Centralblatt,* 1886, Bd. vi. No. 2.

[2] *E.g. Naturforscher,* 1886, No. 1. Eimer, *Deutscher Litteraturzeitung,* 15th May 1886. V. v. Ebner, *Ueber Vererbung,* Lecture delivered to the Society of Physicians of Steiermark, 22d March 1886 (*Betz Memorabilien,* 2 Heft, 1886). Kölliker, *Die Bedeutung der Zellkerne für die Vorgänge der Vererbung, Zeitsch. f. w. Zool.* 1885 ; and *Das Karyoplasma und die Vererbung, eine Kritik der Weismanns'chen Theorie von der Kontinuität des Keimplasma, ibid.* 1886. Virchow, *Descendenz und Pathologie, Arch. f. patholog. Anat.* Bd. ciii. Kollmann, *Biolog. Centralblatt,* Bd. v. 1886. Virchow, *Verhandlungen der Strassburger Naturforscherversammlung,* 1885, Address delivered at the second general meeting of this Congress.

one or more generations, *e.g.* the parents, being passed over (atavism).

Weismann's attempted explanation rests on the assumption that the germ cells, by means of which inheritance must be transmitted, pass over unchanged from the body of the ancestor into that of the descendant; that they form a whole which so far contrasts strongly with the rest of the body; that it takes no kind of part in the changes which the latter experiences during life; that they can therefore be transmitted from generation to generation unchanged. G. Jäger [1] and Nussbaum [2] had previously already sought to explain such an immediate connection between the germ-cells of parent and child, by the assumption that the germ-cells of the offspring separated themselves at the very beginning of embryonic development, or at all events before any histological differentiation, from the rest of the developing ovum.

According to Weismann—the assumption of Jäger and Nussbaum comes ultimately to the same thing—inheritance would be due to the fact that in every reproduction a part of the germ-plasm of the parental egg-cell is not used up in the construction of the offspring, but remains unchanged to serve for the formation of the germ-cells of the following generation. The inheriting substance lies, according to Weismann, in the contents of the nucleus of the germ-cell.

"The germ-cells," says Weismann, "arise in their essential and distinctive substance, not by any means from the body of the individual, but directly from the parental germ-cells." "Inheritance," he continues, "takes place wholly and solely because a substance of definite chemical, and above all, molecular composition passes over from the germ-cells of one generation to those of the next. This substance, the

---

[1] G. Jäger, *Lehrbuch der allg. Zoologie*, Leipzig, 1878, Bd. ii.
[2] M. Nussbaum, *Die Differenzirung des Geschlechts im Thierreich*, Arch. f. mik Anat., 1880, Bd. xviii.

germ-plasm, is located in the cell-nucleus, and possesses, by virtue of its extraordinary complexity of structure, the capacity to develop into a very complex organism."

According to Weismann's explanation of heredity, therefore, the germ-cells appear no longer as the product of the body, but rather as something in contrast with the totality of the body-cells. "The germ-cells of successive generations are related in the same way as a series of generations of unicellular beings which are derived one from another by continued division."

The necessary consequence of such a conception is the proposition that acquired characters are never inherited. If he is right, then no external influences whatever can ever have contributed to the moulding of species. The permanence of forms would be explained by that proposition, but their variability would remain, unless other agencies were brought forward for its explanation, a darker mystery than ever.

Weismann seeks these agencies in sexual reproduction.

By the mixture of characters which sexual reproduction brings about, according to him, are supplied at once—and exclusively—the materials for the origin of new species; among the new forms produced by the mixture, the struggle for existence chooses the fittest for survival and for renewed reproduction.[1]

The great importance of the sexual mixing of characters in modification can be ignored by no one, and was indeed never ignored. But if this were the only cause of modification it would be necessary to assume, supposing the same laws to hold for lower as for higher organisms, that all plants and animals, including the lowest, now reproduce sexually, and always have done so,—an assumption which is known to be

---

[1] A. Weismann, *Die Bedeutung der geschlechtlichen Fortpflanzung für die Selektionstheorie*, Address to the Congress of Naturalists at Strassburg, 1885, and Jena, G. Fischer, 1886.

unjustifiable on the basis of the facts offered to us by the organic world now living, and which is contradicted by the general conception of the gradual development and perfection of organic differentiation. But Weismann compares the increase of the continually unchanged germ-plasm of the multicellular organism with the reproduction of the unicellular; in the latter also the same substance grows continually, and new individuals arise only by its division from time to time. The germ-plasm of the multicellular organisms is to correspond therefore with the whole body of the unicellular. Weismann does not deny that the unicellular beings change in consequence of the direct influence of external agencies; on the contrary, he ascribes to them explicitly this property. From this inheritable individual diversity once produced he derives "that of the Metazoa and Metaphyta,[1] in that this diversity, through sexual reproduction which has in the meantime become general, is preserved for ever, increased and ever produced in new combinations." Thus, according to Weismann, the difference between the unicellular organism and the germ-plasm of the multicellular is a fundamental one—for the former other laws hold good than for the latter.

But against Weismann's conception of the importance of the germ-plasm in heredity are to be set the following facts.

It is known that even highly organised animals and plants have the power of multiplying, by simple division in the case of the former, by cuttings in that of the latter.[2] The new complete individual produced by this method has the same characters as the animal or plant produced at another time from a germ-cell—a proof that the substance possessing the property of heredity is not confined to the germ-plasm, and

---

[1] *I.e.* of the multicellular animals and plants.
[2] Cf. V. v. Ebner, *op. cit.*

that it cannot be something altogether different from other parts of the organism.

When Weismann further expresses the opinion that sexual differentiation itself finds its explanation in his theory alone, he is to be met with the fact that such an explanation is certainly afforded by (1) the importance of preventing in-and-in breeding; (2) the importance of division of labour.

Weismann's argument against the inheritance of acquired characters seems to me to be abandoned by himself in his acknowledgment of the inheritance of the tendencies (predispositions) to new characters acquired during life. In an article on this question[1] I have remarked that any such tendency certainly implies a corresponding molecular peculiarity of the germ-plasm.

"Let us assume," I said, "that all living beings have evolved one out of another, then, on the basis of Weismann's conception itself, it must be acknowledged that the predisposition to the acquisition of characters, and with that the molecular structure of the germ-plasm, has in the course of time experienced great changes; even with predispositions it is a question of characters acquired in the course of time and inherited."

Further, Weismann grants a *slight degree* of inheritable external influence on the germ-plasm.

The greatest stress, however, is laid by him on the fact that no inheritance of injuries incurred during life has been proved.

That injuries incurred during life are but seldom transmitted to the offspring does not appear to me wonderful: the inheritance of the complete form and complete activities of the organism, which took root such enormously long periods of time ago, and has been strengthened at each generation, will as a rule counterbalance in the offspring any

[1] *Deutsche Litteraturzeitung*, 15th May 1886.

such injuries incurred only once and not repeated. It is true there are injuries which, although they have been always repeated constantly, are yet never inherited. Among these is the rupture of the hymen in women.

In such cases we must presume a specially effective power of correlative activity,[1] directed to the part affected and residing in the whole organism — the same compensating power which leads in lower animals during the life of the individual to the regeneration of parts which have been lost or artificially removed.[2] But these cases do not prove the general proposition that injuries are not inherited; they do not prove that even injuries which have been repeated during a considerable period are not inherited. Hitherto little importance has been attached to the demonstration of the inheritance of injuries. Yet single cases of the inheritance of injuries only once incurred seem to me to be thoroughly authenticated. To these I shall refer subsequently.

For the rest, it can I believe be proved as a fact that acquired characters are inherited.

The germ-plasm cannot possibly, in my view, remain untouched by the influences which are at work on the whole organism during its life. Such an immunity would be a physiological miracle, merely on account of the morphological relations of the animal ovum and spermatozoon, and their dependence on the nutritive processes of the body,[3]—a miracle

---

[1] Correlation = the principle that the characters of the living being are so connected with one another that one determines the other.

[2] The more imperfect the structure of animals, *i.e.* the lower their organisation, the less degree to which division of labour is carried out, so much the greater the facility with which injuries are repaired, so that many of the lower animals can be even cut into pieces with the result that each piece grows again into a perfect animal, just as in many plants. (Cf. on this point the Appendix: On the Idea of the Animal Individual.) The inheritance of injuries is thus of but slight importance in the discussion of heredity in general.

[3] I have myself described arrangements of quite surprising character for the nutrition of animal ova. Cf. my memoirs on the ova of Reptiles, Birds, and Fishes, in *Arch. f. mik. Anat.* Bd. viii. 1872.

which would be no less inexplicable than atavism, apart from Weismann's theory, seems to be.

According to my ideas, it is not that sexual mixing of characters, together with adaptation, determines the modification of forms, but rather that sexual differentiation itself is due to acquired and inherited characters.

But it is not my purpose at this point to bring forward all the arguments on which I rest my opposition to Weismann's theory; these arguments will be further developed in subsequent pages.

A theory essentially opposed to Weismann's regarding the causes of the modification of organic forms, in other words, the origin of species, has been set forth by Nägeli under the title "Mechanico-physiological Theory of the Doctrine of Descent."[1]

According to him, internal causes depending on the nature of the organic substance effect the transformation of the "strains" (individuals, species, families, etc.) in definite directions. Such "internal causes" must necessarily be supposed merely on the ground that the modifications or variations of the strains do actually take place in definite directions and are not irregular. The internal causes effect a constant alteration of the strains in definite directions "towards greater perfection, that is, towards greater complexity." The strains grow, as it were, towards greater perfection. Accordingly, Nägeli describes his theory of internal causes as the principle of "improvement." "Superficial reasoners," he says, "have pretended to discover mysticism in this. But the principle is one of mechanical nature, and constitutes the law of persistence of motion in the field of organic evolution." "Once the motion of evolution is started it cannot cease, but must persist in its original direction."

By greater perfection, then, Nägeli understands more com-

[1] C. v. Nägeli, *Mechanisch-physiologische Abstammungslehre*, München und Leipzig, 1884.

plex structure, greater division of labour. Distinct from this perfection of organisation there is a perfection of adaptation, "which is repeated at every degree of organisation, and which consists in the organism having, in relation to the external conditions at the time, the most advantageous structure which is compatible with its degree of complexity and the division of its functions."

The improvement of structure and the division of labour progress according to this conception within certain lines uninterruptedly, on account of the mechanical activity, following certain laws, of the living organic substance; and this course of things is only so far affected by other influences, that the external conditions of each particular period by means of adaptation favour those arrangements of structure which are most useful in the struggle for existence.

Continual advance of its own accord towards perfection is thus, according to Nägeli, a property of living organic substance.

But as Weismann's germ-plasm, unaffected by all external influence, would have been gradually used up, so it seems to me, according to Nägeli's principle of improvement, since it has been constantly at work ever since there have been living beings at all, the lowest creatures which existed millions of years ago must long ago have changed into higher—there could no longer be any lowly-organised simple living things at all. Nägeli explains their existence by the assumption that abiogenesis still takes place at the present day. The origin of the organic from the inorganic, he says, is a conclusion to be drawn from observation and experiment, a fact which follows from the law of the indestructibility of matter and energy (a sentence which, excepting the words "is a fact," shall not be contradicted). To deny spontaneous generation, he says, would be to proclaim a miracle. "In later times, and now still, abiogenesis must everywhere occur where the conditions are the same as in the primitive time... Among known living beings there are none which could have arisen by abiogenesis"—the lowest living plants have already a cell-membrane and the Monera cannot live independently, *i.e.* without the products of decomposition of other organisms, as must be the case with organisms produced by spontaneous generation. "The beings which are capable of arising by a spontaneous origin are therefore not yet known to us. They must have a still simpler structure than the lowest organisms which the microscope shows us" —they may be "beneath the lowest size which is visible with the microscope." The organic being due to spontaneous generation can be but a minute drop of homogeneous plasma " which consists of albuminates without admixture of other organic compounds than food materials, without external form and without distinction of parts, and which grows and feeds on the inorganic or simple organic compounds from which itself arose." "Abiogenesis thus presupposes a spontaneous formation of albuminates." Probably the spontaneous albumen production takes place still at the present time "in the moist superficial layer of some porous material (loam, sand), where the molecular forces of solid, liquid, and gaseous bodies are at work together," favoured by a definite degree of warmth, "so that it may still take place in warmer climates as well as in the warmer part of the year in colder regions."

Nägeli advocates with Weismann the view that external influences—that, in particular, climatic conditions and changes of nutrition—have no effect on the transformation of species. He supports his argument with experiments which he made upon plants; placing them under such changed conditions and finding that these had no relation whatever to the appearance of varieties.[1]

---

[1] C. Nägeli, *Ueber den Einfluss der äussern Verhältnisse auf die Varietätenbildung im Pflanzenreiche. Sitzuugsb. d. math.-phys. Klasse d. k. bayer. Akad. d. Wissensch. zu München*, 18th November 1865; and *Das gesellschaftliche Entstehen neuer Species, ibid.* 1st February 1873.

Thus while, according to Nägeli, the influences of nutrition have no power to produce change in higher organisms in even very long periods of time, he nevertheless assumes that such a change is produced in the lowest creatures. Since the lowest living beings arise by spontaneous generation under different conditions and in different places, therefore they must be differently constituted. Thus the further conclusion follows, that "the organic kingdoms take their origin not from a single organism but from several, which, however, differ but little from one another." The explanation of this important difference (between higher organisms, and those which are being produced from inorganic matter), by the assumption of which Nägeli not merely limits his statement of the non-inheritance of acquired characters but obviously shakes it to its foundations, is sought in a further supposition, namely, that the idioplasm of higher forms "obtains stability of structure from improvement during long geological periods, which structure has a relation to those different influences; while in the introductory period of spontaneous generation the definite order is being sought for the first time, and therefore is determined by the combined effects of all things which modify the molecular attractions and movements."[1]

Nägeli concludes in fact, like Weismann, that it must be a definitely formed firm substance which determines the special character and the specific evolution of an organism, and that this substance impresses its form upon the passive nutritive plasma which is dependent upon it and which forms the principal bulk of the body. To this substance he gives the name "idioplasm." He does not, however, like Weismann with

[1] [If this last quotation from Nägeli seems somewhat obscure in its English rendering, I can only say that it is not less so in the original German. This is not to be wondered at very much, for Nägeli is describing things which neither he nor any one else can really understand. It may be questioned whether such speculations are legitimate in science; they are scarcely due to a scientific use of imagination. Trans.]

his germ-plasm, pack it away in the cell-nucleus alone, but conceives it as a network or framework which permeates the whole body from cell to cell. It is the idioplasm which changes from generation to generation from *internal causes*, and thus, without being essentially influenced by accidental variation and selection, produces quite new forms.

Nägeli thus fundamentally takes us back to the conception of the vital force, applying it to the modification of forms, to the doctrine of evolution, and attempts to divest the principle of utility of any considerable importance.

His principle of improvement therefore effects greater complexity and greater division of labour in organisms in definite directions. When this metamorphosis is once in progress "it continues by mechanical necessity in the direction in which it started. For when, in virtue of the beginning made, one generation produces offspring which in one respect are in advance of the parents, then by the law of persistence of motion the offspring of these offspring must be altered to a farther degree, and the evolution must go on as far as the nature of the conditions allows." Thereupon adaptation comes into action.

We have thus, according to Nägeli, as the mechanical causes of the evolution of the organic kingdoms "the persistent advance towards perfection from the simpler to the more complex, and further, the definite action of the external conditions in adaptation."

"Competition, with its consequent crowding out among living beings, has, within the definitely directed changes towards perfection and adaptation, its effect in separating and in defining, but not in forming, the strains : not a single phylogenetic pedigree owes to competition its existence, but the several pedigrees through the extermination of intermediate forms stand forth more clearly and more characteristically.

". . . . Still better may we compare the vegetable kingdom to a great tree branching from its base upwards, of which the ends of the twigs represent the plant forms living at one time. This tree has an enormous power of sprouting, and it would, if it could develop without hindrance, form an inextricable bush-like confusion of innumerable branchings. Extermination in the struggle for existence, like a gardener, prunes the tree continually, takes twigs and branches away, and produces an orderly arrangement with clearly distinguishable parts. Children who see the gardener daily at his task may well suppose that he is the cause of the formation of the branches and twigs. Yet the tree, without the constant pruning of the gardener, would have been much greater, not in height, but in extent, and in the richness and complexity of its branching.

"In the perfecting process (progression) and adaptation lie the mechanical impulses which lead to the abundance of forms; in competition and extermination, or in Darwinism proper, only the mechanical cause of the formation of gaps in the two organic kingdoms."

Thus Nägeli's theory attempts principally to explain two points which that of Weismann leaves unexplained, namely, the beginning of the formation of characters and the fact of variation in definite directions. But Nägeli's conception seems to me to rest so much more on assumptions acutely thought out than on facts, that it deserves rather to be described as a materialistic-philosophical than as a mechanico-physiological theory.

SECTION II

THE ORGANIC GROWTH OF THE LIVING WORLD

For many years I have myself, from researches on the variation of single species of animals, especially with respect to markings of colour, acquired and expressed[1] views upon the ultimate causes of the origin of species. These views are essentially in contradiction to those of Weismann, and to those of Nägeli, while in details they agree with those of each.

The agreement consists in this, that I have from the zoological standpoint, as already remarked, pointed out and emphatically maintained that the variation of species takes place not in all kinds of directions irregularly, but always in definite directions, and indeed in each species in a given time in only a few directions.

Further, I opposed, as mentioned likewise in the Introduc-

---

[1] Th. Eimer, *Zoologische Studien auf Capri*, ii. *Lacerta muralis cœrulea, ein Beitrag zur Darwinschen Lehre;* Leipzig, Engelmann, 1874. *Untersuchungen über das Variiren der Mauereidechse, ein Beitrag zur Theorie von der Entwicklung aus constitutionellen Ursachen; Archiv für Naturgeschichte* (and separately), Berlin, Nicolai, 1881. *Ueber den Begriff des thierischen Individuum*, an Address delivered at the second general meeting of the "Versammlung deutscher Naturforscher und Aerzte," at Freiburg i. B., 1883. Further, my articles: *Ueber die Zeichnung der Thiere*, in the *Zoolog. Anzeiger*, 1882, 1883, 1884; and (with figures) in the journal *Humboldt*, 1885, 1886. Also *Ueber die Zeichnung der Vögel und Säugethiere*, Lecture delivered at the meeting of the "Verein für vaterländische Naturkunde in Württemberg" at Nagold, 1882, in *Württb. naturw. Jahreshefte*, 1883.

tion, the earlier view of Darwin, afterwards abandoned by that great naturalist himself, that every character occurring in an organism must either be now useful, or must at some time or other have been useful. I brought forward in opposition to this the great importance of indifferent characters. I said then,[1] at a time when the Darwinian principle of utility still exclusively prevailed among zoologists in Germany : (1) "From internal causes conditions of organisation may arise, may as it were crystalise out, which are just as useful to the organism as if they had been due to the struggle for existence. In this case the claims of the principle of utility are accidentally satisfied by the results of evolution from internal causes, and the importance of that principle remains therefore undiminished.. (2) From internal causes characters which are indifferent for the success of the organism, and (3) even harmful characters may arise. . . . But organisms burdened with harmful characters can only maintain themselves, and only transmit their peculiarities through future generations when such characters are inconsiderable in comparison with the useful ones also present, or when such characters stand in correlation with others whose usefulness is greater than their harmfulness."

With these words I already at that time (1874) assumed as important agencies in the modification of species causes independent of the Darwinian principle of utility. At the same time I emphatically stated that it was self-evident that no form could continue to exist which unconditionally contradicted that principle.

Instead of the expression "internal causes" I subsequently employed "constitutional causes," in order to indicate that I sought the causes of the modification of forms not in some fundamental impulse corresponding to vital force, but rather in physical and chemical processes depending on the material composition of the body.

[1] *Lacerta muralis cœrulea*, 1874.

The causes I formerly called "internal," therefore, have nothing to do with the internal causes of Nägeli, a fact which is proved without a doubt by his latest exposition, and for the future I shall avoid the expression.

Fundamental Causes of the Manifold Variety of Organic Forms

According to my conception, the physical and chemical changes which organisms experience during life through the action of the environment, through light or want of light, air, warmth, cold, water, moisture, food, etc., and which they transmit by heredity, are the primary elements in the production of the manifold variety of the organic world, and in the origin of species. From the materials thus supplied the struggle for existence makes its selection. These changes, however, express themselves simply as *growth*.

As individuals grow so the whole world of organic forms has grown up from simple beginnings.

To separate the continuous growing whole into parts, into species, required, and still requires, special means afterwards to be discussed. Let us first of all in what follows exclude the necessity of this separation.

Warmth, air, light, moisture, food, condition the growth of the individual being—appear before our eyes as the mightiest impulses which determine the manifold variety of the forms of living beings. They condition growth through physical and chemical change of the living organic mass, the plasma, through the formation of new and more complex compounds.

Since the external conditions have not always remained the same, but have changed in the course of ages upon our earth, and since they are even now locally different, so that one and the same organism at different parts of the earth, in different dwelling-places even of the same region of the earth, is exposed to different external influences, therefore there results, as a quite self-evident consequence of physico-chemical change in the organism, difference of growth, and therewith, of the shape and form.

Just as in inorganic nature from different mother lyes different crystals separate, as even simple mechanical shock can produce dimorphous crystallisation, so crystallise, if I may so express myself, in the course of ages, organic forms, to a certain degree different, out of the same original mass. Only the organism works with much more complex material, with a much greater variety of compounds. The greater delicacy and multiplicity of the organic processes determine other and more manifold forms in the organic world.

But just because the form of the organism depends upon physico-chemical processes, it is like the form of the inorganic crystal, a definite one, and can, when modification takes place, only change in certain definite directions.

When such new characters, which are simply to be referred to changed growth, become persistent in a group of individuals by constant and still continuing inheritance, and when this group in any way whatever has lost its connection with the rest of its relatives, by the loss of the intermediate forms, then we speak of a species.

New, changed form-constituents appear therefore in the variety and species as the expression of changed growth.

In other words, the origin of species obeys exactly the same laws as simple growth; it is the consequence of unending dissimilar growth of the organic world taking place under changed conditions, with the postulate of permanent separation of dissimilar links of the growing chain of this organic world.

Reproduction and individual development likewise depend upon the laws of growth.

Reproduction as Organic Growth

Sexual and Asexual Reproduction

The peculiarity of reproduction as compared with individual growth consists only in this, that parts separated from the whole under certain conditions continue to grow.

It is a fundamental property of the organism that as a whole it wears out, that as a whole it must perish, die ; another, that during its life it must constantly be nourished, repaired, and in part renewed. It continues its race, it transmits its life by division, or by separation of parts which go on growing.

Thus reproduction is a fundamental property of the organism.

The necessary repair and renewal finds its most complete expression in sexual combination. For the beginnings of sexual combination must be sought in conjugation[1] as it occurs in lower organisms. And these beginnings indicate that each of the organisms in question simply supplies something which makes up a deficiency in the condition of nutrition in the other.

Gradually from such conjugation, which originally took place between two individuals sexually quite equivalent, sexual separation has developed on the basis of the advantage of division of labour, and the prevention of in-and-in breeding.

The difference between sexual and asexual reproduction is not more profound than this.

Indeed, the union of egg and sperm is to be itself regarded as a kind of conjugation, for egg and sperm are parts which separate themselves in an asexual manner from the organism,

---

[1] Temporary contact of two unicellular, not externally dissimilar beings—obviously accompanied by exchange of material—or permanent coalescence of two such beings, in each case followed by reproduction.

to be regarded physiologically as buds from it. The sexual union and reproduction which follow exhibit the whole process of so-called sexual reproduction in this light as a sort of alternation of generations. The sperm-cell as a micro-organism conjugates with the egg-cell as a micro-organism. And indeed the processes which take place in this conjugation are essentially quite similar to those which are known in the conjugation, for example, of Infusoria. Here also the male element (nucleolus = male nucleus, sperm nucleus) unites with the female element (nucleus = female nucleus, egg nucleus) to form a new nucleus, a part of the female nucleus being extruded. This renovation of the Infusorian nucleus, like that of the egg nucleus in sexually reproducing animals and plants, leads to reproduction. For it is now known that fertilisation universally depends upon the renovation of the egg nucleus by its union with the sperm nucleus, accompanied as in Infusoria by the extrusion of a part of the former.

The essential character of egg and sperm is that each contains in homœopathic dilution the sum of the properties of the body of the female or male respectively which produces it, or, as it might be expressed, contains an extract of that body—just as two cells which in simple asexual division arise from a single one contain the material of the latter divided between them, or as the bud which separates from a Hydra, or the bud which separates from a plant for the propagation of its species, is of exactly the same material composition as the organism that produces it.

Individual Development as Organic Growth

Individual development, or ontogeny, is also an abbreviated phylogenetic growth, taking place under special conditions. At the present time the highest living beings still develop from simple cells, grow out of them, repeat in their development the stages of growth of the organic world. The repetition in the development of the individual, in ontogeny, of the ancestral development, of phylogeny, consists at the same time in the compressed abbreviated exhibition of the characters acquired by the whole series of ancestors and transmitted to the developing individual. Phylogeny is the mechanical cause of ontogeny.

If the organic world is a connected whole, the same fundamental laws must hold for all the members as for the whole—therefore also the law of growth.

If all members of the organic world are either directly or indirectly connected by affinity, have been derived one from another,—if, according to the biogenetic law, the development of each single being consists in its growing on through stages which represent the series of its forefathers (stages of growth from the cell to the vertebrate, for example),—then, on the assumption that this individual development morphologically and physiologically forms a repetition of the ancestral evolution, the proof is afforded by the biogenetic law alone that even the highest organisms, that the organic world in general, in the course of ages, has grown up from cells.

If the organic world is thus a connected whole, as biological research now assumes, and if it has grown up as I am here attempting to prove, then two further important questions have to be faced, namely: (1) What causes have brought about a separation of this organic world (whose forms of their own accord would be in uninterrupted connection, would be united by imperceptible transitions) into different members, into kinships—into species, genera, etc.? (2) To what causes is it due that any given highest species in a group of related species—assuming divergent branching in the tree of descent—is a stage farther evolved than those next to it?

I will first endeavour to answer the first of these questions. Neither Darwin nor any other inquirer has done this in a satisfactory fashion, as I remarked in the Introduction. Indeed, this question has hitherto scarcely been seriously taken in hand, and yet it is entitled to as much importance as the general one of the causes of evolution—with its solution alone have we an explanation of the origin of species in the proper sense of the words.

For the rest, it will be seen from what has been said that my *Theory of the Organic Growth of the living World* stands in complete accord with other facts and fundamental considerations. Another circumstance in favour of it is that it is as simple as a true principle must be. This simplicity is probably the very reason that hitherto no advocate of the doctrine of evolution has found words to express it. A question of so extraordinarily wide a bearing seemed to demand extraordinary means for its solution. Very likely, too, my views, after they have overcome the usual first stage of criticism, the assertion that they are unfounded, will be passed on to the second usual stage, and it will be said they have been long ago expressed and known, and then into the third, that they are self-evident. I hope some day to rejoice in the third stage, and shall rejoice even if through their "self-evidence" the recognition of my own work should be diminished. The second stage, however, I may help to shorten, by expressly bringing forward the labours of others, and at the same time discussing the differences between their views and mine.[1]

SEPARATION OF THE WORLD OF ORGANIC FORMS INTO SPECIES—GENEPISTASIS

I have handled this subject in various zoological technical papers, particularly in my memoir "On the Variation of the

---

[1] Compare in this respect especially the section on Lamarckism.

Wall-lizard," also in the addresses "On the Markings of Birds and Mammals," and "On the Conception of the Animal Individual," and have shown principally—

1. That the progressive evolution of a character in a definite direction—let us take as an example the origin of beautiful ocelli on the skin of an animal[1]—exhibits perfectly regular stages. In the case referred to we have (a) longitudinal striping, (b) black spots, (c) formation of black rings, (d) the appearance of the coloured nucleus. These stages succeed one another during the growth of the animal. In other words the whole series of modifications is repeated in the development of every individual.

2. That where new characters appear, the males, and indeed the vigorous old males, acquire them first, that the females on the contrary remain always at a more juvenile lower stage, and that the males transmit these new characters to the species. (Law of male preponderance.)

3. That the appearance of new characters always takes place at definite parts of the body, usually the posterior end, and during development—with age—passes forwards, while still newer characters follow after from behind. Thus during life, e.g. in lizards, a series of markings pass in succession over the body from behind forwards, just as one wave follows another, and the anterior ones vanish while new ones

[1] I have here in my mind an example which by no means serves for all cases. It is to be understood that it is not at all essential that I should mention the origin of ocelli rather than of other characters. Of course all, even the most splendid, ocelli can be proved to arise from the simplest markings, the simplest spots. (Compare my remarks on this point at the Berlin Naturalists' Congress, 1886, in the published reports of that Congress, and also especially my observations on butterflies now in course of publication.) That the gradual development can be demonstrated even on the feathers of one and the same adult bird has been already shown by Darwin for the Argus-pheasant. Herr Stud. Kerschner demonstrated very beautifully the same thing in the peacock, showing in accordance with my theory how in that particular case the development can be followed gradually step by step in definite parts of the body. (Compare his paper in the *Zeitschr. f. Wissenchaftl. Zool.*, 1886.) Unfortunately he has in other respects almost completely misunderstood the laws established by me.

appear behind. (Law of wave-like evolution, or law of undulation.)

4. That the whole of the varieties and variations of a species represent nothing else but stages of the course of development passed through by the individuals of the species —unless they represent new characters usually appearing first in males. One variety will thus stand at stage a, another at stage b, and so on; but one or another is possibly distinguished by a new character, e.

But in the same direction as the peculiarities of the varieties lie those which characterise the related species. The species most nearly related may, for example, present the varieties f, g, h, i. The character e, which occasionally appears in the previous species, is the most characteristic peculiarity of the second species—the variation of the first towards e showed already the direction in which evolution proceeds farther. In the second species appears occasionally a character k, which has the same significance as e had before, and so on.

So, if one examines numerous individuals of a species with regard to its variation, he can in the clearest manner in many cases show from this variation its relation to other species.[1] As I said in my Freiburg address, the evolution—the growth— of species one from another proceeds onwards as though following a plan drawn out beforehand. In the case supposed if all the stages of evolution from a to i were of equal importance, we should unite them all together as varieties of a single species; but if between a and i one stage is altogether wanting, or if one of them, say in the second half of the series, appears suddenly with special prominence, dominant, then we have,

[1] For the clearer comprehension of this compare the details in my paper already cited, *Ueber das Variiren*, etc., with the figures and the last sections. Also my work with coloured plates shortly to be published by G. Fischer in Jena, *Die Schmetterlinge nach ihrer auf die Zeichnung begründete Verwandtschaft dargestellt und beschrieben*. Lastly, my illustrated articles, *Ueber die Zeichnung der Säugethiere*, in the last three years of the journal *Humboldt*.

granted at the same time the impossibility of sexual intercourse, two species.

The law of undulation, and the other laws 1 to 4 which I have formulated, hold therefore as much for the origin or evolution of the variety and the species as for the individual.

One after another, like waves, the new characters appear as stages of growth in the series of organisms which represent the evolution of species. Or as we may conversely express it, the higher species in the ontogeny of the individuals belonging to them, which represents the evolution of those individuals, briefly repeat the characters (stages of growth) of the lower. The highest animals briefly repeat in their ontogeny the whole series of their ancestors (biogenetic law) as stages of growth. And indeed, I have to add, in general the more briefly and the more incompletely they repeat the individual members of the ancestral series, the older these are, the less important they were, and the shorter the duration they had.

Every older stage of the phyletic growth is abbreviated for the benefit of the newer—a proposition which may be added as a fifth to the four already formulated.

Thus the facts established by me afford at the same time a new and complete confirmation of the biogenetic law. Varieties and species are therefore in reality nothing but groups of forms standing at different stages of evolution, that is, at different stages of phyletic growth, whether it be that they outstripped their fellows or their fellows them in the progress of evolution, so that connection by intermediate forms was lost, or that separation in space favoured separation in character.

Isolation in space or separation of habitats is obviously of great importance in species formation, but it is not absolutely necessary.

The same explanation evidently holds good for genera as groups of species, and in general for all divisions of a natural system to which we may with Nägeli give the general name kinships (*Sippen*). It would therefore be more appropriate in general to speak of the origin of kinships than of the origin of species.

We have then before us a graduated evolution, and the essential cause of the separation of species is seen to be the persistence of a number of individuals at a definite lower grade of this evolution, while the rest advance farther in modification. This mode of origin of varieties or species, as the case may be, I have named Genepistasis (γένος race, ἐπίστασις stand still).

With the facts and laws set forth by me in the previous pages as the foundations of my theory of the organic growth of the species, compare Würtenberger: *A New Contribution to the Zoological Proof of the Darwinian Theory*, Ausland, 1873, Nos. 1 and 2, and *Studies on the History of the Descent of the Ammonites. A Zoological Proof of the Darwinian Theory.* Leipzig, 1880.

Würtenberger finds that in Ammonites all structural changes show themselves first on the last (the outer) whorl of the shell—as in living animals, *e.g.* in my lizards[1] at the tail—and that then such a change in the following generations is pushed farther and farther towards the beginning of the spiral—as *e.g.* in my lizards towards the head—until it prevails in the greater number of the whorls.

Then still newer characters may arise again on the most external whorl—just as in lizards at the tail—and drive back the former, and so on. The Ammonites also only at an advanced age, only when they have as exactly as possible gone through the course of development inherited from their parents, acquire the power to vary in a new direction; but this power can be inherited in such a way that in following

---

[1] See my memoirs, *Ueber das Variiren der Mauereidechse*, and on the markings of mammals and birds of prey.

generations it appears always a little earlier, until it itself characterises the greatest part of the period of growth. This "law of precocious inheritance" as Würtenberger calls it, also holds—for it obviously coincides with the law of abbreviated development—for living animals.[1]

"The agreement of the palæontological facts afforded us by the Ammonites with those supplied by living animals," I have said,[2] certainly with justice, "is in the highest degree interesting. It places the perfectly general application of the law beyond all doubt."[3] For not only the markings, but all other characters of animals also thus follow the same law.

THE PARTICULAR CAUSES WHICH DETERMINE THE DIFFERENCE OF THE DIRECTIONS OF EVOLUTION, AND WHICH ALSO CONTRIBUTE TO BRING ABOUT THE DIVISION INTO SPECIES

The ancestral tree of organic forms does not rise in a straight line, but has ramose (forked) branchings. This branching, which is also an important factor in the separation of species, is due to (1) the direct influence of external conditions, which, differing at each locality, act upon each stage of evolution, and are able to divert the farther evolution from the former direction; (2) the functional activity of the organism in relation to the external world, which directly strengthens characters in process of development by the exercise of them

[1] Compare *Ueber das Variiren*, etc., p. 454, separate edition, p. 218, where examples are given.
[2] *Loc. cit.*
[3] Compare also the researches of Weismann, subsequently to be more closely considered, on the markings of the caterpillars of the Sphingidæ in his *Studien zur Descendenztheorie*, ii. : *Ueber die letzten Ursachen der Transmutationen*. Leipzig, 1876. With regard to plants the force of the same law is shown—apart from other facts—by the changes in the form of the leaves on one and the same plant from the lower to the upper part. The upper leaves are often at a new stage of evolution, the lower remaining at or passing through a lower stage— the former indicate the new species, the latter ancestral species.

(herein lies the importance of use and disuse); (3) the struggle for existence, which will indirectly have a different effect according to the difference of the external conditions; (4) the sudden appearance of new formations through correlation (evolution *per saltum*); (5) the principle that an organism continually exposed to the same conditions, under the uninterrupted action of the same influences, will after many generations, in consequence of "constitutional impregnation" ("conservative adaptation"), become different in structure, and have a different relation to the external world than before; (6) sexual mixing, which may, without any influence of adaptation, lead to the formation of quite new material combinations, that is, to the production of new forms.

The expression adaptation is often misapplied. Frequently it is erroneously used in an active sense; people speak of adaptation as of a spontaneously acting power, and in cases in which there can be no question of the use of a character either for the organism which possesses it or for the external world. Adaptation is, however, always something which has been brought about—indeed, implies a character which is in some way useful to the organism which possesses it in relation to the external world, or which is necessary or useful to other creatures or to the generality of things. The first is a special adaptation, the last can be called general (cosmic). In the latter sense, of course, ultimately everything is adapted, and we ought, as will be done in the following pages, only to speak of adaptation without the expressed qualification "general" when special adaptation is meant.

On the other hand, I cannot agree to the limitation of the Darwinian idea of adaptation which Weismann makes, when he will only recognise it as applied to characters gained in the life of the species, not to those gained in the life of the individual. Individuals during their life also adapt themselves to the external world—consider only the variety of the experience which individual animals during life, according to their surroundings and their intelligence, meet with and benefit by, or the special strength of body or any other useful qualities which they acquire in consequence of the external demands upon them. There is a personal adaptation in the Darwinian sense, for Darwin has recognised, as we shall see, the inheritance of characters acquired during the individual life, particularly in emphasising the use and disuse of organs as important in the modification of forms. Kollman certainly goes too far when he supposes that by adaptation we have only to understand the acquisition of definite characters during the individual life under the pressure of external agents.[1] For each character which seems adapted in an organism is not necessarily acquired during the individual life. But the closer consideration of this question is contained in my discussion of the whole theory of evolution. According to the foregoing, we have in any case to distinguish, besides general (cosmic) adaptation, a special adaptation in the usual sense of the word, and in this again personal (individual) and race (or species) adaptation.

The characters above classed under (1), those called forth directly by the influence of external conditions on the organism, without other aid from the latter than physiological reaction, I have previously described as arising by impression. Example: the production of a darker skin by light and warmth. Light and warmth are here the *causæ efficientes* (O. Schmidt). Here belong a number of characters which cannot be described as adaptations, which are rather unessential (indifferent). Also, however, those which are due to the composition of the organism which has been produced by the action of external conditions, and which may be accidentally useful, but cannot have been increased by selection. A host of characters which seem to have arisen through selection will certainly come into this group of the accidentally useful. If the brilliant mother-of-pearl of the internal surface of the musselshell, which is completely hidden under the mantle, were visible on the outer surface, it would certainly be indicated as useful. To the same category belong the black back and bright silvery belly of fishes, etc. Probably also many colours due simply to interference, such as the splendid blue of the wings of the male Calopteryx virgo and others. There can be nothing, indeed, more splendid in colour than the iridescence of Labradorite, and is this useful to the stone? And are colour and brightness useful to gold and countless other minerals? Are they useful to the soap-bubble?

Among the characters described under (2) I place those which have been produced with the aid of the external action of the parts of the organism concerned. Example: the formation of the hard skin of the heel, of the hard skin of the sole of the foot in the barefooted African

---

[1] Compare Kollmann, *loc. cit.*, and Weismann, *loc. cit.*, *Biolog. Centralblatt.*

negro (cf. Livingstone's *Travels*, etc.), of the nails and hoofs of men and animals. Here selection may play a part, but no more necessarily so than in the first case. However, most of the characters so produced are certainly useful, are to be described as adaptations. The action of the foot is here the *causa agens*.

The first case is direct, the second indirect adaptation; in the third case, the struggle for existence helps in the acquisition of useful characters; the fourth, correlation, may call forth useful or hurtful or indifferent characters as it may chance to happen; the fifth, change of the organism, through long continuance under the same conditions, will in like manner produce partly indifferent, partly useful characters; the sixth, sexual combination, likewise.

Even in the fifth case, therefore, there need not be necessarily any question of adaptation, and the usual expression "conservative adaptation" is only to be applied to the cases in which some advantageous character has been added to the organism through the constancy of relations. Accordingly, I have both in former works and in the present called the change which takes place on account of persistence of the same conditions constitutional impregnation, without regard to advantage or disadvantage.

Sexual Combination—One-sided Inheritance

I take in hand, first, the discussion of the sixth of the factors in modification just enumerated, on account of the exclusive importance which Weismann ascribes to sexual combination, assisted by selection, in the modification of organic forms in general, not merely in the separation of the forms into species.

My view, that sexual combination can lead to the production of quite new forms without the assistance of selection, is based upon the fact that from the union of two beings intermediate forms are not as a rule, as is generally assumed, produced, but very frequently forms differing from either of the two. For the characters of the parents and the ancestors either strengthen or cancel one another (to this, indeed, is due the harm of close breeding, the danger of marriages between relations), or possibly form in similar manner, either with one another or with characters of ancestors present in a latent condition, quite new combinations, as in the union of different chemical substances or elements.

But on the other hand, to avoid exaggeration of the effects of sexual combinations, it must be pointed out how often the offspring retain unaltered the characters either of the father or the mother, how often one-sided inheritance takes place.

Every nursery, especially in South Germany, where blonde and black hair together with blue or gray and black eyes are seen side by side, will show this. I mention, for simplicity's sake, hair and eyes only. But it is often obviously a question of a whole series of correlative characters, colour of the skin, strength of the bone structures, shape of head—in brief, of more Germanic or more Romance race.

Dark and fair parents together do not usually produce children which in colour are intermediate between them, but fair and dark again.[1] Only when the offspring continue this

[1] A remarkable confirmation of this is afforded by squirrels. Black and red young ones of the same litter are often found in a nest, but intermediate forms not so often. A similar assertion might be made of black and white sheep, although here artificial selection may contribute to the prevention of mixture; but I am assured by farmers that the colour of the wool is not so very important, because it is usually, at least in Germany, dyed of a dark colour. In any case, intermediate forms between black and white are very seldom found among sheep. I was particularly struck by this in Bulgaria, and in the Balkan peninsula generally, also in Italy, where one sees, besides the pure white, very many pure black sheep, but scarcely ever a mixed form in the pasturing herds. Such mixed forms are piebald. Landowners skilled in breeding horses tell me that from the union of black and white horses, as a rule, not intermediately coloured, or piebald foals result, but black or white again.

Very important in this question is of course the result of various interbreedings of races of men. Here also it would seem that a complete mean is not the rule. Thus Levaillant, 1793, speaks of the number of white slaves at the Cape of Good Hope, which were derived from the crossing between Dutch soldiers and slave negresses (mostly from Madagascar and Mozambique), and which in colour completely resembled Europeans. Likewise the hybrids between Hottentots and Europeans (in which case the Europeans mostly if not exclusively have furnished the male element) resemble the Europeans more than the Hottentots (male preponderance). Dr. Bernard Schwarz similarly speaks in his account

intermixture for a long time does an intermediate race gradually arise.

Dark, however, according to my observations, has a preponderance over fair.[1] When it is once there it is not easily eradicated from the blood; it necessarily has this preponderance simply on the ground that it is a positive quality as compared with the mere deficiency of pigment in the fair. Darkness of complexion, therefore, among us under the present conditions will always rather increase than diminish in its ratio to fairness.

The length of time for which the dark and the blonde type can constantly recur separately in the children of light and dark complexioned parents is particularly well shown by several south German villages, like those in the immediate neighbourhood of Tübingen, where a thoroughly dark, almost Romance, and a purely Germanic race of people appear often enough sharply distinguished in the children of one and the same family—notwithstanding the fact that in these little villages intermixture is continually taking place, for the people usually marry among one another, and seldom outside the village. The dark type will here too gradually become predominant, simply in consequence of the preponderance which it has over the blonde and blue.

Undoubtedly, in our temperate climate, sexual intermixture has been the essential cause of the extension of darkness of complexion, not the sun. But whether sexual selection gives a preference to darkness is at least doubtful—taste in this matter is very varied. It is possible that the intermixture

of his journey to the Cameroons of the white child of a Dutch agent and his black wife. I should be very grateful for additional accurate facts.

[1] I first expressed this opinion in 1881 (*Variiren*, etc.), and am rejoiced to find that M. A. de Candolle in his book, *Histoire des sciences et des savants depuis deux siècles, précédée et suivie d'autres études sur des sujets scientifiques, en particulier sur l'hérédité et la sélection*, published in 1885 (Genève-Bale, H. Georg), likewise defends the view at p. 81 with regard to the dark colour of the eyes. He promises to discuss the subject more minutely in a later work.

of dark and fair, that is to say, the formation therewith connected of other characters, has an advantage in the struggle for existence over the pure original Germanic blonde. In this last case only would the increase of the darker type appear as an adaptation. But I am of the opinion that the simple physiological preponderance of the dark colour principally determines the result; that the dark complexion in Germany is accordingly to be taken as an example of the principle that modifications resulting from sexual intermixture are not necessarily useful to the organism in which they appear, but may be indifferent.

Wallace, it is true, ascribes to pigment an importance in the production of greater acuteness of the senses.[1] Nevertheless, according to him, the blonde possess greater intelligence; they have, he believes, acquired this because in the struggle for existence, in consequence of their deficiency in acuteness of sense, they were driven to rely upon their intelligence. Perhaps it might be urged on another side in favour of adaptation as explaining the predominance of darkness of complexion, that intermixture of blood is within certain limits an advantage. In Germany, as a matter of fact, the purely Germanic blondes of the north show no less vigour in the struggle for existence than the mixed race of the south.

That darkness of hair and eyes is among us something newly ingrafted and is on the increase is proved by the universally known fact that the children of dark German parents are as a rule, in the earlier years of life, fair, and have blue or gray eyes: here also characters which were dominant in the ancestors are repeated in youth. This biogenetic fact struck me first and very forcibly in the Upper Engadine, where obviously an intermixture of blonde Germans with dark Romans has taken place, and where it is the more surprising because the climate in that region would

[1] Wallace, *Tropical Nature*, German translation, Braunschweig, Vieweg, 1879.

favour rather lightness of colour. For example, in the neighbourhood of Sils-Maria, in villages where the adults are all perfectly Romance in darkness of complexion, one sees children of such dark parents with quite fair flaxen hair and blue eyes. I shall subsequently consider more facts of the kind, especially from North Italy.

Accordingly, statistics intended to serve any really scientific end, in dealing with the relative numbers of fair and dark in Germany, would have to give, besides the age of the children enumerated, the colour of the hair and eyes of their brothers and sisters and parents, and, wherever possible, that of the grandparents.

It is also to be considered whether male preponderance does not also play a part in the extension of the dark type in our territory, for, in any case, it has been for the most part men who have first introduced it among us. We ought, then, to find, if this preponderance has been at work, that there are more dark men than women among adults. But this must (a further point of view for the statistician) have been the result according to laws already explained in any case, because the dark colour is for us a new character, and female individuals always remain nearer to the stage which corresponds to the condition of the young, that is, nearer to the original condition.[1]

Moreover, male preponderance has an important bearing on the questions before us from another point of view. It is

---

[1] [The author here states that supposing there are more men than women possessing the dark complexion in Germany, this effect might be due (1) to the fact that the type was introduced chiefly by men, not by women, from abroad ; (2) to the mere fact that the dark complexion is a new character, however produced, and therefore appears first in the adult males among Germans. The first supposed explanation is said to be a case of male preponderance ; but if we refer to p. 28, where the law of male preponderance is formulated, we shall see that it is the second possible explanation which exemplifies that law. In the statement of the law, p. 28, nothing whatever is said about the effect of introducing a few males of a somewhat new type among the members of a given race. Trans.]

certainly also the cause of the fact that we are able to recognise family likeness in so great a degree among the earliest male ancestors of a family, as they appear in many a gallery of family portraits ; and further, it also explains why the male line of descent is considered of such great importance, while the female is scarcely regarded. If this prepotency of the male did not exist, if the female element in all unions had exactly the same value as the male, then, after comparatively few generations, by the combination of equally potent male and female types, all similarity to ancestors, excluding cases of one-sided reversion, would be completely effaced.

A very remarkable example of male prepotency in this connection is afforded by the large under lip of the Habsburgers. Portraits of Rudolf I. of Habsburg already show it. It was inherited by his descendants up till the last of them, the Emperor Charles VI. (1740), that is, for 500 years. With Charles VI. the male line of the Habsburgers became extinct. Its place was supplied by the Thuringian line, derived from the marriage of Francis of Thuringia with the Habsburger Maria Theresia. In the male descendants of this couple the great under lip appeared again, and has been transmitted up to the present day, although the wives of the Habsburgers, coming from various families, could not possibly by chance have generally possessed it, and did not possess it.[1]

At all events, the fact of male prepotency in itself shows that a complete mingling of characters is no more the usual consequence of sexual union than the production of hermaphrodites that of the coalescence of sperm and ovum in animals of separate sexes. Atavism is no greater miracle than these facts which must be explained with it.

Another factor contributing to the separation of forms into species without selection, to be added to those already mentioned

---

[1] Cf. *Pinacotheca principum Austriae*, by Marquard Hergott and R. Heer (Benedictines in St. Blasien), Freiburg, 1770.

as connected with sexual reproduction, is that any new characters may produce such a correlative change in the composition, or even in the form of the sexual products, that sexual crossing between the forms concerned is no longer possible.[1]

Often, for example, species quite nearly related show surprising differences in their spermatozoa (Rana temporaria and esculenta).

I have already, years ago, pointed out how inadmissible it is for this reason to base the argument that species are to be considered as independent and isolated from the first, as groups created in their present condition, on the fact that true species are not fertile when crossed with one another. On the contrary, the impossibility, or at least the difficulty, of fertile union, and therewith the separation into species, can be brought about or favoured by any change indirectly or correlatively produced in the sexual products, *e.g.* the spermatozoa, affecting their morphological character, their motion, and material composition.

The sperm-nucleus and the egg-nucleus must in their physico-chemical properties form with perfect exactness the complements of one another, if through their union a new organism is to be produced.[2] The form and motion of the

[1] Compare on this the remarks in my publication, *Ueber den Bau und die Bewegung der Samenfaden* in *Verhandlungen der physikalisch-medicinischen Gesellschaft zu Würzburg*, N. Folge, Bd. vi., and Würzburg, Stahel, 1874.

[2] I must refer in this connection to a conception to which much importance has been attached as a support of Weismann's theory of heredity. The latest researches on fertilisation having shown that sperm-nucleus and egg-nucleus (= germinal vesicle) do not dissolve in the process, that it is effected not by liquid but by solid substance, therefore the inheritance of acquired characters has been declared impossible, and fertilisation a morphological, not a physico-chemical, combination. On this view we have before us, to consider more fully what we have previously touched upon, an unchangeable, never wearing out, eternally living organic substance (changeable only by sexual combination or by disease), which cannot even be nourished like other parts of the body, for if it were it would necessarily be affected by the constitution of the body. And whence did it originally obtain its peculiar properties ? How and whence has it come to be ? Before we are obliged to ask ourselves these riddles, and before we

spermatozoon must be adapted with perfect accuracy to the morphological peculiarities of the egg-envelope if there are to be no morphological hindrances, even the minutest, to fertilisation. For the formation of new species in this way it is therefore only necessary that the sperms and the eggs of a given number of individuals should have peculiarities of structure corresponding with one another, and not corresponding with those of others. Complete isolation of the varying individuals is thus on these grounds also not absolutely necessary for the formation of new species.[1]

The correlative alteration of the sexual products leads to the consideration of a fact which seems to me to cry loudly against Weismann's assumption that those products remain essentially unaffected by the condition of the body from time to time. I refer to the great correlative influence which conversely the condition of the sexual products, namely, their maturity (puberty) and their artificial removal (castration), as well as the extinction of their powers in old age, has upon the

go so far as to doubt that the character of the process of fertilisation is that of a physico-chemical, *i.e.* physiological, process, let us rather, I say, leave the problem of atavism for the present unsolved. What if the solid nuclear substance had the power to reproduce the condition of the whole body at any given moment because it was definitely and peculiarly adapted to contain an extract, as it were, of the whole body? Who will undertake to prove that it does not take up such an extract somewhat like a delicate sponge, so that the inheritance of acquired characters, as in the older view, can be physiologically explained in spite of its firm structure? In any case, such a hypothesis is not in contradiction to general physiological principles, as the conception of fertilisation as a purely morphological process seems to me to be.

Weismann, moreover, speaks in his more recent paper (*Bedentung der sexuellen Fortpflanzung*) of an increase of the germ-plasm by assimilation ; but such assimilation merely on the ground of the laws of physiology obviously implies that the germ-plasm is influenced by the general nourishment—in other words, by the condition of the body.

[1] Apart from this I have shown in reference to this question in a very carefully-studied example—in the wall-lizard—that varieties differing to any important degree fight with one another when they live together, as if there was profound antipathy between them ; that, therefore, they do not sexually mingle, while the similar individuals breed together, and so establish their characters more firmly.

condition of the whole body. This relation is so well known that it is unnecessary for me here to give any proof of it. But it is at the same time one of the most striking examples of correlation, and certainly affords most valuable support to my conclusion that changes in the characters of the body appearing suddenly through correlative growth may lead to the formation of new species without the assistance of selection.

Finally, some importance is to be attached, as supporting my conclusions in opposition to Weismann's, to the evidence which tends to prove that not only the bodily, but the mental, condition of the parents at the moment of procreation has an influence on the offspring. This evidence is so abundant and comes from so many sides, that it will be difficult to doubt at least its partial trustworthiness.[1] But if it is really based on fact, a further important proof is afforded of the inheritance of acquired characters, and of course also of the part played by sexual combination in the alteration of the characters of living beings. But it is sufficiently evident from the foregoing that I place a high value on the latter influence. The divergence of my interpretation from that of Weismann consists only in this, that the latter pronounces natural selection to be the indispensable handmaid of sexual combination in the transformation of species; whilst I, although recognising in the present work and always the importance of selection,

---

[1] In this connection may be mentioned the fact that, according to the belief of breeders, a pure-bred mare or bitch if only once covered by a worthless male is spoilt for all subsequent breeding—indeed, it is asserted that such females subsequently, even when they were covered by thorough-bred males, produced young which possessed characters of the mongrel with whom they had once bred. (Probably this is essentially a case of nervous influence [trophic nerves].) Compare my later observations on the influence of the age of the parents on the offspring. It is generally asserted that procreation during drunkenness causes inferiority of mental powers or even idiocy in the child—in this case general influences of nourishment are the probable explanation. Compare with respect to the importance for the offspring of the condition of the parents during procreation, A. de Candolle, *op. cit.* p. 49, *et seq.*, and the literature there referred to.

regard it as by no means absolutely necessary to that transformation.

Lastly, although I ascribe to sexual combination, with and without the help of selection, an important influence in the moulding of species, yet it cannot be granted that this influence is essentially paramount, or even that it is thoroughly independent in its action, if we start from the assumption that sexual differentiation is itself at first an acquired character.

In reality its own beginnings are to be traced to external influences, and obviously it continues to serve as the means for the transmission and distribution of other characters acquired by the body in different localities under different conditions. In that it unites, combines different elements, it can engender a new third element (crossing). Without this mixture of what is foreign—different, however, through the continued union of the cognate, it leads ultimately on the contrary to equalisation, not to progress.

Thus sexual combination, like the principle of utility, can only work with the material which phyletic growth offers to it; universally something new must first exist before either can commence to operate.

I years ago expressed my belief that the original directions of evolution are shown by the facts to have been followed with an extraordinary persistence; let the external influences be what they will, the new directions never seem to radiate from the starting-point of modification—the new lines pass off at an acute angle, at first near the original ones—only the intermittent deviation caused by correlation forms an exception. But the fact of evolution in a definite direction is explained neither by selection nor by sexual combination by itself.

I proceed now to the closer consideration of one of the most important of the causes to which I have attributed the separation of species, to the just-mentioned action of correlation, to what I have called intermittent evolution.

Intermittent Evolution—Kölliker's Hypothesis

It appears, therefore, according to my views, to be possible, that since one character is followed by one or perhaps several others correlated with it, a new form may suddenly arise, which, if the variation extends to the sexual products, may be no longer capable of crossing with the original form. At the same time, this latter condition is not necessary for the production of a new species. For example, hair, horns, and hoofs have been proved to be connected correlatively with one another, and by simultaneous variations in these three organ-systems a very considerable change in the external appearance of an animal may be effected. If only one of the altered systems is more useful to the animal than that of the original form, then, notwithstanding the possibility of crossing, a new species may arise.

There are of course few examples among living animals which illustrate this mode of origin in so complete a manner as the Axolotl. But if this were the only one it would suffice, and would justify the conclusion that the same metamorphosis has under similar conditions probably occurred in other allied animals. The metamorphosis of the Axolotl into Amblystoma is clearly the consequence of the transition of the animal from an aquatic to a terrestrial existence. With the cessation of branchial and the commencement of exclusively pulmonary respiration went correlatively hand in hand the appearance of a spotted marking on the skin, of the smoothness of the latter in place of the warty character, especially on the head, of that of the Axolotl, the slight change in the form of the body, and lastly the formation of a more cylindrical tail, while also, in consequence of terrestrial habits, the legs became more powerful. It is highly remarkable that the same peculiarities distinguish Salamandra maculata in comparison with its aquatic larva. In fact, it is scarcely possible to distinguish a Salamander larva so long as it retains its gills from a young Siredon. Both have the gray colour without large yellow-spots, the laterally compressed swimming tail, and the feeble legs; and lastly, in comparison with the adult Salamander or Amblystoma, respectively, a somewhat flatter, longer, and narrower head, with less prominent eyes.

Siredon is for the problem of evolution one of the most important of living animals, in that it brings so beautifully before our eyes the transition of a lower sexually mature into a higher sexually mature form, and at the same time shows so clearly the causes of the transformation. We discern these causes simply in the reaction of the organism under external conditions, the increased exertion of an organ already in process of formation (the lung), and the disappearance of another (gills) in consequence of definite demands of the environment, and changes connected with these by correlation.[1]

Amblystoma appears where the Axolotl has too little water to live in, where it is compelled to live on dry land. That this is the case is proved by the possibility of artificially rearing Amblystoma from the Axololt by gradually withdrawing the water. Before this can be possible it must of course be a previous condition that the Axolotl has already a great tendency to transform itself into a land animal. This may be due to the possible fact that it has in its aquatic state already made much use of its lungs; but the transformation which follows its assumption of terrestrial habits is none the less sudden, and evidently connected with correlative changes. When an Amblystoma is formed, then this is the adult animal,

[1] These are the essential causes; I do not pretend to have exhausted the explanation of the metamorphosis.

the Axolotl its larva. If it is not formed, the latter is the adult animal capable of sexual reproduction.

Since, then, in the spotted Salamander and its allies exactly the same metamorphosis in structure occurs in the passage from larval life to the adult condition as in the transition from the Axolotl to the Amblystoma, we have in these amphibia again clearly revealed definite directions of evolution.

Weismann has offered another explanation of the development of Amblystoma out of Siredon, namely, that it is a reversion to an earlier form. But no cogent reason for this hypothesis seems to me to have been given, any more than there is any reason to suppose that the adult Salamander is to be regarded as a reversion to an earlier form. What is true in the one case is reasonable in the other. The only difference I can see in the two cases, besides the sexual immaturity of the Salamander larva, is that in the spotted Salamander and its nearest allies the metamorphosis into an exclusively air-breathing animal has become the rule, while in Siredon it is only commencing, and occurs at first in rare individual cases.

If this be so, then, as I have said, the metamorphosis of Siredon pisciformis into Amblystoma is evidently the most perfect example of the occurrence of the evolution *per saltum* of a species of animal now living, in consequence of the action of external conditions.

My view of the meaning of the metamorphosis of the Axolotl is supported by the experiments of Fraülein von Chauvin. That lady reared larvæ which were the progeny of Amblystoma. These larvæ were kept under conditions in which larvæ begotten by the Axolotl would never have changed into Amblystoma. The larvæ thrived well, and their gills attained in abundantly aerated water an unusually great development. In spite of this they came often to the surface of the water to get air, and remained there for hours—a thing which in an Axolotl usually occurs only at a more advanced age, and in water but slightly aerated. When the larvæ were one year old, the weather continuing warm for some time, a reduction of their gills commenced. Fraülein von Chauvin gave twenty of the larvæ an opportunity to quit the water, and to her surprise some of them immediately crept out into the moss. The metamorphosis began in these after only a few days. One of the larvæ completed its change in ten days, and after twenty-three days all the twenty larvæ had of their own accord quitted the water.

Six other Amblystoma larvæ of the same brood were kept from the beginning in very cool rapidly-flowing water, and their tendency to breathe air was thus checked, and these remained in the water.[1]

The Amblystoma larvæ had therefore inherited the characters acquired by their parents, and the experiments justify the conclusion that in consequence of this inheritance, after a longer duration of the external conditions, the new form Amblystoma would become the permanent species as truly as the perfect spotted Salamander is the species, and not its gill-breathing larva.

It is supposed that the North American genera Menobranchus and Menopoma stand in the same relation to one another as Siredon and Amblystoma.

Further, I may refer to the transformation, shortly to be mentioned, of the crustacean Artemia salina into another species or another genus through the increase or decrease of the saltness of the water in which it lives. Here also a great deal is due to correlative modification.

On a previous occasion I have endeavoured to describe the physiological causes of this fact of correlation which plays so important a part in the modification of forms, and therewith the ultimate causes of evolution *per saltum*, in the following

[1] M. v. Chauvin in *Zeitschr. für wissensch. Zoologie*, xli., p. 385, 1886.

words: "As soon as something or other in the original state, in the original arrangement of the parts of the organism, is changed, other parts also are set in motion, all arranges itself into a new whole, becomes—or forms—a new species,"—just "as in a kaleidoscope, as soon as on turning it one particle falls, the others also are disturbed and arrange themselves in a new figure—as it were recrystallise."[1]

And these considerations, I added, also throw light on the question of the absence of intermediate forms: in the formation of new species the animals have by no means necessarily passed through all conceivable intermediate stages.

These sentences also imply the assertion that through correlation, in other words, through *per saltum* evolution, new species can arise without the aid of selection, however much the latter may facilitate the formation of the first new character, as indeed was probably the case in the Axolotl.

I need scarcely point out that this sudden evolution for which I am arguing has nothing to do with the hypothesis of sudden evolution brought forward by Kölliker.

Kölliker, who has repeatedly argued in opposition to the utility principle, to pure Darwinism, contends for the evolution of forms from "internal causes" on the basis of a "general law of evolution." In this evolution the ova even of the higher and highest of now living animals play a special part as the "original organisms": it is supposed that the ova, or the germ-cells of a given form, in consequence of an altered mode of development due to internal causes, could give rise to new forms. If the newly produced forms are widely separated they belong to a new genus, family, or order; if they are less different from one another, they are related to one another as varieties and species.

Further, according to Kölliker, there is reason to consider

[1] *Variiren der Mauereidechse.*

whether new forms could not be generated by internal germs or external buds. In support of this the phenomena of alternation of generations are adduced.

Thirdly, it would have to be considered whether not only germs and buds, but free-living young forms of animals are not capable of entering upon a development other than the typical.

Lastly, the possibility of a "rapid modification of adult organisms into others" is considered.

In all these cases we should have a sudden evolution, "yet this is essentially to be traced to the embryonic period, and, indeed, to its first stages."

In addition, a more gradual modification in a less degree is recognised as possible, and some effect is ascribed to it, and this likewise is chiefly to fall within the embryonic period.

We should have, therefore, I said in my "Variation of the Wall-Lizard," "an upward evolution towards higher forms, from internal causes. Original organisms—ova—must, according to this, have developed all at once into higher forms, *e.g.* into mammals, and they would have succeeded in this just the same wherever they happened to be, no matter what the external conditions in which they lived or were about to live. Adaptation has nothing to do with the question—whether a part newly arising is useful or harmful is of no consequence; it is as if the whole course which the original organism had to pass through was definitely pre-ordained for it by internal causes, and as if this plan (the words plan of evolution are repeatedly used, and used as if equivalent to 'general law of nature') corresponded from the beginning to the external conditions."

Kölliker also explained, although not in his first paper on this subject, yet later, that his internal causes were physico-chemical.

The facts on which my conception of the modification of species is based show, in harmony with those brought to light by Würtenberger in the field of palæontology, that the last, the highest stages in the development of animals govern the modification of the species. The opposite view advocated by Kölliker is no more supported by facts than his other propositions—it rests only on assumptions, on the conceivable and the possible. Moreover, Kölliker, as already remarked, opposes Darwinism, the utility principle, altogether.[1]

The rest of the agencies previously mentioned as causes of the division of the organic world into species will be examined in subsequent sections. Here I have only to make the following preliminary remarks on

CONSTITUTIONAL IMPREGNATION

(Constitutional Adaptation)

The proposition that a character must establish itself more firmly the longer it remains in an organism exposed to the same conditions, in other words, the longer it is maintained by continually repeated inheritance, scarcely requires any further proof than is given by general physiological considerations. As I have elsewhere expressed it :[2] "If a form remains stationary at a low phyletic stage, then, from purely constitutional causes, the longer it remains at that stage the more does it become different, because its characters stamp themselves more and more deeply on the organism (constitutional impregnation). It will, therefore, after a certain time no longer be the same as it was when its relatives diverged from it. The longer it is able to exist with these characters the more it will change in another way while its relatives change by correlation, but the more also

---

[1] Cf. Kölliker, *Ueber die Darwinsche Schöpfungstheorie, Zeitschr. f. w. Zool.*, Bd. xiv. Also *Morphologie und Entwicklungsgeschichte des Pennatulidenstammes, nebst allgemeinen Betrachtungen zur Descendenzlehre*, Frankfurt (Senkenbergsche Abhandlungen), 1872. *Entwicklungsgeschichte*, Second Edition, 1879, pp. 6, 27.   [2] *Variiren*, etc.

will it be in a condition to maintain the persistent characters in face of the coercive powers of adaptation, and these latter will be thrown on to other characters with greater modifying effect."

Elaboration and Simplification in the Evolution of Species

I come now to the second of the questions previously formulated, to the question, On what grounds each species high in the scale of evolution was able to surpass the one next below it—what are the causes of increased perfection?

This advance in structure in the organic world is most clearly revealed by the examination of the development of some one individual organism—so long as it be not a degenerated form. The individual in this development passes by growth through the series of its ancestors, but it grows a stage farther on to the condition which constitutes its individuality, which distinguishes it from the next lower stage.

Nägeli calls in the aid of a "principle of improvement."

I assume with him that the conditions for a progress towards the more complex and towards division of labour exist in the fact that a higher stage once reached can afford a foundation for one still higher, since the former, the existing stage, will necessarily be the starting-point for further modification.

Nevertheless, my view has certainly nothing to do with a "principle of increasing perfection" in Nägeli's sense. I confess that with respect to the criticism of this foundation of Nägeli's theory I count myself among those whom its author describes as the "less far-sighted," at least in so far as I consider the assumption of such a principle acting in accordance with organic persistence of motion as merely an assumption. Although he explains it as a mechanico-physiological principle, I hold it to be a kind of striving towards a goal, or teleology, in face of which the recognition of a directing power conceived as personal, existing outside material nature and ruling all things, would seem to me fully justified.[1]

The formation of species is not completely comparable to the sprouting of the twigs of a tree. It depends not only on increased perfection, but just as much on deterioration and simplification, and on retrogression in complexity and in division of labour, *i.e.* upon growth, with all the changes which growth can undergo through hindrance and through the modifications which are imposed upon it.

The abundance of the species which have been formed by degeneration, by retrogression, is known to every zoologist. It is self-evident that their origin is to be traced to the action of external conditions, that they have been produced by acquired and inherited changes of growth depending on these conditions.

My own extensive researches on the variation of animals,

---

[1] Here I may be allowed also a remark *pro domo*. Nägeli in the introduction to his book speaks very severely of those who without any justification undertake to express opinions upon the question of the origin and the evolution of organisms. He claims this right exclusively for those who are physiologists by profession, and counts among the non-physiologists both Darwin and Haeckel. Against such a close corporation I protest. Investigators who have enjoyed a thorough physiological training, and whose ideas must in consequence of this training have a physiological basis—and among such I would modestly request to be numbered—are surely not to be excluded. But, as every one knows, outsiders have at times larger ideas than the members of a corporation. Besides, we might in accordance with Nägeli's argument conversely inquire whether the laws of vegetable physiology and the facts at the disposal of this limited branch of science are sufficient to justify the treatment of a matter which concerns the whole morphology and physiology of the animal kingdom? At all events, the animal kingdom, on account of its greater complexity and the active functions of its organs, affords a wide field for the examination of questions which never come to the notice of the botanist at all. For this reason, the impulse towards a more scientific doctrine of evolution has come from zoologists, or at least, from men zoologically trained, not exclusively from botanists.

especially with regard to colours and markings, only the smallest part of which are published, show, however, in the most distinct way, that the directions, not only of variation, due, for example, to parasitism, but, as I may express it, of free variation, in very many cases tend towards simplification and not towards greater complexity. How far it is true in such cases that the partial simplification is connected with simultaneous advance in evolution in another respect, and how much is to be attributed to adaptation, must be ascertained for each individual case; in my examples these questions have been considered. The facts already published by me on the markings of animals, the transformation of longitudinal striping into spots, then into transverse striping, and the final disappearance of the marking, are sufficient to contradict Nägeli's principle of increased perfection.

But my inquiries on the markings of butterflies will furnish still more striking examples of the fact that variations tending to simplicity as well as those to greater complexity take place with wonderful persistence in direction, as if following a path previously ordained. Such variation is often in the same definite direction in species nearly related but completely separated in habitat or even belonging to quite different faunæ, so that sometimes external conditions appear to be undoubtedly the essential controlling factor. Moreover, in many cases we have simplification alone, uncompensated by greater complexity in another part. Thus these examples also show once more that neither the inferior *rôle* ascribed to adaptation by Nägeli nor the commanding one given by Weismann are justified.

The process of evolution is not, as Nägeli believes, to be compared to an eternally growing tree, on which adaptation acts only by cutting branches away and by pruning the whole. Adaptation has also, if only indirectly, a perfecting and strengthening—in any case, a modifying effect. On the tree many shoots are stunted and many are metamorphosed in manifold ways, but others die away. I have on occasion used another simile for evolution :—

"Thus we may," I said, " compare the whole process of the modification of forms to the results of the migration of a people over an extensive foreign territory. Some tribes, not having the strength to follow, soon, others later, remain behind (genepistasis), others again reach a distant goal. Some retain their characters in their new home or strengthen them, even modify them by correlation, others change under the influence of external conditions and adapt themselves to the environment—all that is not sufficiently capable of endurance is left lying by the way and perishes, and if the struggle for existence is at all severe only the toughest of all survive. The sooner the connection between the separate tribes is lost, the sooner each appears as a new species, as a new genus ; but all bear the stamp of common descent."

I willingly confess, however, that this simile, like all similes, is incomplete.

I find, then, the ultimate and the most essential causes of progressive evolution in all the causes of growth in general—therefore in all the influences of the external world upon organisms.

It has been said that it is impossible to comprehend how the varying external conditions should produce a continual advance in evolution (Weismann). In answer to which the first proposition to be emphasised is, that each higher stage of evolution attained is a firmly established condition,—by the previous argument, all the more firmly established, fastening the more upon the new form, the longer it continues; a condition which, therefore, the longer it exists is with the greater difficulty disestablished, but which merely through the continual repetition of stimuli, although these have no unusual strength, will undergo a change in the course of long periods in the direction of more complex (higher) organisation, especially with the aiding action of selection. But such higher development will be in a greater degree experienced when an unusual increase of stimulation occurs —in both cases as a consequence of modification, that is, intensification of growth.

For the action of continuous unvarying stimuli the influence of the use of an organ may serve as an example; further, the effect of cold and other climatic conditions which produce hardiness. For the action of increased stimuli, the influence of warmth on plants as it manifests itself in several of their characters in various warm regions of the earth.

Similarly, permanent cessation or diminution of the stimuli will be followed by simplification of organisation as a result of change of growth.

Whenever the power of growth, be it through persistence of characters, through retrogression, or through progress, has undergone a change in any species under particular external conditions, then the reapplication of a stimulus—even if it be that which was formerly the most usual—will have a new ground for action, will excite a peculiar kind of growth, and lead to new structure.

We have, however, to answer the question, Why male animals for the most part are in advance of the females in organisation, why, again, the older males first possess the new characters, and why this new formation of characters (advance in organisation) takes place on the body in a constant direction, for the most part from behind forwards? And since the changes in markings consist principally in the conversion of a longitudinal striping into spots, and these into transverse stripes, we have to inquire what are the causes of this conversion. What are in general the causes of the modification of the marking in definite directions in the intermediate stages between longitudinal striping and spotting and transverse striping?

I have permitted myself to express the supposition [1] "that the fact of the original prevalence of longitudinal striping might be connected with the original predominance of the monocotyledonous plants, whose linear organs and linear shadows would have corresponded with the linear stripes of the animals; and further, that the conversion of the striping into a spot-marking might be connected with the development of a vegetation which cast spotted shadows.—It is a fact that several indications exist that in earlier periods the animal kingdom contained many more striped forms than is the case to-day."—The American fauna exhibits in many respects a lower stage of evolution in the markings of its members than ours; it gives, therefore, at the present day an idea of the condition of the ancient world referred to. This supposition of mine is also supported somewhat by the fact "that at present strongly spotted forms mostly occur in places with spotted shadows, the longitudinally striped more in grassy regions. Young slugs and caterpillars are often longitudinally striped. Cross-marking is perhaps to be connected with the shadows, for example, of the branches of woody plants,—thus the marking of the wild cat escapes notice among the branches of trees." Others have previously suggested such relations as that of the cross-striping of the tiger with the shadows of the bamboos in which it lives.

Weismann also has sought in the different markings of the caterpillars of Sphingidæ adaptations to their environment. And while he provisionally leaves out of consideration the ultimate causes of the origin of any of these markings he says, " Even the first commencement of striping must have been useful, for it divided at once, to the eye of the observer,

[1] *Variiren*, etc.

the large conspicuous surface of the caterpillar's body into several areas, and so made it less conspicuous."

In fact, one has only to look at such a light green caterpillar with long stripes as it rests lengthwise on a grass-stalk or a pine-needle.

Weismann points out that all caterpillars of the Sphingidæ, in which the longitudinal striping is at the present time the permanent marking, live either among grasses or on conifers.

Oblique striping is regarded by him as adapted to the veins of leaves, the circular or ocellus marking as imitating parts of the plants fed upon, berries, etc., or as means of frightening enemies, or as enabling the possessor to be recognised as unpalatable.

Preponderance of the males, I myself suggested further, might be possibly explained by the fact that the males fight the battle of existence more than the females, and therefore must always be first to respond to new demands. The postero-anterior evolution by the fact that the part of the body farthest from the head is most in need of mimicry, because it is least protected in other ways by the sense-organs, and because it is at a special disadvantage; that it is the last part to be withdrawn from the pursuit of an enemy (some instances of the opposite condition in caterpillars I endeavoured to trace to special adaptations). The fact that in mammals the new marking appears first on the sides was comprehensible, I thought, because the sides were most exposed to the view of an enemy, certainly more than the marking along the back, which always remains at a lower stage.

"The preponderance of age," I said, "is due, in the first place, to the fact that those individuals which are most adapted to the environment as a rule also live longest, and have the longest time to transmit their peculiarities.

"Since, moreover, the individuals provided with the new

marking live longest—in other words, bear the new marking for the longest time, this becomes, as it were, most firmly inoculated in the organism, and is therefore with greater certainty transmitted to the offspring. The longer it is borne by the possessors the more completely from constitutional causes will it be inherited, and as the duration of its existence in the posssessor will be proportional to its usefulness, the degree of its heredity will also be proportional to its usefulness, so that an important share must be ascribed in the process to the conservative adaptation which depends on constitutional causes."

But all this does not explain the first occurrence of the new characters, nor the undeviating course of the evolution in a particular direction.

For when a number of varying individuals are compared it is seen that the variations of all tend to a definite end, and that the majority of the intermediate forms show stages in the development of the characters which are absolutely without use to them.

This cannot be explained except by natural growth, whose operations are changed, intensified, or diminished to a certain extent by the stress of adaptation, and may also at times be entirely restrained.

That this growth proceeds in definite directions and begins at definite spots is not more wonderful than the same processes in individual growth, which are of course more evident in plants than in animals; in the latter, on account of the more active life, they are more altered and obscured by adaptation than in the former.

Some important ultimate causes of the direction of growth must, however, be sought in the general physiological mechanisms of the body—for example, in the distribution of the blood, etc. In particular, the tendency to symmetrical, and even to metameric development in the marking, for instance, depends obviously on such mechanisms.

That the male outstrips the female in phyletic growth is comprehensible apart from the importance of adaptation in this process, from the greater strength developed in the former; and that the growth of new characters shows itself first in the later period of life, in the period of most complete development, at the end of the ordinary and general development, is comprehensible, again apart from the importance of adaptation, merely from my theory of growth.

This view of mine is, however, further supported by the fact that an essential cause of male preponderance consists in the appearance of characters during the breeding period, at the time of maximum vigour, many of which are subsequently transmitted by heredity first as constant male characters, and finally to the whole race. These characters include not merely colour and strongly developed markings,—a long series of male characters could be mentioned which were evidently only present originally as such sexual distinctions in the male. And we know many instances in which such characters are intensified to a surprising degree in the males at the breeding period, as, for instance, the dorsal crest of many water newts (Tritonidæ), the hook-shaped prolongation of the lower jaw in salmon and trout. In this sense it is an interesting fact that the latter formation becomes permanent in old male salmon, so that such hook-jawed specimens have been considered as forming an independent species.

The facts referred to in the preceding: (1) the definite directions of evolution; (2) the appearance of new characters at definite regions of the body, and their progress in an unvarying direction during individual life; (3) the symmetrical or in some cases metameric occurrence of characters; (4) the first occurrence of characters at a late period of life, as it were at the end of the stage of evolution previously attained; (5) their first occurrence at the time of the maximum development of vigour—these facts I might point to as not merely in complete harmony with my theory of growth but as very clear evidence of its truth. They may be compared with processes which can be recognised in the, so to speak, more transparent individual growth of any plant. Such comparison holds especially for the 2d and 4th of these facts; plants grow from the apex; and from the terminal point of the evolution which has previously taken place, that is, of the structure already formed, from the so-called vegetating point, new characters arise.

Thus the proper permanent leaves arise after the appearance of the seed-leaves; after the foliage-leaves, the flowers, with their parts formed of modified leaves, and among the foliage-leaves, again, sometimes various successive forms or even colours.

As in plants, the new characters show themselves also in the shells of Ammonites at the youngest part, as it were at the growing point. For in these shells, as in snail shells, the parts near the aperture are the newest. The growth of these shells takes place, as we may observe still in the shell of the Nautilus, in an intermittent manner; it becomes more vigorous at each period of maximum vigour in the animal, which probably coincides with the most favourable season of the year, again to decrease and come to a standstill, just as the growth whose results are seen so beautifully in the shells of our vineyard snails, and the annual rings of any tree-trunk. That this periodical growth results in the formation of characters appearing successively more strongly or more feebly is not to be wondered at. A high degree of symmetry or metamerism may be produced by it.

These considerations seem to me also to afford an explanation of the origin of the symmetrically, *i.e.* metamerically arranged characters of segmented animals, and to enable them to be included in the same class of phenomena. That in such segmented animals each somite should have the same characters ought not to surprise us. These somites have arisen by the subdivision of a whole, and the first has the same relation to the second as the second to the third, and so on. Now, since in lizards, as in caterpillars (*vid.* subsequent sections), and also in mammals and birds, new peculiarities of marking first arise on the hinder part of the body and extend thence forwards, the question at once occurs whether this fact harmonises with the explanation given for plants and Ammonites, *i.e.* whether the hinder part in these animals is to be regarded as the youngest. It is to be noted that a vertebrate is segmented, and that there is much to be said for the supposition that the vertebrates were once forms similar to segmented worms; on various grounds they are regarded by many as descended from the latter. The Arthropoda, including caterpillars, have similar relations to the segmented worms (Annelids). But in the latter the most anterior segments obviously as a matter of fact are composed of the oldest tissues. This is proved by the fact of ontogeny: so far as special attention has hitherto been paid to the commencement and succession of segmentation in Annelid larvæ, it has shown that the process begins in the anterior part of the body and is continued thence backwards; the worm must therefore grow at the posterior end. Exactly the same thing is observed in our fresh-water worm, Nais probescidea. In the young worms produced by division the new head always consists of the oldest material—growth takes place at the posterior end.

At present, my only intention in referring to this subject is to indicate a definite point of view from which to investigate a question which requires a work to itself. Moreover, I am not at present in a position to explain in a similar way the infero-superior modification of the marking which takes place in many mammals.

SECTION III

INFLUENCE OF ADAPTATION IN THE FORMATION OF SPECIES

Is everything adapted?

FROM the preceding considerations it appears to me, therefore, that leaving aside Kölliker's hypothesis, neither Nägeli's view, which ascribes to the principle of utility an almost infinitesimal effect, nor Weismann's, which regards adaptation as all-powerful, can be unreservedly accepted. The truth lies between them. Weismann in his latest paper explains everything as adaptive. As an example of such complete adaptation he describes the whale.

He alludes to Nägeli in his remark that he can perfectly understand that it is more natural to a botanist than to a zoologist to take refuge in internal forces of evolution. "The relations of form to function, the adaptation of the organism to the internal and external conditions of life, are less conspicuous in plants, less easily observed, indeed—are often only to be discovered by the most careful and acute investigation. The temptation is therefore greater to regard everything as due to causes acting from within." And he says further: "In any case, animal biology cannot point out too emphatically how exactly and to the minutest detail form and function go together, how completely adaptation to definite conditions of life prevails in the animal body. There nothing is indifferent, nothing could be otherwise than it is; every organ, nay, every cell and every part of a cell, is as it were tuned to the part which it has to play in relation to the outer world. True, we are not able to demonstrate all these adaptations in any one species, but wherever we succeed in fathoming the significance of a structural feature, it always turns out to be another adaptation, and any one who has ever attempted to study profoundly the structure of any one species, and to give an account to himself of the relation of its parts to the function of the whole, will be much inclined to say with me, everything depends on adaptation." Of course Weismann adds: "These are convictions, I grant it, not absolutely proved facts."

"If then," he continues, "the organism consists only of adaptations based upon the constitution of its ancestors, it is impossible to see what remains for any phyletic force to do, even if it be conceived in the refined form of Nägeli's idioplasm.".

Weismann then describes the whale in order to show the completeness of adaptation, and concludes: "And now I repeat my former question with regard to this special case, If all that is characteristic in the animal depends on adaptation, what is there left to represent the action of an internal force of evolution?"

I also repudiate any special internal force of evolution. According to my view, everything in evolution is due to perfectly natural processes, to material, physical causes.

I also have investigated the phenomena in question in single forms of animals; but I have arrived at a conclusion which does not agree with Weismann's—" that everything depends on adaptation."

But when Weismann says: Even where we cannot recognise the purpose of a character at present, it. will ultimately prove to be adaptive, then I feel justified in reversing his assertion with respect to his own example, and saying: We do not know the numerous relations and conditions of the life of the whale with anything like sufficient accuracy to enable us to conclude that it is so perfectly adapted to them as Weismann believes. Our own conditions and relations of life we do know accurately, and although it is a matter of taste with us whether we think our adaptation greater or less, it must surely be conceded that in any case this adaptation cannot be called perfect, even in the prime of life, not to speak of childhood and gray old age.

What would be the total, to take only one instance, if we were to add up the number of men in vigorous life who perish miserably every year simply in consequence of the entrance of fruit-stones or similar bodies into the useless vermiform appendix of the cæcum?

Besides, it is self-evident that adaptation must be more perfect in highly-organised forms much exposed to the dangers of life than in many of a lower order; in those which have and had more enemies much greater than in others which have fewer, and whose ancestors had fewer. And I do not dispute the fact that there are forms which seem to be adapted to external conditions in as high a degree as Weismann supposes. But that only proves that those particular forms are possibly so adapted. But with respect to all forms which vary in a high degree, the assumption of perfect adaptation is *à priori* improbable. And there are forms of which we can say with all the required certainty, that their bodily shape must depend not on adaptation, but on a "crystallisation," resulting from the physical and chemical action of external agents on the material of the organism subjected to them.

In my essay on "The Variation of the Wall-Lizard" I expressed, with reference to adaptation, the view that the individual is not necessarily constructed entirely in accordance with its own requirements: "Only an inconceivably gross egoism could seriously make this assumption——the individual is nothing but a wheel in the clock-work of the universe; to the universe its characters must be subordinated, must be adapted; for the wheel, for the individual itself, only such a proportion of advantage is possible as the order of the whole allows, as in this order belongs to it." It is certainly no adaptation in organic beings that they cannot live without nourishment; and to the lamb devoured by the wolf the deficiency of adaptation is particularly palpable, as it is to us in death itself, our gradual extinction just when the highest degree of knowledge and experience has been obtained. But both hunger and death are conditioned by the substance of which we are composed, and this again is conditioned by the circulation between organic and inorganic nature.

"Wherefore," I said, "have the calcareous or siliceous bodies which crystallise out from the cells of a sponge, wherefore have the calcareous spicules of the coral skeleton, wherefore have the spicules of the Holothurian's skin, just this or that exquisite form and no other? Surely on the same grounds on which a crystal has its definite form, and not on the ground of utility. To what end the exquisite shapes of the Radiolaria, the exquisite sculpture, markings, and colours of the mollusc's shell, which latter are, moreover, generally covered throughout life by mud and dirt, and whose beauty of line and colour often only appears after polishing? To what end the black colour of the peritoneum of many vertebrates? To what end the various delicately-wrought patterns of the leaves of our foliage trees? To what end the whitening of the hair and all the other changes of old age in animals and in man? To what end the necessity of material metabolism, of death following the acquirement of the highest degree of capacity and knowledge, of the highest adaptation to the environment? Surely not for the advantage of the individual, nor that of the species—at most their use is to maintain the circulation of life upon the earth ; and to place a higher value upon this than upon the continuance of vigorous, joyful bodily existence must be difficult even for the most consistent advocate of the principle of immediate utility."[1]

Death as an Adaptation

Immortality

Weismann, we know, explains death itself as an advantage, as an adaptation.[2]

The unicellular animals, the Protozoa, which multiply by simple division, are, he says, immortal. The multicellular, however, the Metazoa, are mortal. Originally the latter were also immortal, but after they had differentiated their germ-cells from their body-cells (somatic cells), so that reproduction could only be carried out through the former, they became necessarily more and more imperfect, because they were no longer able to repair the damages incurred during life. Hence the utility of their death.

The direct opposite of this is held by Götte.[3] According to him, all animals are mortal, and reproduction is itself the cause of death. Reproduction in Protozoa is preceded by encystation. In this condition the organism passes into a non-living condition, from which it revives with renewed youth and renewed life; a similar condition occurs in the egg of the Metazoa, during a certain period in which it forms an unorganised, non-living body composed of organic substance. The idea of "death" is here conceived as the stand-still of the organic life of the whole—the idea of a corpse is not included.

---

[1] For botanical instances compare Askenaz, *Beiträge zur Kritik der Darwinschen Lehre*, Leipzig, 1872.

[2] A. Weismann, *Ueber die Dauer des Lebens*, an Address delivered at the Salzburg Naturalists' Congress, 1881 ; and *Ueber Leben und Tod, eine biologische Untersuchung*, Jena, G. Fischer, 1884.

[3] *Ueber den Ursprung des Todes*, Hamburg, Voss, 1883.

Weismann replies, in his latest essay, such an arbitrary definition of the term death is unjustifiable. The process of encystment is by no means comparable with death, and does not even occur in all Protozoa—a view which is certainly correct. Weismann says further: "The gradual evolution of death is to be explained in this way, that at the first differentiation of the body of the Metazoon into somatic and germ-cells, the life of all cells was limited to one generation, that of the somatic, therefore, to a short duration. The somatic cells first began to last for several generations among the higher Metazoa, and life was prolonged. This change was brought about by processes of selection, on the basis of the principle of division of labour. The shorter or longer duration of life depends entirely on adaptation. Death is not due to an original property of living substance, nor is it necessarily connected with reproduction. Reproduction, on the other hand, is an original property of living matter. Life is continuous, not, as Götte would have us believe, interrupted, discontinuous."

With regard to the different views of Weismann and Götte, I have elsewhere [1] expressed myself as follows: "On account of the definite evolution of science, certain ideas are not far to seek. I have for years discussed the immortality of the Protozoa in my lectures,[2] but in consideration of the action of metabolism, and the necessity of the renewal of the organs of even these forms in consequence of wear, would place some limitations on its acceptance.

"In the Metazoa, according to my opinion, the germ-cells are immortal like the Protozoa (with the same limitations as hold for the immortality of these); only the soma dies.

"The latter is not really an end in itself,[3] but rather its

---

[1] In a report in the *Deutsche Literaturzeitung*, 1884, No. 19.

[2] Cf. also Bütschli, *Zool. Anzeiger*, 1882, p. 64.

[3] [It would be just as true to say that the germ-cells are not an end in themselves, but serve only to produce the soma. Trans.]

principal function is to ensure the maintenance of organic life by favouring reproduction, by sheltering the germ-cells till their maturity, and in order to deposit them repeatedly; further, by the dispersal of the same in space, by incubation in the widest sense, and so on. Further, it has the function of strengthening the power of endurance of the species by the inheritance of acquired characters.

"Reproduction is unending growth. Not reproduction is the essential cause of death, but the differentiation of the constituent parts of the soma resulting from division of labour, in the last result division of labour itself, to which the soma owes its very existence: the germ-cells alone are still individually complete elementary organisms, and as such retain still the general complete combination of matter which in them as in Protozoa ensures endless growth. To the soma such growth is impossible because these fundamental conditions in it are wanting; it dies in consequence of the exhaustion of the organs.

"Separation of the sexes is likewise due to division of labour, and sexual reproduction finds its explanation in this very necessity of the complete combination referred to as a condition of endless growth. And thus at the same time parthenogenesis would explain itself."

This conception is in some measure completed and developed in the preceding and following pages. I have to add here at once: there are also Metazoa whose bodies (soma), in that they divide or multiply by budding, may be immortal. Amongst these are instances in which division of labour in the organisation has not yet attained any considerable development (many Zoophytes, *e.g.* Hydra), and others in which the body consists of equivalent parts, into which it may separate (*e.g.* Annelids, like Nais). In both cases such a reproduction (possibly unending) of the soma without the aid of the germ-cells can only occur because the parts have all the properties, because they have the complete combination of the whole. The higher animals, with more advanced division of labour, cannot reproduce by division, because each part does not contain the properties, the complete combination, of the whole. In this sense the germ-cells alone are still individually complete elementary organisms, in relation to the present question.

Further, the Blastozoa,[1] which reproduce only by germ-cells, are only distinguished with respect to immortality from the Protozoa, inasmuch as we are accustomed to regard their soma as the essential part. But let us place ourselves at the point of view indicated above, "that the soma is not really an end in itself, but rather its principal function is to ensure the maintenance of organic life by favouring reproduction,"—let us reflect a moment while we picture to ourselves the unbroken chain of organic nature, its unity, and look upon the individual animals and plants as merely seed-vessels,—then the Blastozoa appear as immortal as the Ablastozoa.[2]

If we philosophise about nature at all, the naturalist who desires to be thorough must not halt half-way.

Thus the above way of regarding the matter is perfectly justifiable—if the multicellular animals have really been evolved little by little from the unicellular, it is a mere self-evident consequence.

Although, however, we may be willing to reason with as much self-abnegation as is demanded by this conception, yet by the above we are not compelled to go so far as to consider death as useful, as an adaptation.

Besides, Weismann himself speaks concerning the relation between the germ-plasm of the multicellular organism on one hand and of the unicellular on the other, and concerning the significance of the body (soma) of the former, as follows:—

---

[1] Animals with germinal layers = Metazoa = multicellular animals.
[2] Animals without germinal layers = Protozoa = unicellular animals.

After he has stated that his notion of the continuity of the germ-plasm reduces heredity to simple growth, and parallels it with the reproduction of unicellular forms in which the same substance grows on and on, and new individuals are only produced by its division, he continues : " The distinction between the unicellular and the multicellular consists only in this—that in the latter every division of the 'germ-substance' is followed by a process of development which leads to the formation of a multicellular individual. This, then, exceeds in mass by an enormous amount the unused remnant of the germ-plasm, but yet it is only a by-product of the eternal germ-substance, is abandoned to death, must die after a time, while the germ-substance, under the shelter and nurture of the multicellular body (soma), continues to grow, increases in mass, and produces new germ-cells, which possess the power of forming another generation of bodies (somata), in which the same process again takes place. The germ-plasm may therefore be conceived under the simile of a long creeping root, from which at intervals separate plants arise, the individuals of the successive generations."

If the body of the multicellular organism is thus, even according to Weismann's ideas, of secondary importance in comparison with the germ-plasm, if the latter corresponds to the unicellular organism, it follows that the multicellular is just as immortal or mortal as the unicellular. And thus it is impossible to see why, between the germ-plasm of the multicellular on the one hand, and that of the unicellular on the other, there should exist this profound difference, that the latter acquire characters during life and transmit them by heredity, the former not,—how the former any more than the latter can nourish itself and grow without being influenced in its nature by its nurture.

FURTHER CONSIDERATIONS ON THE ADAPTATION AND DIRECTION OF EVOLUTION OF THE MARKINGS OF CATERPILLARS. ABSENCE OF SEXUAL COMBINATION IN THIS EVOLUTION

I much regret that Weismann's views and mine with regard to the causes of evolution are now so divergent, after the great agreement between us, which I felt myself entitled to assume in my publication on the variation of the wall-lizard, judging from Weismann's researches on the markings of the Sphingidæ caterpillars, which demonstrate definite directions in the modifications of these markings.

I was rejoiced to be able to appeal to the fact that the results of these researches of Weismann admitted of an explanation in accordance with my views, and thought I had good grounds, therefore, for believing that we agreed. I was the more pleased at this supposed harmony because I had arrived at my results quite independently of Weismann's paper.

Weismann was one of the first who insisted that variation takes place in constant directions; he also recognised the truth that simply from the fact of the given constitution of the body it could not occur in any direction at random. Formerly, moreover, he did not allow such exclusive dominion to adaptation as now. Thus he says in his work on season-dimorphism:[1] "Little as I am inclined to talk of an unknown transmutation force, I would as emphatically insist here again that the modification of a species depends only in part on external conditions, and in the other part on the specific constitution of the species."

[1] A. Weismann, *Studien zur Descendenz-Theorie.* I. *Ueber den Saison Dimorphismus der Schmetterlinge,* Leipzig, Engelmann, 1875.

Weismann discovered in his caterpillars a regular series of steps in the marking from longitudinal lines to transverse lines and spots, and came to the conclusion: "That among species which are ornamented with oblique streaks or with spots there are many whose young stages are streaked lengthwise, but the converse never occurs; never do the young caterpillars exhibit spots or oblique streaks when the adult caterpillar has only longitudinal streaks. The first and most ancient marking of the Sphingidæ caterpillars was therefore the longitudinal striping." A derivation of the oblique from the longitudinal striping is not demonstrated; both occur together, but the former appears later than the latter, and remains when this has vanished.

I thought it possible that the longitudinal streaks broke up first into small spots and dots, and that these spots and dots formed the oblique streaks. If so, complete agreement with my laws of marking would be proved in this case.

In caterpillars also, as already mentioned, the new characters arise mostly at the posterior end of the body.

Weismann endeavours, as before remarked, to explain the markings of caterpillars as adaptations to the conditions of the environment. Nevertheless, he comes to the conclusion that these markings have phyletically developed extremely gradually, according to certain laws, and in quite definite directions. And he further says: "The evolution of the species of Deilephila shows that the evolution of the marking follows throughout a certain law, that it proceeds in all species in the same manner. All the species seem to steer towards the same point, and this gives the impression that there is an internal law of evolution which, like an impelling force, determines the future phyletic modification of the species." This conclusion seems to be supported also by the fact that in caterpillars there is a tendency for the same characters to be repeated in succession on all segments; and further, that characters which have newly arisen appear afterwards at an earlier and earlier age, although no advantage is to be discovered in this.

Although Weismann, even at that period, showed the greatest repugnance to recognising a special force of evolution, and, in my opinion, was quite right in his opposition, yet how can he explain by his theory of the action of sexual reproduction the facts established by himself concerning caterpillars? And how is it possible to explain definite directions at all in evolution by sexual reproduction, unless every step in the process of modification is demanded by adaptation?

Characters which are inessential (indifferent) to the Life of the Organism

I once devoted myself for a long time to the study of the siliceous sponges. As is well known, we find in these a degree of variation which is extremely troublesome to the investigator. One form passes into the other, and almost innumerable, often in the highest degree insignificant, divergences are met with in the parts on which diagnosis depends—the siliceous spicules and particles of the skeleton. It cannot possibly be of importance for the success of a given variety whether the spicules have this or that minute character or not. That this is true is proved by the occurrence of many such varieties side by side. As there can be therefore no question here of useful adaptation, the variations under the circumstances can be only ascribed to an extreme sensitiveness of the protoplasm to external influences. The "crystallisation" in these sponges of the siliceous parts, often so graceful in shape, from the protoplasm is, to an extent most favourable to my ideas, comparable to processes of form-production in inorganic nature, where there can be no question of special adaptation. It is known, for example, that in calcareous sponges the variation and disappearance of one axis of the spicule actually forms a transition from one of the principal divisions of the order into another, a subject which will subsequently be discussed fully in connection with a particular case.

And of what use was the gradual modification of the Ammonite's shell in such a way that new characters, some of a very beautiful kind, arose always round the aperture, and extended in the descendants more and more over the whole structure, so as to form new species, as Würtenberger has shown? How can these characters, especially at their first origin, have been useful?

And what demand upon the organism could entirely exclude the development of such indifferent characters? Certain it is that this one or that among such characters was once useful to the ancestors—as, for example, possibly the traces still occurring of marking in Canidæ[1] (dog, wolf, jackal), or possibly the stripes on the shell of our garden snails, *Helix hortensis* and *nemoralis*. But it is equally certain that this does not hold for all.

I mentioned the garden snails as an example intentionally. It is the less easy to discern any use in the striping of the shells of these snails; not only does the striping vary greatly, but it is about as often absent as present. The striping might be regarded as an ornament which acted as an advantage in sexual selection. But such an assumption is inconsistent with the following facts : I have observed for years in my garden that striped and unstriped individuals of *Helix hortensis* unite without any selection. And what makes this case particularly noteworthy as a support for my view of the comparatively slight effect of sexual crossing in the production of intermediate forms (one-sided heredity) is this, that the

[1] Cf. my articles in *Humboldt*.

offspring of these striped and unstriped parents are again striped or unstriped—in spite of constant crossing (pammixis) these two forms appear everywhere side by side with no connecting forms between them.

If everything were adapted, there would be no characters at present useless, representing either rudiments of formerly useful or the beginnings of new characters. There would be no gradual transitions, and, more particularly, no change of function, and also there would be no correlative characters — there could be no striking variability in forms at all.

The universal dominion of adaptation is finally also contradicted by the reflection, that from the given materials only instruments of a certain quality can be produced in organic nature[1]—that these instruments might be many times better for their purpose, more perfect, less easily damaged—that they are not in all cases completely adapted to the requirements seems to me beyond a doubt.

Nevertheless, what great importance, in spite of my refusal to recognise its exclusive dominion, I ascribe to adaptation in the modification of forms will be best gathered from the section on the adaptation of lizards.

It was not my purpose in this place to exhaust the question of non-essential characters, as it is discussed elsewhere in this work, both in preceding and following pages. I wished merely to mention here the existence of such characters as an argument against the assumption that adaptation rules exclusively in all cases.

If this exclusive dominion did belong to it, if everything

---

[1] Compare here my remarks, reproduced in a later passage, on the formation of sense-organs, in my monograph *Die Medusen, anatomisch und physiologisch auf ihr Nerven-system untersucht*, Tübingen, Laupp, 1878. I there showed how clearly the structure of sense-organs, especially of the optic and auditory organs of various animals, indicated that the given material permits only the formation of instruments constructed on certain few principles, which instruments satisfy the requirements, but are obviously often very imperfect.

were adapted, all evolution of the living world would be also excluded—there would be stagnation.

The chief evidence for my view of the organic growth of the living world as yet withheld is now to be given : (1) The evidence that external conditions modify organisms, and (2) that characters so acquired are inherited.

SECTION IV

ACQUIRED CHARACTERS

Methods of Investigation—The Period of Time to be claimed for Evolution.

NÄGELI says in the introduction to his book, when he is objecting to the discussion of the doctrine of descent from the standpoint of descriptive natural history, and arguing for the "exact physiological method": "As the 'descriptive naturalist' is accustomed to attain in his progress to only disputable hypotheses, not to certain laws, he regards everything, even the results of exact observation and rigid criticism, not as matter of fact, but as matter of opinion. This was the case, for instance, with the fact of the promiscuous evolution of species of plants, and with that of the insignificance of the influence of climate and nutrition in the production of varieties, both of which I believe I have adequately established, and which an unprejudiced and conscientious observer can easily test and confirm. These facts inflict a very heavy blow on the whole fabric of the doctrine of descent as it now exists, and could not therefore be considered unless the existing doctrine were abandoned. . . . 'I cannot make use of that,' they say."

With regard to the promiscuous evolution of species of

plants, all that need be said from my standpoint is, that it is by no means in contradiction to my view, but rather helps to confirm the correctness thereof : Nägeli shows, for instance, that the crossing of nearly related plants living together (Hieracium species) does not, as might *à priori* be expected, lead to the production of intermediate forms, but rather of varieties which possess the characteristics of the one or the other parent forms intensified—a result which harmonises completely with the process described by me as one-sided heredity, and with the points of view indicated in that description.[1]

Against Nägeli's supposition that the influences of climate and nutrition are of no consequence in the formation of varieties, it is, however, to be urged that isolated negative examples prove nothing. This is especially the case when the observations are made only on one distinct kind of objects, only to a limited extent, and only during a limited period of time. Such experiments prove only that in the given cases the result was negative. A single positive example on the other side is, however, sufficient for complete proof. I have adduced here such examples, and will bring forward more. Nägeli's arbitrary general application of his results, be it remarked in passing, is sufficient to show his misuse of the word fact, and also to prove the injustice of his charges against the " descriptive naturalist."

When, therefore, this inquirer believes he has by his experiments quite settled in the negative the question of the influence of external conditions on the permanent modification of species in general, and that he has shattered the whole structure of the doctrine of descent hitherto held, I cannot agree with him, actively as I am endeavouring myself to alter that doctrine.

---

[1] Cf. *ante* p. 39, and Nägeli, *Das gesellschaftliche Entstehung neuer Species, Sitzungsberichte der Münchener Akademie*, 1873.

Against the validity of the evidence afforded by Nägeli's experiments I would further, above all, bring into the field the circumstance that they are artificial throughout, and that they cannot therefore claim to be held as complete evidence for the processes which occur under unrestricted natural conditions. Not that I would attach no value at all to artificial experiments, but they must certainly be employed with great caution. Their relatively slight importance depends in my opinion on the fact that definite directions of evolution become fixed in living beings by extremely long continuance, and these can only be changed by continuous external influences acting for a long time. Thus an organism is restricted long beforehand to a certain direction of modification. If an artificial experiment chance to promote change in this direction, it soon results in the appearance of permanent modifications: these will persist even after the artificial external conditions have ceased to act. But not so when art is in opposition to nature—then the organism will rather return to its former condition immediately after the cessation of the artificial action, or its descendants will do so. This last seems to have been the case in Nägeli's experiments with plants, which he placed under different and more favourable conditions of nutrition, and which were modified by the change, but whose seeds sown in poorer soil again produced plants of the old kind.

An Alpine plant, for example, is necessarily covered with snow for a long time in winter, and in summer is exposed to strong heat from the sun and to intense light, and has acquired through its peculiar conditions of life the power of very rapid development and also special morphological characters. If this plant is placed where these conditions are intensified, in the far North for instance, these characters ought to be still more strongly developed, and the advance ought to remain permanent even in the offspring which are reared from their seeds in the Alps again—this on the additional ground that the advance would be useful also in the Alps. But it is otherwise when, as was done by Nägeli, plants growing on the Alps (Hieracium) are transplanted to the Botanical Garden of Munich, and their offspring are reared in gravelly soil (as a substitute for the conditions of their Alpine home). That these offspring, deprived of the rich garden soil, revert to the original form seems to me less surprising than the reversion of all the garden plants everywhere artificially cultivated for a much longer time as soon as they grow in a wild condition; and thus Nägeli's experiments, on which he lays so much stress, seem to me to prove nothing new against the modification of forms by external influences.

The effect of the gradual withdrawal of water upon the Axolotl is a case in which art hits upon and hastens the movement of evolution already in progress. Likewise, undoubtedly, the action of water of greater or less saltness on the crustacean Artemia salina, shortly to be considered more closely. It is of course also possible that one part of this latter case is an instance of reversion, for examples show that when animals are transferred to the life-conditions of their nearest ancestral form, or retained in those conditions from the young state onwards, they retain the characters of the ancestral form which they possess as larvæ. Thus we are able to prevent Salamanders from losing their gills and giving up branchial respiration by compelling them to remain in water all their lives, and so cause them to remain similar to their ancestors the Perennibranchiata, because their other characters also, through correlation, remain permanently at the lower stage.

Very pretty experiments in relation to this point have been made by Fraulein von Chauvin [1] upon the Alpine Salamander, Salamandra atra. The gill-bearing larvæ of this animal were taken out of the oviduct of the mother and put into

[1] *Zeitschr. für wiss. Zoologie*, Bd. xxix.

water. It is well known that these larvæ under the usual conditions at a later time, while still within the mother's body, lose their gills, and, what is very rare among Amphibians, are born as completely terrestrial animals. The gills of the larvæ placed in water at so early an age were disproportionately large, and hindered the animals in their movements, and in some cases they were cast off, whereupon new smaller organs arose in their place. These new gills persisted in one case for a surprisingly long time (fourteen weeks), and then atrophied. This larva, like the rest, ultimately developed into a land animal. But the remarkable fact remains that on account of the peculiar conditions of life artificially produced, after the original gills, which were unadapted for use in a free state of life, had perished, new and suitable gills were formed, not in the struggle for existence against competitors with the cumulative effect of selection, but, as I believe, directly from purely physiological causes. These causes must, to my thinking, be sought in this, that as the pulmonary respiration was not allowed to develop, and the original relations of the branchial circulation therefore continued, new outgrowths of the skin, *i.e.* gills, were formed in consequence of the unchanged distribution of nutrition at the place which that distribution made most favourable; yet in the end the phylogenetic tendency which had been for a long time established got the upper hand.

In thousands of cases, on the other hand, we are unable by changing the external conditions to bring about any change at all in the organism, even of temporary duration—the animals or plants perish rather than adapt themselves; we succeed by experiments only in killing them very rapidly. It is clear that from this most simple physiological effect to those of Nägeli's experiments an unbroken chain of transitions must exist, and that therefore the latter are in reality as little surprising as the former.

But in addition to this an important consideration has been entirely left out of consideration in Nägeli's argument—the length of time necessary for the production of important modifications.

When the external change favours completely the process of evolution in progress, then the modification may—apart from the possibility of the aid of adaptation—be brought about very rapidly. In most cases, on the contrary, even when the external change is perfectly natural, the result must evidently be very gradual. Leaving aside the very gradual disappearance of unused organs, *e.g.* parts of the skeleton, this is proved by the obstinacy with which, as I have pointed out, traces of primitive markings are inherited in animals.

These facts prove the extreme tenacity of heredity. In other words, they show how difficult it is to divert forms from their usual direction of evolution, to conquer their *vis inertiæ*, and with what difficulty they can in general be led into new paths of evolution.

And this whole argument of mine shows also, to recur again to the subject of the previous section, how unjustifiable it necessarily is to talk of universal adaptation, for this very tenacity of heredity is sufficient surety for the existence in animals and plants of innumerable inessential characters which have gone out of use or which belong to a deserted path of evolution.

Even natural science has found it hard to recognise the effects of long periods of time. Only a few decades back the 5000 years of the Bible were accepted even in Geology. Now Geology reckons merely in the history of living beings with endless time, and Darwin required a long period of time to explain the modification of a single form. But that the conviction is even at the present time far from forming part of the mental constitution of the naturalist, is shown by the fact that so eminent and talented an inquirer as Nägeli can forget its claims, and believe himself entitled to draw conclusions as to the evolution of organic nature from cultivation experiments, carried out during a few years with negative result, and to deduce from them a "fixed law."

Yet more: it is only a few years since we have reached the conviction that the Egyptian civilisation stretches back more than 6000 years behind us. A few years ago, after Darwin had proposed his theory of the origin of species, his conservative opponents believed themselves justified in urging against the doctrine of the modifiability of species, that the species of cereals and other plants and also animals known from the ancient Egyptian times had not changed up to the present day, for the period of time in question seemed to them eternal. The Darwinians explained in reply that demonstrably the external conditions have not changed in Egypt since that time, and that therefore in the particular case no inducement, no necessity has existed for the new adaptation of the living beings, and therewith for their modification. But quite recently Weismann employs the fact cited by Darwin's opponents to prove that the germ-plasm is a substance of extreme power of persistence, a substance, he says, "which is nourished and grows to an enormous extent without changing thereby in the slightest degree its complicated molecular structure. We ought," continues Weismann, "to maintain this with all certainty, like Nägeli, although we can get no direct perception of this structure. But when we see that many species have reproduced themselves for thousands of years without change—I mention merely the sacred animals of ancient Egypt, whose embalmed bodies must in some cases be 4000 years old—this proves to us that their germ-plasm possesses still to-day the same molecular structure which it had 4000 years ago. And since, further, the amount of germ-plasm which is contained in a single germ-cell must be supposed very small, and since of this again only a very small fragment can remain unchanged when the germ-cell develops into an animal, therefore, even within a single individual, a quite enormous growth of germ-plasm from this fragment must occur. In every individual, as a rule, thousands of germ-cells are produced. It is thus not too much to say that the growth of the germ-plasm in the Egyptian Ibis or the crocodile in those 4000 years must have been utterly immeasurable. In the plants and animals, however, which inhabit equally the Alps and the Polar Regions, we have examples of species which have continued unchanged for much longer periods, namely, since the ice age, in which, therefore, the growth of germ-plasm must have been much greater.

"Since, nevertheless, the molecular structure of the germ-plasm has remained exactly the same, it must be very hard to modify, and there is little prospect that the slight temporary differences in nutrition which will of course happen to the germ-cells as well as to every other part of the organism, will bring about any change, however small, in its molecular structure. Its growth will proceed now faster now more slowly, but its structure will be affected thereby all the less that these influences are mostly of an alternating character, and act now in one direction now in another."

The hereditary individual differences Weismann infers must therefore have another root—they are to be derived from sexual reproduction.[1]

Against the above exposition, which reproduces once more Weismann's whole theory of the continuity of the germ-plasm, I must repeat that I cannot conceive how the germ-plasm grows, grows to an enormous extent, without being influenced by the conditions of nutrition, by the composition of the body. What is the cause of its growth, and why is it sometimes faster sometimes slower?

[1] *Die Bedentung der sexuellen Fortpflanzung*, etc., p. 27 *et seq.*

The objection against the influence of external conditions, that these are of variable nature, tending now in this now in that direction, is discussed in the preceding pages. The view, however, that the variations in nutrition to which the body, and as Weismann acknowledges the germ-cells also, are subjected, must be regarded as transitory and minute, I cannot admit.

The fundamental difference between Weismann's view and mine seems to me to lie just in this point. I grant entirely that the permanent action of external conditions on the body of the organism in most cases is not immediately perceptible. From physiological principles this is not in general possible. The question essentially depends on what ideas we possess of the time occupied by organic evolution. It is my opinion that we must accustom ourselves to a much less limited conception than even that introduced by Darwin.

My theory of the progressive growth of the living world and of the origin of species demands for the modification of a form, according to physiological principles, to the extent shown in any given case, enormous periods of time—periods compared with which the few thousand years of the history of Egyptian civilisation may be but as a moment in comparison with the individual growth of a plant or an animal.

My evidence of the importance of external conditions for the origin of species requires that this demand shall be borne in mind. None the less, I am able, as shown already in preceding pages, to bring forward cases in which influences show themselves in their effects in a short time. I proceed now to produce further evidence.

Every character which must have been formed through the activity of the organism, is an acquired character. All characters, therefore, which have been developed by exertion are acquired, and these characters are inherited from generation to generation. The same holds for all organs atrophied through disuse—the degree of atrophy is acquired and inherited. In the first class we see especially the action of direct adaptation, in the second the results of the cessation of this action. A third class of acquired characters are to be traced simply to the immediate action of the environment on the organism, and originally, at the commencement of their appearance, all characters must have belonged to this class.

Let us take first an example of the last class.

Acquired Characters due to Direct External Action

The formation of pigment is universally subject to the influence of light and warmth. Numerous species of animals living in the dark have become completely colourless; the deficiency of pigment due to external influences has been inherited and becomes at length a constant character. Everyone who like myself has undertaken, with his eyes open on such questions, a journey from Germany directly to the South, to Africa, as far as the tropics, will acknowledge the gradually increased darkness of colouring even in one and the same race of men, and must ascribe it to the gradual increase in the power of the sun. The result of this action of the sun, however, has been inherited and has become a constant character. We meet near and in the tropical regions of Africa tribes of men who are almost as black as ebony, and who, as for instance the nomadic Bischari, have features, skulls, and general form of body of perfectly Caucasian type; and the negresses give birth to light skinned children—a proof, for those who take their stand upon the biogenetic law, that their forefathers were light-skinned.

That I allow special importance to sexual admixture in relation to the distribution of dark colouring of the skin, hair, and eyes, I have already shown, since I pointed out that dark hair and eyes in Germany must be principally referred to this process. But that the dark colour of the inhabitants of hot countries cannot be explained by sexual admixture, any more than the light colour of cave animals (see below), needs no elaborate argument.

To me it is utterly incomprehensible, not so much that views opposed to the relation of the sun to the dark colouring of the skin are expressed, but that, without further investigation and reflection, any scientific importance is given to them. To form a decisive judgment on the question one must travel through a region like the Nile valley, which forms a uniform, continuous, isolated, whole—not over mountain and valley, which not only separate different tribes, but also afford sudden transitions in climatic conditions, and with them in the habits of life of the inhabitants. The perfectly gradual transition in the colour of the inhabitants from brownish yellow to black in the Nile valley in passing from the Delta to the Soudan is particularly conclusive as evidence for my contention, for the very reason that various races originally of various colours dwell there. The Berbers, who live from the First Cataract southwards to Nubia, are as a race much darker than the Egyptians. The anthropological separation of the two races is at the present day perfectly distinct, and the difference between the languages, except for one or two interchanges of words, is well-marked. The traveller accustomed to the sound of Arabian, who has learnt to make himself understood in that tongue, beyond Assouan no longer understands a word of the language of the natives. And yet the colour of the neighbouring Egyptians passes into that of the Berbers, that is, the former at Assouan are scarcely lighter than the latter, whose colour again towards Wadi-Halfa is still blacker than at Assouan. Crossing between the two peoples has of course to be considered, but it is not the most important cause of the facts described. On this subject I agree entirely with R. Hartmann, who says, in his book on *The Races of Africa* (Internat. Bibliothek, 1879, p. 9): "Our travellers make too much of the contrast between the light-coloured Egyptians and the dark Nubians. It always seemed to me as if these gentlemen slept away the time of the journey between Kene and Syene (Assouan). In this district one sees transitions enough between the two types of people. This is not merely a result of the immigration and settling of Nubian families in Said (Upper Egypt), but the inhabitant of Said as he gradually approaches the tropic becomes darker, darkened by the sun, but also in consequence of marriages with Berbers. So also the Nubian settlers in the Nile valley possibly become gradually lighter under the mild sun of Middle Egypt, partly also of course in consequence of marriages with people originally lighter. But that in such processes a certain adaptation to the physical and climatic conditions of the district takes place, seems to me a fact in natural history which cannot be denied."

That the black Berbers under the slightly greater mildness of the Egyptian sun become lighter I should consider not very probable. But I have traced step by step the quite gradual increase in darkness in one and the same race, in the Nile valley up to Dongola, as clearly as it can be traced from Germany to Calabria.

Such observations on dark peoples, and on the influence of the sun on colour, bring the significance of the peculiarity of our Germanic race prominently into the foreground. There is no race among all the peoples of the earth which is nearly as peculiar in the absence of dark colour in the body-surface, and this to me is evidence that we Germanic people had our original home in a very moderate climate, indeed in quite northern regions.[1]

---

[1] It is true that the greater action of light in the north has favoured the development of pigment in plants and in many animals, but this does not apply

The physical cause of the dark colouring of the integument is obviously that in consequence of the greater flow of blood to the skin under the action of light and warmth, of their stimulation, pigment is deposited there. Possibly in some cases greater moisture has contributed to the result. An increased supply of blood can effect the deposition of pigment in particular parts of the skin without the influence of light. This happens, as we know, in woman round the nipples, in man on the scrotum, in both sexes in the arm-pits and round the anus. Dr. Passavant, known for his travels in the Cameroons, states in his doctorate dissertation that in the negroes just those parts which are least exposed to light are darkest, *e.g.* the arm-pits, and he argues from this fact against the conclusion that light is the cause of darkness of colour in man. This apparent contradiction finds a very simple explanation in the above considerations.

Like the pigmentation of the human skin, and pigmentation in general, the formation of leaf-green (chlorophyll), the green colour of plants, depends also on the influence of light. But here the colour is not inherited by the offspring, for when these are reared in the dark they remain colourless.

The profound action of light on the whole physical constitution of the plant-body, on the whole physiology of plants, is shown by the fact that many tropical plants, as for example the South American species of Bougainvillea, in European hothouses either do not bloom at all or only incompletely in spite of all application of warmth, on account of the deficiency of light.[1]

And who would deny that action when he only thinks of the influence which light exercises on the direction of growth of plants, and therewith on their whole form?

to northern races of mankind, who are not constantly exposed to this action.

[1] Cf. Charles Martin's edition of the *Philosophie Zoologique* of Lamarck. German, Jena, 1876. (In the biographical introduction by Martin.)

To this cause I also ascribe the first commencement of the vivid colours in the skin of many animals, *e.g.* of reptiles, which, clearly in consequence of the increased circulation of blood, appear under a warm sun, and disappear under a low temperature, and which with the aid of selection have developed into the permanent brilliant colours in the skin of southern animals. All the animals that pass their lives in darkness must have become colourless in consequence of the absence of light, since their nearest relatives which live in the daylight are coloured. Absence of pigment is indisputably an acquired and inherited character. No one will seriously maintain that male and female individuals which chanced to be slightly pigmented have everywhere crept into caves, and there, in consequence of sexual mixture and selection, have given rise to colourless races.

Weismann advocates the view that the absence of colour in cave animals is due to pammixis, to indiscriminate sexual mixture consequent upon the cessation of selection. But how can selection have contributed to the evolution of the original dark colouring—what use can this have had? Sexual selection on the ground of beauty is out of the question in the gloom of the caves. And the principal question is, Why is a colourless race evolved everywhere in darkness by pammixis, and not one of some other colour, green, yellow, blue, or red? Since we know that light is favourable to pigmentation as it is to the development of chlorophyll, since we know that pigmentation like chlorophyll universally vanishes in the absence of light, we need, it seems to me, no other explanation than that which first offers for the absence of pigment in cave animals. Animals are in general more or less colourless in the early stages of development, and very many are born light-coloured or colourless, and only subsequently develop a dark colouring. This circumstance evidently facilitates the loss of colour in cave animals : since the first offspring of the individuals which have betaken themselves to caves are born with light skins and are no longer exposed to the action of light, the later descendants will very soon remain permanently light coloured.

Further, how immense is the direct effect of warmth on the vegetable world, and how obvious the inheritance of this effect. Consider merely the characters of the flora in the different zones of the earth. Quite different forms of plants have been developed in the tropics than with us—yet obviously in part essentially through the direct influence of the climate. Warmth and moisture have always, as at present, stimulated the growth (*i.e.* the multiplication) of cells beyond the limit which it reaches in colder climates, and this growth necessarily produces new and more luxuriant developments.

The following example shows to what a degree increased growth results directly in change of form : the luxuriant shoots which spring from the stumps of trees, *e.g.* elms and other forest trees, which have been cut down, often bear leaves which have a quite different shape from those the tree usually bears, so that they have quite a strange appearance. This is the immediate effect of unusually increased nutrition; the same roots which previously supplied nourishment to the whole tree have now only to provide for a few shoots. If such a shoot grows into a tree, this again bears the ordinary leaves. And if most of the roots of the stump were cut away so that the shoot was reduced again to the usual supply of nutrition, then likewise the new leaves would revert to the ordinary form. Certain characters which are directly conditioned by better nutrition, and others connected with them by correlation, must obviously again disappear with the cessation of the condition—thus certainly hunger causes leanness and feebleness, over-feeding fat and laziness. In any case, experiments continued for a year or two prove nothing. It is another question whether specially abundant nutrition continued through several thousands of years does not produce constant characters in a species of plant, so that the species could no longer exist if brought again under conditions more unfavourable to nutrition.

The directly modifying influence of climate is exhibited most distinctly in our common plants. The resistance, for example, of the plants of a single species to the influence of cold or warmth is very variable. It is very surprising at what different times trees of the same species in a wild state under the same conditions develop their foliage in spring—for instance, beeches growing close together in a wood. Years ago this struck me particularly in the garden of the castle of Veitshöchheim, near Würzburg, which is laid out in the rococo style. The straight walks are there enclosed between hedges of cut beeches like walls. In the spring, parts of the hedges formed by certain beeches are already green, while parts formed of others which stand between the former, and are even mingled with them, have not begun to open their buds.

Every gardener knows that individual plants are more able to endure cold than others of the same species under the same conditions. Darwinism is satisfied with making use of such differences to explain how they render selection possible, and how they are increased by its means, but does not inquire into their causes. Yet these differences cannot be due to chance. There must be a peculiar condition of the tissues underlying them, which peculiarity must ultimately depend on external influences which have acted on the plants themselves or their ancestors. For that such powers of endurance are inherited no one, in the face of so many facts known to every fruit-grower, will deny.

Perhaps the above difference in beeches, considering that nearly every beech-wood with us is more or less artificially planted, is due to the fact that the ancestors of the trees which come into leaf at different times were derived from different stations in which they had accustomed themselves to the prevailing external conditions. That the trees themselves—even in the garden of Veitshöchheim—came from different stations is a less probable supposition.

We can observe every spring in our gardens that the shrubs which belong to the south begin to bud later than our own—they are so accustomed to a certain slight degree of warmth that it has no effect upon them; their tissue is indifferent towards this degree of warmth, is not stimulated by it. It may thus be assumed that plants whose ancestors grew in cold stations have become accustomed to cold, which means in physiological terms that their tissues have gradually been modified simply by the influence of climate.

In this manner many of our cultivated plants have acclimatised themselves in a high degree. There cannot well be any doubt that the different kinds of summer and winter corn are forms which, originating from one and the same species, have not only gradually accustomed themselves to ripen at different times, but have also acquired new morphological characters. Very remarkable instances of such acclimatisation are described by F. C. Schübeler in Scandinavian plants, especially species of cereals.[1] Schübeler finds that—

1. When various cereals in Scandinavia (Norway and Sweden) are gradually transplanted from the plains to mountain districts, they can be accustomed not only to develop in the same, or even in a shorter time than in their native region, but even at a lower average temperature. When such grain after it has been grown for several years in the mountain regions is again sown in its native soil, it ripens at first earlier

---

[1] F. C. Schübeler, *Viridarium Norvegicum.—Norges Växtrige. Et Bidrag til Nord-Europas Natur- og Kulturhistorie.* Bd. i. Universitets-Programm. With numerous woodcuts in the text, and four maps. Christiania (Dybwad) 1885. The above extracts according to Foslie (Tromsö), *Botanisches Centralblatt*, 1886. Bd. xxviii. p. 205.

than other grain of the same kind which has been cultivated uninterruptedly in the plains.

2. The same thing occurs when species of cereals are gradually transplanted from south to north, even when the warmth is less and the sky more cloudy than in the native region.

3. The seeds of various plants increase up to a certain degree in size and weight when the plants are transplanted to the north, provided that they have attained their full development. But they return to their original size when the plants are again grown on their native southern soil. Similar relations hold for the leaves of several trees and other plants.

4. Seeds which have ripened in northern regions produce larger and more vigorous plants, able to endure severe weather better than those of the same species or forms which are reared from seeds brought from southern districts.

5. The formation of pigment in flowers, leaves, and seeds is greater (at least up to a certain degree) in the same species and varieties in northern than in southern latitudes.

6. In plants in which certain organs are aromatic this property increases as we go northward, provided that the plants attain their full development, while the proportion of sugar up to a certain degree decreases.

From these facts, of which 3 and 4 are probably to be explained by natural selection, it follows amongst other things that my view, namely, that an organism is altered by the continued action of definite external conditions, is fully justified; for the longer it is exposed to this action the less will be its tendency to reversion. Thus also the Egyptian varieties of wheat are certainly not exactly the same as they were 4000 years ago, even although the change is not expressed in their external form but only in their constitution and their powers of life. For it is a proposition of special importance to my argument, as well as in itself self-evident, that physiological changes must always precede morphological changes of structure in the organic world, because the former determine the latter.

I would draw particular attention to two results of Schübeler's experiments.

It is generally known that the formation of pigment, *i.e.* the brilliancy of colour in plants, is greater in elevated regions and in the north. Alpine plants afford evidence of this. This phenomenon is to be explained in part by selection, for it is clear that in the brief flowering period those plants will be soonest fertilised by insects which have the most striking colours with which to attract them. But in consequence of the shortness of the nights and the clearness of the sky, sunlight has a more continuous and more powerful action on the heights of mountains and in the north, notwithstanding the shorter duration of the summer, and there can be no doubt that this directly contributes to the development of the more vivid colouring. It has always struck me how much the garden and window-flowers of the houses in mountain districts surpass the same kinds among ourselves in splendour of colour, and this is a difference not due to insect selection.

But it is also a known fact that many species of animals, especially of insects, which are found at a high level on mountains have a darker colouring than their allies at a lower level. Thus there are remarkably dark species and varieties of beetles occurring at high levels.

The great variation in colour of our common wayside slug, Arion empiricorum, is universally known. The colour varies from light yellowish red to deep black. In my publication on the variation of the wall-lizard, in discussing the question of the causes of darkness of colouring in lizards, I have minutely considered this peculiarity, and I reproduce the passage here because it contains references to other conditions which produce this character and further examples of the inheritance of acquired characters in general.

"Leydig has pointed out that variation towards greater darkness of colouring, the tendency to become black, is connected with the action of moisture in the environment, having observed this connection in the case of the dark variety of Lacerta vivipara, of Amphibia,[1] and above all, of Arion empiricorum.[2] He observed that, besides Arion empiricorum, other molluscs, *e.g.* Helix arbustorum, Succinea Pfeifferi, Helix circinata, become darker than usual in moist localities. In Arion the colour even changes in this way according to the dampness or dryness of the weather: 'In early spring, when the ground and the air are still very moist, all the slugs, in places where later on in the year only reddish-yellow specimens are seen, have a dark brown colour.' For instance, in the cool and rainy May of 1873 and in June with cold weather and heavy rains the slugs in the saturated forest of the Spitzberg (near Tübingen) were of a deep black. Numerous other instances of the dark colour of Arion empiricorum in moist places are given by Leydig. I can confirm his statement that this slug under otherwise similar conditions exhibits a dark colour in moist localities; I found that in the extremely damp summer of 1879 all specimens were dark in places where at other times they were rather light."

"Some years ago, however, I pointed out a relation between elevation of habitat and darkness of colour in Arion which is inconsistent with the supposition that moisture has an absolute influence in producing this character."[3] I observed, for instance, that Arion empiricorum on the heights of the

---

[1] Leydig, *Die anuren Batrachier*, Bonn, 1877.

[2] Leydig, *Die Hautdecke und Schale der Gasteropoden*, Arch. f. *Naturgeschichte von Troschel*, 1876.

[3] *Württembergische naturw. Jahreshefte, Vortrag gehalten im Verein für vaterl. Naturkunde zu Tübingen*, 1878.

Rauhe Alb (*e.g.* above Urach), where there is little water, was generally darker than at its base in the well-watered valleys. Down the descent into the latter the animals became lighter and lighter; and I found afterwards that on all mountains on which I examined this species the greater number of the specimens, or even all, were dark, almost black, *e.g.* on the Schwarzwald, on the Harz, on the Rigi, etc.

Leydig states, on the contrary, that he found almost exclusively Arion rufus on the dry heights of the Alb,[1] but my observations are confirmed by another excellent observer of molluscs—Weinland, who found that Arion on the heights of the Alb near his own home was usually dark, and never reddish yellow as it so often is in the valley. "It might be said," adds Weinland, "that darker pigment is always produced on mountains as in Vipera prester, the black mountain variety of Vipera berus, as in the black rattlesnake of the White Mountains in North America." "But," he continues, "the law does not always hold: near the Hohe Neuffen I found almost exclusively light-red specimens." He attributes, therefore, the difference of colour to adaptation, in which I am unable to agree with him.[2] The explanation of the contradictions is probably that both moisture and elevation produce darkness of colour—both causes may act together, and probably often do on mountains, or great dryness may partially counteract the influence of elevation. More dark slugs ought to be found on the heights in wet than in dry summers.[3] It must also be pointed out that in many localities all colours occur side by side, which can only be explained by the individuality, and

---

[1] *Op. cit.* separate copy, p. 60.

[2] Weinland, *Zur Weichthierfauna der Schwäbischen Alb, Jahreshefte des Vereins für vaterl. Naturkunde in Württemberg*, 1876.

[3] Only black specimens of Arion are said to occur at Hamburg. Perhaps the moisture always produces this colour on the sea-coast. Compare the following.

the particular habits, the special lurking-places of the different specimens.

It seems to me, therefore, certain that moisture and elevation promote dark colouring directly and without the aid of adaptation. Adaptation, in my opinion, in the case of our native slugs is out of the question, for they are to be numbered among the animals which, creeping over road and path, expose themselves openly and freely to the view of all the world, as if exposing themselves intentionally in order to proclaim : I am an uneatable, loathsome thing, touch me not.

But what is it that causes darkness of colour at high levels ? Only two causes, apart from moisture, seem to me possible, either light or decreased atmospheric pressure. Since the latter facilitates the flow of blood to the skin, it might also promote the deposition of dark pigment.

In any case, elevation and moisture in various animals have determined a permanent change of colour, which serves as one of the characters by which species are distinguished.[1]

A second point to which I wish to refer in connection with Schübeler's results is the increase in the aromatic flavour of fruits towards the north. Pleasant to the taste as oranges and figs are in Southern Europe, they do not, in my opinion, by any means compensate for the loss in aromatic flavour which our native fruits undergo in the south. The wild strawberries, equally with the cherries and apples, are in South Italy almost absolutely tasteless. This is certainly a change brought about simply by climate and soil,—just as there is no wine in the world which equals good Rhine wine in fine aroma.

With regard to the acclimatisation of plants transferred to new regions, and the modifications shown by Schübeler to be caused by the process, similar facts, as I have already mentioned, are known in abundance to every fruit-grower,

---

[1] Additional proofs of this are given farther on.

indeed to every agriculturist. Cultivated plants of the most different species are gradually accustomed to new stations, to rich or poor soil, by being transplanted themselves or by the planting of their progeny. The fruit-grower of a district where the climate is severe will import his trees, no matter of what species, not from a warm, but from a similar region, if he wishes to make sure of their success.

Hundreds of examples supporting Schübeler's conclusions could be obtained in all directions.

The objection will be urged that in these artificial experiments there is no formation of new species. In the preceding pages I have already recognised the justice of this criticism as a statement of fact. But I do not acknowledge it as a valid basis for the proposition: that the fact that external influences, that artificial cultivation, effect changes in animals and plants which last as long as the influences themselves proves nothing with regard to the action of external conditions, to the inheritance of acquired characters, to the modification of species.

All the results of cultivation which man successfully produces in plants and animals, and for thousands of years has produced, prove rather most incontestably the fact that acquired characters are hereditary.

Why permanent species have not been produced by such results of cultivation, is a question by itself which I have attempted to answer in the preceding.

In my opinion, to expect that effects which have been obtained and maintained by artificial conditions continued during a relatively short period should persist after the sudden cessation of those conditions is to expect what is perfectly unnatural and unphysiological.

Only by gradually bringing the organism slowly, step by step, during a very long period of time, into new conditions —if moreover the new conditions harmonised with the direction given to the evolution—could the artificially produced characters possibly be maintained. In Schübeler's experiments, by attention to the necessary requirements, artificial characters were to a certain degree maintained. But nature alone can completely satisfy these requirements. If we had in our artificial experiments command of the measureless periods of time which have been employed by nature, we should be able perhaps in many cases to imitate artificially, or even to surpass, her success in producing permanent modification.

The facts, however, which are brought to light by study of the gradual transformation in plants and animals seen in passing from north to south, or from high to low elevations, afford the clearest proof that climatic conditions have promoted, *i.e.* have helped to produce that transformation; and the gradual transformation of extinct plant and animal forms also points in the most definite way to such external causes.

This gradual transformation is in many ways so obvious that its explanation becomes of secondary importance compared with the problem of the causes of the separation of the organic chain into species. That the chain was originally continuous is proved by the fact that it can be actually traced without interruption between widely separated limits among extinct organic forms.

The tropical climate, however, certainly has exercised, sometimes in a very short time, a direct influence on particular characters of animals, especially on the integument, on the thickness of the coat of hair, on the nature of the wool, etc. This is demonstrated by authentic instances.

In India our races of dogs after only two generations lose their distinguishing characteristics; thus, for example, in pointers the nostrils become more contracted, the snout more pointed, the size smaller, the limbs more slender (Everest). On the Guinea coast the ears of dogs become long and stiff, as in foxes, to which they also approximate in colouring, so that in three or four years they deteriorate into very ugly creatures, and after three or four generations their bark becomes a howl (Bosmann). In Paraguay the domestic cat has become one-fourth smaller, its body is slender, its hair short, shiny, thin, and pressed closely to the skin, especially on the tail, which is almost naked (Rengger).

In the Malay Archipelago, and in Further India, the cats have a stumpy tail of only half the usual length, and often a kind of knob at the end of it (Crawfurd).

Also dogs in the tropics, according to several authorities, often become thin haired.

The regular summer and winter changes of the hair in mammals are supposed by Darwinian reasoning to have been gradually acquired by selection, no thought being given to the ultimate causes of the phenomenon. But the facts above stated seem to indicate that these causes are to be found in the direct influence of climate, or rather in the changes in the physiological condition of the skin produced by climate.

The fact that our mammals acquire a thinner coat of hair in spring, a thicker coat in autumn, might also be due to another cause, namely, that they are as a rule in a better condition of nutrition in autumn than in spring. In that case the changes of hair in our domesticated animals would be an inherited acquired character.

In Porto-Santo the rabbits which have there run wild have on the back red hair only occasionally mixed with black, or black tipped hair. The throat and certain parts of the lower surface are, instead of pure white, pale gray or lead colour, the upper side of the tail reddish brown instead of dark gray, the ears without a blackish border. In less than four years a specimen imported into England from Porto-Santo almost entirely lost the peculiarities of the race (Darwin).

In New Zealand the climate is said directly to favour the formation of a longer and stronger wool.

In more southern regions animals and plants have in general better conditions of nutrition than in the north, and to these also, not alone to the effect of climate, must be attributed the changes of form which they undergo when transferred from north to south. Of the importance of the influence of nutrition on growth and change of form I have already spoken.[1]

I have only to add here that the increase in size of animals of the same genus, and even of the same species, in passing from north to south is sometimes in the highest degree conspicuous. It is often then most clearly evident that where this increase in size has been accompanied by the development of other characters, the formation of new species has become possible, and it is often a debated question in such cases whether the changed forms are to be called new species or not. I may mention in illustration the species of the genus Scorpio, which in its smallest form as Scorpio germanus occurs as far north as the Tyrol. I may mention further the wonderful increase in size of the species of Julus and Scolopendra towards the Equator.

I have minutely investigated and described an example of such increase in size and change in colour, and of the characters connected with these by correlation, in the common lizard (compare below).

The acquired and inherited characters depending on nutrition in our domestic animals are so well known and so obvious that nothing further need be said about them.

A little careful consideration, however, will also show that a large number of these characters must have arisen without

---

[1] With reference to the effect of nutrition in the modification of forms, compare especially what is said below on bees as an example of the importance of acquired and inherited characters.

any assistance from selection: ultimately, in fact, all of these characters owe their origin either to the direct influence of changed conditions of life, or to correlation.

I will mention some examples in which selection is excluded. A particularly striking one has recently been communicated to me by a German landowner who is a large breeder of cattle and sheep. He assures me that by feeding the lambs with powdered bones he has obtained in a few years a race of sheep of much greater weight, more massive skeleton, and larger in size than the original form. In this case selection was not employed.

Another similar example is this, that the feathers of domesticated ostriches, according to the statements of experts, are heavier than the wild, and for this reason that the quills are thicker. In consequence of this a given weight of the feathers of domesticated ostriches fetches a smaller price than of the wild.

Other examples of characters acquired and inherited through cultivation without the assistance of adaptation I shall bring forward in the next section.

We have to consider next the already mentioned case of the crustacean Artemia salina, as a proof of the change of one species into another in a state of nature through a change in the saltness of the water.

The little crustacean Artemia salina, which occurs in our salt inland waters, acquires, through diminution of the saltness of the water in which it lives, the characters of the fresh-water genus Branchipus, among them a nine-jointed instead of an eight-jointed abdomen; but through increase in the saltness it becomes transformed into the species Artemia Milhausenii, which lives in the Crimea, and this on diminution of the saltness conversely into Artemia salina.

As with the Axolotl this single example alone speaks forcibly in favour of the view I am advocating.

Herr Schmankewitsch[1] observed that the transformation into A. Milhausenii, M. Edw., in a lake in South Russia, in consequence of the increase of saltness, was completed after an interval of three years. The latter animal differs from A. salina chiefly by the absence of the caudal lobes. Schmankewitsch has also produced the A. Milhausenii through artificial cultivation, *i.e.* by rearing several generations of A. salina in water of increasing saltness.

The species of Branchipus into which Artemia salina is transformed in consequence of the dilution of the salt water in which it lives is Branchipus spinosus Grb. We have here, therefore, a transformation not into a form which is described merely as a distinct species, but into one which is ascribed to another genus.

Branchipus has more somites than Artemia. It must be regarded, compared with Artemia, as the phylogenetically older form, and thus the transformation from Artemia into Branchipus can be considered as a reversion. But not so the transformation from Artemia salina into A. Milhausenii, for the latter stands higher in evolution than the former. Moreover, it is most remarkable that the saltness of the water has also an effect on the duration of the development: a higher proportion of salt retards, a lower accelerates it.

In other crustacea also, in minute Daphniæ, Schmankewitsch observed similar changes as a consequence of the increase or decrease of saltness. In my opinion, there can be no doubt that, among other cases, the nature of the waters in which they live has governed the evolution of new forms among our fresh-water fishes. The species of Salmo in our fresh waters, which are so nearly related to one another, are evidently in part local forms, or they are obviously influenced by the character of the water in which they live. Especially noteworthy in this connection is the lake trout,

---

[1] *Zeitsch. f. wiss. Zoologie*, Bd. xxv. Suppl. and xxix.

Salmo lacustris, as compared with the brook trout, Salmo fario. The epicure easily distinguishes the one from the other. The lake trout has coarser flesh than the brook trout, and also tastes somewhat peaty. The two forms also show differences in the colour and marking of the skin, but the specific distinction is generally founded principally on the dentition of the vomer. Quite recently, Professor Klunzinger has, certainly with reason, publicly embraced the view long ago put forward by me, that both fishes are but one species. From the influence of different habitats, the one living in lakes, the other in the rapid waters of mountain brooks, they have become so different that they have been taken for distinct species, notwithstanding that it has ever been of the greatest difficulty to discover any really definite distinction between them.

In a similarly slight degree does the salmon-trout Salmo trutta differ from the lake and brook trout, so that we must regard, with Klunzinger, this also as a "biological species."

The comparison of the American fauna with the European shows a large number of curious parallel forms, *i.e.* several of our species of animals have representatives in North America which are very similar to them, but which have so many peculiar characters that they are described as distinct species, or at least form distinct varieties. Thus the reindeer: Cervus tarandus and C. caribou, Aud. and Bachm.; Canis lupus, of which orientalis as our, occidentalis as the American wolf are distinguished; Ursus arctos our brown bear, and Ursus americanus and ferox in North America; Cervus elaphus, our stag, and C. canadensis, the Wapiti stag; Bison europæus and B. americanus, etc.

These relations of forms can only be explained on the ground that the faunæ of America and Europe were once united by the connection of the two continents, and that they have gradually formed two distinct groups from the difference of the environment. It is also certain, as I have before mentioned, that the American fauna, especially in regard to markings, has remained at an earlier stage of evolution; thus the differentiation has been partly accomplished by genepistasis.

Species representing one another in this way (vicarious species) exist also, as is well known, in abundance in the faunæ of various parts of the world; only both the separation and the relationship are not as a rule so clearly expressed as they are between the European and American forms.

The most striking example, however, of the importance of climate, character of the soil, and isolation in the moulding of the organic world, is supplied by the very peculiar fauna and flora of Australia. Australia affords us an enclosed area of evolution in nature on a large scale than which no better could be wished. Marsupial animals represent there the most diverse groups of our mammals in a series of parallel forms. There are marsupials which are proxies for our Rodents, others for our Ruminants, and so forth, and these have a dentition similar to that of the Rodents and Ruminants respectively; nay more, the representatives of the Ruminants have a multiple stomach.

Darwinism of course says this similarity has arisen through selection based on the same requirements: since in Australia no other animals except marsupials have developed out of the lower forms, therefore the marsupials have adapted themselves to the diverse possibilities of the country. In other countries, however, the evolution of mammals proceeded in various directions from the original marsupials, and so arose the diversity of the mammalian kingdom. But, without doubt, the direct influence of the uniformity of the Australian area ought to be held responsible for the uniformity of the animals and plants belonging to it, and the fact that the marsupials have assumed in their organisation similar relationships to those of the other mammalian groups outside Australia is undoubtedly to be attributed partly to the immediate influence of the external world—for the beginnings of the various organs appeared originally within and without Australia independently, and have evolved themselves into completely analogous structures.

Besides, we cannot say that other higher forms of life could not have arisen in Australia, for new plants and animals transported thither from other parts of the world get the upper hand of the Australian forms and crowd them out.

One of my assistants, Dr. Vosseler, has made an observation which, if the result is really due to the causes to which he ascribes it, forms a noteworthy contribution to the view advocated by me as to the causes of the transformation of forms.

Dr. Vosseler had at the beginning of the year 1886 (6th Feb., that is now one and a quarter years ago) a number of fully developed young Salamanders (Salamandra maculata) taken from the oviduct of the mother. These he put into a spacious aquarium for subsequent use, where he left some which he completely forgot, and which therefore were not fed. After the little creatures had remained thus over a year without being properly fed (for the aquarium contained only Algae and a few Infusoria) they were again found. In the whole of this time they had only grown from three centimetres to five centimetres, and had not undergone any metamorphosis. They had not quitted the water, although it was possible for them to do so, and up till the present day they still remain in the water; they have retained the large gills, the long tail, and in general all the characters of the larvæ, and they lead an active existence as aquatic animals. Investigation showed that these creatures, which usually feed on worms, all kinds of larvæ, etc., had nourished themselves with Algae together with Infusoria. They had thus become almost complete vegetarians.

In other cases, Amphibians, which under natural conditions lose their gills and change into exclusively air-breathing terrestrial animals, have been compelled to retain their gills and remain aquatic by being prevented from leaving the water, or at least from breathing much air directly, as for instance when the vessel in which they were contained was covered by a cloth. The success of such experiments shows by itself the important part which the direct influence of external conditions plays in the modification of organisms.

The fact, if proved, that insufficient nourishment can cause an organism to remain at a low stage of its normal development, *i.e.* can retard its phyletic growth, just as bad nourishment hinders the individual growth, would go very far to justify the explanation of the evolution of forms as growth.

Indeed, nature affords instances which make it *à priori* probable that inadequate nutrition may have this effect. It is well known that the cockchafer takes three years to develop in the south of Germany, four years in the north. In South Germany every third year is a cockchafer year, in the north every fourth. But it happens in South Germany, after a long inclement winter, that in one region or another even there the development is retarded by a year; and accordingly in localities which are adjacent but sheltered in different degrees the cockchafer years are not the same. In exceptionally warm summers even in the north the cockchafers appear a year earlier; these are the beetles which fly in August and September.

The last period of its life beneath the earth the larva passes in the chrysalis state. The change into the chrysalis takes place usually in the early summer of the year which precedes the appearance of the beetle. The variations of temperature accelerate or retard the chrysalis stage. As the variation of temperature, however, shortens or prolongs the feeding-time of the larva, it may be assumed that the change of the voracious larva into the chrysalis stage is conditioned partly by nutrition.

Now, the larval stage of the cockchafer certainly repeats a stage in the ancestral evolution. And this example shows that the ancestral evolution must have been retarded under unfavourable conditions of temperature or nutrition—in other words, that the evolution of the type was, like the individual growth itself, conditioned by warmth and nutrition.

Moreover, it seems self-evident that a corresponding effect must be ascribed to warmth alone, as is shown by the influence of temperature on the period of development, in the most various species, not only of larvæ, but also of chrysalids, which take no nourishment. I shall shortly have to speak at greater length on the importance of warmth in the modification of species. Previously I have to point out some facts which show still further the influence of nutrition in the process.

Every boy knows that the butterfly called the "brown bear," Euprepia caja, can be reared in various varieties, according to the different nourishment which is supplied to the caterpillar. Here again the varieties which arise through change of food are not fortuitous but perfectly definite. For the differences of the markings which numerous such varieties in my collection exhibit are not irregular, but reveal, as in other cases, quite definite directions of modification.

Other butterflies are also known on whose colouring and markings the food of the caterpillars has great influence. I have special grounds for the conviction that many new species have arisen through the caterpillars having been at one time or another forced to accommodate themselves to a change of food. In evidence of this is the fact that numerous very slightly differing species of Vanessa, *e.g.* V. polychloros, xantho melas, l. album, and urticæ, lay their eggs on different plants. It is natural to infer that in such cases the difference in the quality of the food has been the cause of the origin of different characters.

I have to consider now a special example, which I have

carefully followed out, of the effect of a general change in the external conditions, and particularly in the nourishment, on the modification of organisms, another proof of the inheritance of acquired characters.

Our domestic cat is certainly derived from the fawn-coloured cat (Felis maniculata). The two cannot be distinguished, as I have proved in the journal *Humboldt*, 1886, either by the skeleton or by any other really decisive criterion. Still the fawn-coloured cat, apart from her yellow-gray colour, seems to be somewhat more slenderly built and also to have somewhat shorter and smoother hair than our domestic cat usually has. The latter, moreover, in consequence of the protection which it enjoys in the house, has gradually varied much in colour and marking, while the wild (although among the Niam-Niam, according to Schweinfurth, it is half tamed and replaces the domestic cat) fawn-coloured cat in Africa is adapted to the colour of the desert, and has retained the original transverse stripes, which so often occur even in the domestic cat on a gray ground. This transverse striping and gray colour is also evident in the wild cat, which, in my opinion, is likewise to be regarded, not as a proper species, but as a variety of the Felis maniculata domestica, which is developing into a species. Special breeding, selection, has evidently not been much applied to the markings of the cat in Egypt, where it has been domesticated from the most ancient times. The pleasure of reproduction has been freely permitted there to all cats; scarcely any cats were killed, for the cat was sacred. Thus we have in the variety of marking in the house cat, not something produced by selection, but something which has arisen from the effect of external environment, firstly from the absence of selection, and ultimately from some direct causes of modification.

I have pointed out that this variation also is not altogether irregular, but exhibits certain fundamental lines.

In still higher degree is this true, as I have shown, for our domestic dogs. "The apparently innumerable varieties in the markings of these animals—spots, dots, and streaks—are by no means fortuitous, irregular, but can be traced back to a perfectly definite fundamental plan."[1]

This fundamental plan of marking in the domestic dog and the domestic cat of course approximates to the original marking, from which it is derived by the strengthening and coalescence of parts, but as a whole it is something new. There is evident in it (at least in the dog) a new definite direction of evolution, which can only be due to causes connected with the domesticated condition. It is true that human selection in the breeding of dogs must be directed to a certain symmetry, even in markings. Dogs which have less of this symmetry are less propagated, but in valuable races of dogs, *e.g.* pointers, even this degree of selection is out of the question; and as the regularity of which I now speak, and which I have demonstrated by figures in *Humboldt*, has hitherto been noticed by no one, and was unknown, therefore it could not be an object to the breeder. I may remark in addition that according to my observation the apparently irregular spots in our cattle can be referred to a perfectly definite law. For the dog and the cat, however, the remarkable fact comes to notice, that the new plan in the markings has a certain agreement in both, in both arose independently, through modification of the same parts of the original transverse striping common to both. This striping can still be recognised as a token of the original blood relationship between the dog and the cat.

I have traced in detail this modification of the marking in its beginnings in the street dogs of Constantinople, and have already published some remarks on it in the journals already named. These dogs are evidently the immediate descendants

---

[1] Cf. *Humboldt*, and *ZoologischerAnzeiger, loc. cit.*

of jackals. They were originally jackals which, under the protection of the noble Mohammedan custom not to harm animals but to cherish them and feed them, established themselves in the neighbourhood of men in villages and towns, although men did not domesticate them or assume ownership in them. They are thus in perfect liberty. These dogs supply the best proof of the truth of what I have before said, viz. that the cessation of natural selection through the struggle for existence in a natural state, and the positive influence of the environment of civilised life, are the causes which have produced the new characters in these dogs. In place of the old jackal colour, the yellow of the desert, and the traces of definite inherited markings upon this general colour, appear the beginnings of another colouring and of a new marking of spots, the latter following the law which, with the help of the explanation given by me in *Humboldt*, can always be recognised more or less distinctly in our domestic dog. Each dog is on the occurrence of these changes apparently quite different from every other, but in reality the dark spots are constantly on definite portions of the skin, the light or white spots on the intermediate portions. No man, no kind of selection, has contributed to this result: for the dogs' sense of beauty cannot be adduced in favour of sexual selection (in which, in dogs, the sense of beauty scarcely plays a part) in this modification: because the new characters, making their appearance for the first time in spite of the long continued inheritance of the old, are always irregular in comparison with these, and often have a positively ugly irregularity. The ears of these Constantinople street-dogs occasionally begin to show signs of drooping at the tips, a change which is so complete in many of our races of dogs, *e.g.* in pointers, that they have lost much in the sense of hearing by domestication. This change is obviously due to the absence of the need to use their ears, and to the cessation of selection in relation to hearing—an acute perception of sound is no longer so necessary as in the natural state. But the reason why these dogs begin to erect the tail and carry it upright, while the ancestral jackal, like the wolf, carries it hanging down, is not so easy to discover, although the fact could scarcely be explained as a case of adaptation.[1] As a rule, the Constantinople dogs retain not merely the exact shape of the jackal, the pointed head, the narrow body, etc., but also the brownish yellow ground colour, although they are spotted with black, or black and white. Occasionally, however, the black or white becomes preponderant—dogs uniformly black have come under my notice, especially on the Asiatic side of the Bosphorus, in Scutari.

In a journey through Roumelia and Bulgaria over the Balkans—from Constantinople to Adrianople, Philippopolis, Sophia, and thence to Lom-Palanka on the Danube—I was able to trace, I may say step by step, a transformation of the brown, somewhat spotted jackal-dog into an ordinary domestic dog, which is at first like the Pomeranian breed, with a *short* compressed body and tail rolled up, and which, in consequence of better nourishment, is large and strong, and more uniformly coloured, most often white. The farther one passes into Christian regions—that is, the more the dog becomes a house animal, the more does he show this transformation, of which a great part is undoubtedly due to better nourishment—in other words, to directly acquired and inherited characters.

I have to remark further that through these observations, and through the study of ancient monuments, particularly of those of the tombs in Athens, where the departed are constantly represented with their favourite dogs, I have reached the conviction that the Pomeranian is a breed directly derived from the jackal, and must be considered one of the most ancient breeds. All the dogs on those Greek monuments are

---

[1] See a subsequent passage.

Pomeranians. A breed equally ancient is the Eskimo dog, which probably comes directly from the wolf. Both breeds have retained the pointed, erect ears of their ancestors.

Apart from all this there is, it must be allowed, scarcely an animal which affords a more perfect example of pammixis than the street dog of the East—*e.g.* in Constantinople—and the house cat. In the former pammixis, which in our dogs is troublesome, is unchecked. The animals carry on their family life on road and path in sight of every one. The mother drops her pups in the most crowded streets of the city, and lies suckling and warming them in the middle of the path, without avoiding man, who, if he be a Mohammedan, disturbs her not, but habitually steps aside and leaves her in peace.

In spite of this pammixis in these dogs definite new markings, formed according to a law and identical with those of our domestic dogs, have appeared, obviously in consequence of new conditions of life.

Why a definite shape of body has not been impressed on domestic dogs I will not decide. Reasons are not far to seek. In any case, a continual struggle still occurs between the old shape and the new. The external conditions in the regions in which the dog has been domesticated are very various. He has been modified in very different degrees in these different regions. Crosses between the separate stages of modification are always taking place, and man constantly by selection seizes upon the result of these unions. In the Eastern dog also this struggle between old and new still occurs, but man does not interfere in it; in the Eastern dog, therefore, the new characters are least obscured in their beginnings: through this dog I was first led to the discovery of the law which the markings follow.

Dogs thus show, in my opinion, that new characters are due to external conditions and an internal direction of evolution, and can be acquired and inherited in spite of all pammixis.

Experiments on the Influence of Temperature on Lepidoptera

The direct influence of warmth on the modification of animals is further shown by the experiments of Dorfmeister and Weismann on butterflies.[1]

Since the fourth decade of this century it has been known that the two butterflies Vanessa Levana and Vanessa Prorsa, formerly considered as different species, are really one and the same. And, indeed, in these two forms of the same butterfly we have two generations developed at different seasons of the year. V. Levana is the winter form, V. Prorsa the summer form. The chrysalis of Levana remains dormant during the winter, the butterfly emerges in the spring, breeds immediately, and its progeny go through their whole development in the summer; from their chrysalids emerge the V. Prorsa, whose progeny then pass the winter as chrysalids, and in the spring produce the Levana. These two forms of butterflies are differently coloured and marked, and it is in complete agreement with many other examples of the effect of warmth on the formation of pigment in the integument, that the summer form (exposed to warmth), Prorsa, is much more deeply coloured than the winter form (exposed to cold), Levana. The former is deep black, the latter brown-yellow in its ground colour. The characters of the two also afford a very strong proof of the correctness of my theory of the importance of correlation in the formation

---

[1] G. Dorfmeister, "On the Effect of Different Degrees of Warmth applied during the Period of Development on the Colouring and Markings of Butterflies," *Mitt. des naturwiss. Vereins für Steiermark*, 1864; and A. Weismann, *Studien zur Descendenztheorie*, I. Ueber den Saison-Dimorphismus der Schmetterlinge, 1875.

of new species; for the transition from one form to the other is sudden and immediate, inasmuch as not one but several new characters separate the one from the other.

The original form is obviously the Levana, which approximates in colour and markings to related species of Vanessa, especially to Vanessa polychloros and c. album. The summer form, Prorsa, shows nothing of this marking; it has on the two black wings a transverse row of light spots, and on the anterior wing in addition to this row some white dots; its whole marking appears to connect it very closely with the genus Limenitis.

If in the whole habitat of Vanessa Prorsa such climatic changes were to take place that the form produced by cold were no longer developed, then the Prorsa would alone remain as a new species, which would be classed according to its marking rather among the species of Limenitis than among the species of Vanessa above named, or would in any case take a position completely separate from the latter.

Weismann has himself attempted to prove that their colouring is not protective to butterflies during their flight, "because the colour of the background against which they are seen is continually changing, and because their fluttering movement would betray them to their enemies in spite of the most complete adaptation to the background." He declared, on the basis of his own observations, that butterflies are most liable to be seized by their enemies when at rest with their wings folded together, and especially at night; in the latter case, mostly by spiders. For this reason butterflies so often have colourings and markings on the underside of their wings which mimic the surface on which they settle.

In fact, who has not, in the attempt to catch butterflies, met with the experience that the insect suddenly vanished from his sight? It had settled somewhere, and even with the keenest sight could no more be found. Even the common thistle butterfly (Vanessa cardui) shows something of the kind, although not so completely as other species. As soon as it has settled on the ground, as it does in play, with wings folded together, it can only be discerned if it has been seen to settle, or if, as it generally does, it flaps its wings.

It may be held, indeed, that butterflies are protected from the beaks of birds which pursue them in flight by the size of their wings, since a bird is more likely to bite a piece of the wing than to grasp the body.

If I mistake not, another naturalist has already expressed this view somewhere. Some years ago I came across a peculiar proof of its correctness. On a hot summer's day I was on the high plateau of the Swabian Alp; far and wide no water was visible, but at one spot in the field-path there ran over it the outflow of a little spring, forming a shallow, clear pool in the track. Here sat hundreds of butterflies, all whites and blues, closely crowded together, drinking thirstily. On my approach a number of birds (stone-chats) flew from the spot, and when I came up I found a number of maimed butterflies lying fluttering on the ground; pieces had been bitten from the wings of most of them—indeed the wings were often torn to pieces before the birds succeeded in getting the bodies of the butterflies, although these were sitting quietly on the ground. And only because they were sitting on the ground had the birds been able to get their bodies.

Thus the colour and marking of the upper side of the butterfly's wing cannot be regarded as an adaptation, as a protection against enemies. But there remains another adaptation, that due to sexual selection: the colours and the markings may be a sexual attraction. I am of opinion that this is generally the case. Details of colour and marking, however, as their sudden development shows, and as numerous facts besides have convinced me, are due to physiological causes.

Such causes were assumed by Weismann for the origin of season-dimorphism, *i.e.* the formation of winter and summer forms of animals.[1] He tried accordingly, by raising the temperature in winter, to rear Prorsa directly from the offspring of Prorsa, and conversely, by lowering the temperature in summer, to rear Levana directly from the brood of Levana. Similar experiments had already previously been made by the Styrian entomologist Georg Dorfmeister. It seems, however, that such experiments have been even more widely made, for an acquaintance told me that as a boy he had practised the same thing with his companions successfully at his home in Darmstadt, and indeed on Vanessa Levana and Prorsa. Weismann put chrysalids of Vanessa Prorsa, which he had reared from the caterpillars of butterflies which emerged in April, into an ice-box with a temperature of $10°$-$12°.5$ C. This temperature, however, was not low enough to produce Levana again; but the experiment succeeded so far "that instead of the Prorsa form to be expected under ordinary conditions most of the butterflies emerged as the so-called Porima, *i.e.* as an intermediate form between Prorsa and Levana, occasionally observed in nature, which possesses more or less of the markings of Prorsa, but still mingled with much of the yellow of Levana."

As already remarked, the Levana, not the Prorsa, is the original form. Weismann's figures of this single case again seem to give support to my view that the males, as a rule, are a stage in advance in evolution, and that therefore their characters indicate those of the variety or species which follows next in evolution. The male in the Levana figured by Weismann in Fig. 1 has the greatest likeness to the female Porima which passes through its male (Fig. 3) into the

---

[1] I would suggest instead of the word season-dimorphism, first used by Wallace, which is in every respect monstrous, the word hora-dimorphism, or the shorter expression season-variation.

female of Prorsa (Fig. 6). But as there are many more intermediate forms than those figured by Weismann, this explanation requires further confirmation. Afterwards Prorsa chrysalids were placed by Weismann in a temperature of 0°-1°.25 C. in the ice-chamber: out of twenty butterflies, fifteen now changed into Porima, and among these were three which were almost indistinguishable from Levana. Five remained unaffected by the cold and emerged as the summer form Prorsa.

Dorfmeister had never applied such low temperatures, and had only succeeded in obtaining Porima.

The converse experiment of Weismann to rear Prorsa from Prorsa by raising the temperature succeeded, among forty chrysalids which were kept in a forcing-house at 15°-31° C., only with four, of which three were Prorsa and one Porima; all the rest produced Levana in the next spring.

The majority of the species of our white butterflies (Pieridæ) show very strikingly a winter and a summer form. In the winter form of Pieris Napi the roots of the wings are much powdered on the upper side. In this case Weismann, by keeping the chrysalids of the progeny of the winter form three months in the ice-cellar and allowing them to emerge in the hothouse, obtained perfect winter forms. But there is a variety of P. napi which lives in the Swiss Alps, in the Jura, and in the polar regions, and which can be described as a very dark form of the winter variety of Napi: P. Bryoniæ. The male of Bryoniæ is almost exactly similar to the ordinary winter form of Napi, from which the female differs by the gray-brown powdering of the entire upper surface. In the polar regions Bryoniæ is the only form of Napi; in the Alps (excepting some isolated regions) it is mixed with the ordinary Napi. I agree with Weismann that Bryoniæ must be the ancestral form of Napi, basing my opinion on the fact that, as Weismann's figures show, a series exists from the female of the former as starting-point through its male to the female of the ordinary winter form of Napi, through the male of the latter to the female, and so to the male, of the summer form. Here also the males always show the beginning of the new characters which distinguish the next variety, which is one degree higher. The common Napi must therefore, according to this interpretation, be regarded as a variety which has arisen from Bryoniæ through the influence of a warmer climate, as the commencement of a new, with us the only, species.

Weismann placed the chrysalids of Bryoniæ in a hot-house at 15°-30° C., but only Bryoniæ emerged; he failed to convert Bryoniæ into Napi, as he had conversely, by the application of cold, changed the summer form of Napi into the winter form.

Thus in Levana, as in Napi, the progeny of the cold form are converted by cold again into the cold form, but only exceptionally the progeny of the warm form into the warm form. In other words, the application of cold had an effect, the application of warmth generally none.

But in summer two generations of Prorsa occur. The first takes flight in July, the second in August. The latter supplies the chrysalids which form the Levana after the winter, and which Weismann could only exceptionally, through warmth, change again into Prorsa. Weismann concludes from this that different reactions follow the same stimulus, and that this difference can only lie in the physical nature of the generation operated on, not outside it. According to him, cold and warmth are not the direct causes of the origin of the Levana and Prorsa forms, but only the indirect. The change of the progeny of the cold form back again into the cold form through cold he refers to reversion to the ancestral form.

The origin of Prorsa is explained as having occurred gradually through the gradual rise of the warmth of the climate, *i.e.* of the summer. As the Levana is the ancestral form of Prorsa it cannot revert to Prorsa, and therefore the latter is not produced artificially. Weismann further argues that the different Prorsa generations must have different physical constitutions, because the later of the two cannot be converted by warmth into Prorsa.

In answer to this, it is to be remarked that it was this *second* generation of Prorsa which in one case on the application of warmth again produced some Prorsa (Weismann's experiment, 10 A). And thus the hypothesis of reversion breaks down and that of the direct influence of warmth is proved.

Dorfmeister, however, has, as will be discussed in the following, by the application of warmth produced Prorsa out of Prorsa.

If warmth had less influence than cold there might be an explanation for it. Weismann goes on the assumption that artificially applied warmth affords the same stimulus as the natural. I am of opinion that this is not to be expected. At least in artificial experiments the effect of the light of summer is wanting. Further, the duration of the period of development in them has always been longer than in the case of the warm forms under natural conditions. Besides, the natural rise and fall of temperature, and in general the naturally effective degrees of warmth as well as the effect of radiant heat, cannot be imitated. So that the stimulus of artificial warmth is not the same as that of natural. In the case of cold the conditions are much simpler.

In actual fact, however, Dorfmeister's experiments show that artificial warmth has the same effect as natural.

Taking all in all, from the facts already detailed, in my opinion, the simple conclusion follows, that it is obviously the action of warmth which in the natural state originally produced one form of butterfly out of another, and that adaptation had nothing to do with the matter.

There are still other butterflies known which have a summer and a winter form which were formerly believed to be distinct species, *e.g.* the two blue forms, very different in marking and in size, Lycæna Polysperchon and L. Amyntas, as shown by P. C. Zeller[1] by rearing experiments; further, the white forms belonging to the Mediterranean shores, Anthocharis Belia and A. Ausonia, as shown by Dr. Staudinger.[2]

Weismann speaks of five such species. But there are more whose winter and summer forms were formerly described, not as species, but as varieties. Weismann mentions twelve. Among these are several of our commonest white butterflies, and also the commonest of our blue species, Lycæna Alexis, in which the differences between winter and summer form are very small. I may add that the sail-butterfly (Papilio Podalirius) forms a southern darker variety, P. Feisthamelii Dup. (South Europe, North Africa, West Asia), and that this produces in Algiers a second peculiarly marked generation, the P. Letteri, Const.

The American butterfly Papilio Ajax wherever it occurs appears in three varieties, viz. var. Telamonides, Walshii, and Marcellus. The American entomologist Edwards has shown by breeding experiments that all three belong to one cycle of development, and that Telamonides and Walshii emerge only in spring, while Marcellus occurs only in summer, and in three successive generations. Marcellus is therefore the form produced by summer, that is, by warmth. Telamonides and Walshii, according to Weismann, are to be regarded as the ancestral forms. Of these Telamonides is considered to be due to "imperfect reversion," like Porima, Walshii as the original form of the species. Edwards records that out of fifty pupæ of the second summer generation of

---

[1] *Stett. entomolog. Zeitsch.* 1849.
[2] *Ibid.* 1862: "The Species of the Lepidopterous genus Ino, with Remarks on Local Varieties."

Marcellus forty-five produced Marcellus imagines after fourteen days, and five did not change until April of the next year, and then into Telamonides; of the pupæ of each of the three summer generations only a portion produced butterflies after a short period (fourteen days); another much smaller portion did not change till the following spring, and then into the winter form. Weismann concludes from this that "here there is no doubt that, not external conditions of various kinds, but purely internal causes determine the retention of the long-inherited direction of development, for all the larvæ and pupæ of the many different broods were exposed simultaneously to the same external influences."

This statement, in my opinion, is probably to be thus altered: The external conditions cause, as a rule, the reproduction of the parent form, but on some individuals the summer warmth has not sufficient influence; they do not develop before the winter, and the winter cold, *i.e.* the absence of warmth and light, then determine the emergence of the cold form. It is a question, therefore, obviously of the different sensitiveness of the individuals in regard to external influences, a difference which is equally striking in the metamorphosis of the Axolotl, in which also some individuals are more prone than others to change into Amblystoma.

In the highest degree noteworthy for my views, however, is the fact already pointed out, that as in the Axolotl, so in these butterflies, there are not a large number of transitional forms in the metamorphosis side by side, but that in every case when transitional forms do occur they form distinctly separate stages. Only when the ancestral and the new form live side by side and interbreed, as with Bryoniæ and Napi in the Alps and the Jura, do the varieties show intermediate forms. In Lapland, on the contrary, Bryoniæ is constant. The intermediate form Porima is a great rarity in nature as in experimental breeding.

Weismann, regarding the reappearance of the cold variety as a reversion, not recognising the direct influence of cold, but rather postulating internal causes, attempts to discover other causes which might likewise determine this reversion. And he believes he can point out warmth and mechanical movement as such additional causes.

In the summer of 1869 great heat prevailed at the time of the development of the second summer generation of Levana-Prorsa. From a batch of this generation an unusually large proportion of Porima were developed (out of sixty to seventy, eight to ten.)

As Porima is half way to the cold variety, the fact is to me very incomprehensible, and I should suppose that other causes than the extra warmth determined the formation of Porima, for to assume this as the cause would be to completely contradict all the facts previously detailed.

In another case, a number of Pieris Napi, although they had been reared in a heated room, all emerged in the following spring as the winter form. But the pupæ had made in the summer a seven hours' railway journey; and so the shaking on the railway is taken as the cause of the reversion in this case.

Unable as I am to grant that causes so completely opposite as warmth and cold have the same effect in the modification of forms, I am equally unable to agree that such different causes as warmth and mechanical shaking produce the same result. Since warmth has obviously, under natural conditions, determined the transformation into new forms in the cases previously discussed, I shall hold for the present that only cold has the power of causing a reversion to the original form. That effect of warmth, under natural conditions, which consists in the formation of new varieties which are so distinct that they have been in many cases long regarded as species, has been evidently a direct effect. And it would need, as I before pointed out, only a slight change in the external conditions, or a change in the sexual products arising by correlation, to produce from such varieties absolutely separate species, or to exterminate one of them.

It is in complete accord with this hypothesis, and with my whole view of the origin of species, that the variety Bryoniæ is in the far north a constant sharply defined species. There are other such cases of great interest. Anthocharis Belia and Ausonia were first recognised as the summer and winter form of the same species by Staudinger. They live on the shores of the Mediterranean, extending to the middle of France. In the mountains of the Valais in the neighbourhood of the Simplon Pass the winter form is alone produced. A second generation is not developed in the short summer, but all the pupæ remain throughout the winter. Here also, as in Bryoniæ, my claims as to the effect of external conditions are completely fulfilled, whether Ausonia and Bryoniæ, as Weismann, supposes, are or are not survivals from the ice-age in the localities in question.

Polyommatus Phlæas L., the blue butterfly so common with us, which ranges from Lapland to Sicily, has in Lapland only one generation in the year, in Germany two. But only in South Europe do these two generations differ from one another, in Germany they are still similar.

Another blue form, Lycæna Agestis, has a twofold season variation: "The butterfly occurs in three forms: A and B alternate in Germany as winter and summer form, B and C are the winter and summer form in Italy. Thus the form B occurs in both climates, but appears in Germany as the summer in Italy as the winter form. The German winter form A is completely wanting in Italy, while the Italian summer form (var. Allous) does not occur in Germany." Here we have plainly a little chain of modifications obviously produced by climatic conditions, in which adaptation plays no part, and for which only the views on the causes of modification which I maintain give a satisfactory explanation. My works on the affinities of butterflies will bring to light numbers of such species evidently produced by climatic influences, which are obviously connected but distinct from one another. Here I have intentionally discussed the examples brought forward by Weismann, firstly, in order to use his results for my argument as those of an impartial witness; and secondly, in order to introduce the agreements and the differences between my explanations and his. To these examples of the effect of climate on the formation of butterflies, may be added that of Pararge Egeria, which in South Europe appears in the variety Meione. Meione, however, is connected with Egeria by an intermediate form of the Ligurian coast, which does not produce a summer and winter generation.

In a separate chapter on "The Nature of the Causes of Change" Weismann lays stress on the case of Polyommatus l'hlæas as showing that climate, and not the duration of development, determines the formation of climatic varieties. But, he proceeds, the nature of the change essentially depends on the organism itself, not on the warmth to which it is exposed. It is not the quantity of the black pigment produced which distinguishes the summer from the winter form, but the mode of its distribution on the wings. Under the influence of warmth arise quite different markings: starting from the markings already present, quite new ones are developed, or, "as I may express it in other words, the direction of evolution of the species is quite altered. The complicated chemico-physical processes in the metabolism of the dormant pupæ are gradually altered until they result in a new marking and colouring of the butterfly. That in these processes the constitution of the species plays the chief part, and not the external agent warmth; that the latter rather performs the function of the spark which, as Darwin appropriately expresses it, sets fire to the explosive substance, while the nature and extent of the explosion caused depend on the quality of the explosive material—this is proved by many other facts. Were it not so, increased warmth would necessarily in all butterflies change a given colour always in the same way, always into the same other colour. This is not the case; for while Polyommatus Phlæas becomes black in the south, Vanessa Urticæ, which is likewise red, becomes blacker in the north; and many other examples well known to entomologists can be adduced as evidence. On the other hand, we find conversely that species of similar physical constitution, that is to say, species nearly related, under the same climatic conditions change in an analogous way." Of this our white butterflies are mentioned as examples. "But nothing can more strikingly prove that here everything depends on physical constitution than the fact that in some species the male individuals change differently to the female." The females of Bryoniæ have undergone through climate a greater change than the males,[1] the converse is the case in Phlæas.

"The external influences were exactly the same, but the reaction of the organism was different. . . . I bring this specially forward because, in my opinion, Darwin ascribes too great an influence to his sexual selection when he refers to that alone the formation of secondary sexual differences. The case of Bryoniæ teaches us that the latter can appear from purely internal causes, and, until experiment has supplied some fixed point as to the extent of the influence of sexual selection, the view is justifiable that the sexual dimor-

---

[1] In answer to this I must point out that the warmth variety, P. Napi, is the newer, Bryoniæ the ancestral form. On my view of male preponderance the female Bryoniæ must be the farthest removed from Napi, and the male Napi the farthest evolved. So it is true that the female of Bryoniæ and the male of Napi are at the greatest distance, but it is the latter, not the former, which has been most changed by climate.

phism of butterflies has its causes in great part in the differences in physical constitution of the sexes. It is quite another matter in the case of those sexual characters which, like the voice of male grasshoppers, have an undoubted importance in the relations of the sexes. These can certainly with greater probability be derived from sexual selection."

Regretting as I do that my views are so much in opposition to those of Weismann, it is some consolation to me that this opposition essentially applies only to his new views, not to his old, which latter in very important points agree with mine. From the quotation just given this is sufficiently evident, and I myself could not find better cases in support of my views than those there brought forward; nor could I more distinctly oppose the excessive estimation of sexual selection than Weismann formerly did when criticising Darwin. And yet Darwin attributed to that process an importance but very slight in comparison to the omnipotence with which Weismann to-day endows it, notwithstanding that the experiments desired by Weismann in the above passage are yet to be made. The great agreement between many propositions of Weismann's earlier theories and my view strikes me now most forcibly as I read his treatise again for the purpose of this discussion of the question of the effect of climate on the modification of forms. In the sixth section of this treatise, under the heading of general conclusions, Weismann even makes the statement that differences of specific value can arise entirely through the influence of external conditions of life. He believes that species-formation, at least in Lepidoptera, has taken place to a large extent in this way, and among them more than other animals, because the conspicuous colours and markings of the wings and body are without biological importance, that is, are of no use for the preservation of the individual, and therefore of the species, and thus they cannot have come under the operation

K

of natural selection. He says that it is on these grounds that Darwin sought to derive the markings of butterflies, not from ordinary, but from sexual selection; but that sexual selection in the origin of the colours of butterflies can be dispensed with.

He then raises the question whether species of Lepidoptera, in so far as they arise through changes of climate, only alternate between two forms—a cold-form and a warmth-form—or whether each new change of climate does not rather, if it be pronounced enough to cause a modification, give rise to a new form. In the previous pages I have contended against the view which Weismann now holds on this important question, that the alternation of external influences must hinder the modification of species, forgetting that Weismann himself has formerly upheld my view at least in respect of the changes of climate. He says: " I believe that the old forms are never produced again by alternation of climate, but always still newer ones; that therefore a periodically repeated change of climate is by itself sufficient in the course of a long period of time to produce a series of successive new species. . . . The climate, when it has acted in the same manner on several generations one after another, will gradually bring about an alteration in the physical constitution of the species which will make itself visible by an alteration in colour and marking. Now, when this newly-acquired physical constitution of the species, which we will suppose has been firmly established by inheritance through a long series of generations, is subjected again to a permanent change of climate, this latter, even if it be exactly the same as at the time of the first form of the species, cannot possibly bring back that first form. The nature of the external influence is the same, it is true, but the constitution of the species is by no means the same. Just as a white butterfly, as was shown previously, exhibits quite other alterations than a blue or one of the Satyridæ under the influence of the same change of climate, so must the modification which is produced from the transformed species of our example after the return of the first climate be different—though perhaps in a less degree—from the original form of the species. In other words, if in succeeding geological periods only two different climates alternated on the earth, yet from every species of butterfly subjected to this alternation would be derived an endless series of different species. In reality the variety of climates must have been much greater, and climatic changes for a given species have taken place, not only through periodical variations of the ecliptic which may be assumed to occur, but through geological transformations, and also through migrations of the species itself. Thus from this single cause—variation of climate—a continual change of species must have taken place. When we reflect that species extinct elsewhere must have survived locally, and add to these those local forms which owe their origin to amixis,[1] we cease to be astonished at the enormous number of species of Lepidoptera which we find on the earth at the present day."

It is unnecessary for me specially to point out how distinctly the above passage, which so exactly agrees with my own views, implies the recognition of the inheritance of acquired characters.

Dorfmeister has, he tells us, ever since 1845 made experiments on the effect of temperature on the colour and marking of butterflies. He explains first of all that through many years' experience in rearing caterpillars he has been convinced that the production of varieties of butterflies depends much more on climatic conditions, in which temperature is a chief factor, than on either food or hybridisation. Dorfmeister learns from his experiments that temperature exercises the greatest influence on the colour and marking of butterflies when it

[1] Cessation of interbreeding. Pammixis = indiscriminate interbreeding.

acts upon them during the change into the pupa, or shortly afterwards. In many a rise of temperature produces a lighter, more brilliant ground-colour, a fall a darker or less brilliant; for example, in Vanessa Io, Utricæ, etc. In Euprepia Caja the red-yellow ground colour of the posterior wings is changed by a rise of temperature into vermilion-red ; by a fall, into ochre-yellow.

Dorfmeister considered the Vanessa Prorsa, on account of the slighter definition of its marking, to be the lower form. The marking of Levana is much more complex, and much more finely delineated. But as the simple Prorsa is in fact the higher form, it affords a perfect example of the proposition sustained by me in opposition to Nägeli, that simplification of characters can appear in new forms.

The chief result of Dorfmeister's experiments is just this, that he has obtained Prorsa, with its simple colours and marking, by means of warmth from Prorsa, and in fact from the summer generation of Prorsa which takes flight in August. Thus in his hands the experiment, which in Weismann's with few exceptions failed, repeatedly and abundantly succeeded. On the other hand, he was unable to rear the cold form from the cold form.

Dorfmeister draws no particular inferences from his results. He refers finally to the necessity of completing his experiments, and emphasises his inability to decide whether the modifications obtained be the direct consequences of the rise of temperature, or only indirect, depending on the shortening of the period of development caused by the increased temperature.

But it seems to me to follow decisively from the cases recorded by him that the greater the warmth he applied, and consequently the quicker the development, the darker the colour and the simpler the marking of the Prorsa forms he obtained, and therefore the farther removed they were from the Levana. In the case in which he got the darkest and the simplest Prorsa he applied a still higher temperature than Weismann, although only for a short time: the caterpillars, as soon as they had hung themselves up to change into the pupa stage, "were exposed in the morning on a part of the stove to a maximum temperature of $32°.5$ C. and towards midday had become pupæ;" they afterwards developed in the room.

Since the pupa, which under natural conditions produces Prorsa, has a dormant period of only a few days, while that which produces Levana has one of six months, Dorfmeister, by increase of temperature and the consequent abbreviation of the period of development, has simply imitated the effects of summer.

But the fact which appears to me of the utmost importance is that Dorfmeister, from different temperatures and different periods of development, has obtained a whole series of stages of modification between Levana and Prorsa, while such intermediate stages under natural conditions very rarely occur. He himself remarks that during more than forty years' collecting he met with only one specimen of such an intermediate stage in the wild state, in places where Levana and Prorsa were quite common. The rarity of the intermediate forms (called Porima) is evidently the reason that Levana and Prorsa have been considered distinct species. Under natural conditions, therefore, there is obviously as a rule only one average of summer temperature which produces the ordinary Prorsa. But Dorfmeister has by the application of temperatures beneath this average in a few experiments artificially produced a whole series of such transition stages. These stages can be divided into two groups—(1) those which must be described as transition forms between Levana and Prorsa; (2) those which are different stages of Prorsa. Thus, while the latter were produced by the highest temperature and the shortest period of development, the former were reared after the action of moderate warmth from the same second generation of Prorsa. Dorfmeister kept caterpillars in process of changing into pupæ, or the pupæ themselves, twenty-two days at a temperature of 12°.5 C., and then let them develop in the room. A portion of these pupæ, which belonged to the second Prorsa generation, emerged the same autumn, another portion, like Levana, in the following spring. The former were intermediate between Prorsa and Levana (Porima) in two stages, the latter were Levana partly approximating to Porima.

The Prorsa obtained agree with forms which also occur in natural conditions. I find them all in my collection. Whether this is true for the various Levana or Porima forms I cannot say. The possibility that artificial varieties are produced which do not occur under natural conditions is meanwhile not excluded.

If I interpret Dorfmeister's experiments rightly, it is possible to satisfy the greatest demand that can be made as to the influence of external conditions on the modification of organisms; it is possible to call forth, thermometer in hand, definite varieties which are possibly absolutely new, not occurring under natural conditions.

I cannot sufficiently emphasise the fact that the warmth and cold forms of Prorsa and Levana, differing as they do, not by one, but by several new characters, together afford the most conspicuous example of the correlative, "kaleidoscopic" origin of new forms.

This fact leads at once to a question closely connected with it, namely, whether such discontinuous modifications depending on external influences are not only apparently discontinuous, whether they are not stages in the course of a modification of the species which once proceeded gradually, so that they are merely reversions to forms formerly dominant. I am quite of opinion that this explanation applies in many cases, and equally so that in others it does not—that rather in these other cases we have the formation of new characters by the influence of external conditions. This is proved by the conversion of Levana into Prorsa, and that of Bryoniæ into Napi. In both cases the cold-form is the original, the warmth-form the new. Warmth has evidently in Levana-Prorsa gradually produced a darker, more simply marked form, which in its extreme stages hitherto occurs but rarely. Artificial warmth can easily produce this extreme stage. Retarded development depending on cold can also produce the cold-form —nothing indicates any necessity to explain the origin of either form by reversion.

Further Remarks on the Causes which Change the General Colouring of Animals

I must here make some further remarks on another question, namely, that of the effect of external agents, particularly of light and warmth, on the total coloration.

Dorfmeister would of course say that in Lepidoptera also warmth produces more vivid, more brilliant colours, while dull shades predominate under the influence of cold. This agrees perfectly with my assumptions, and does not contradict the facts brought forward by Weismann, provided that we consider the part that light plays in certain cases (*e.g.* in high regions, Pieris Bryoniæ), and when we further reflect that lighter general coloration can be obtained by distribution of the dark pigment without any necessary diminution thereof.

But the case also occurs in which the colours under the influence of warmth and light undoubtedly become lighter. Thus by rearing the caterpillar of Xanthia Cerago in a higher temperature up to the pupa stage, Dorfmeister obtained the variety flavescens Esp., which occurs also in natural conditions, and which is lighter coloured than the principal form. Exceptional cases of this kind, in which warmth actually calls forth lighter coloration, can only be explained by a peculiarity in the composition of the pigment.

More frequently light causes paleness. Indeed, in animals after death, light causes bleaching of the same colours which, united with warmth, it helped to produce during life. The care with which collections of insects have to be shielded from the light is sufficient proof of this; during the life of the animals the brilliant and dark colour was called forth by active metabolism under the influence of warmth and light, after death this colouring matter undergoes changes in the light which cause it to bleach.

But there are also colours which are bleached by the effect of light during life. In birds this bleaching appears to play a great part in the change of colours. It is clear also that in life both processes may compete together, and I have entered upon these questions once more in order to show with what caution they must be handled, and how little can be inferred from single examples.

Exactly the same antagonism exists, as has been previously mentioned, in the influences of moisture and other agencies. I may be allowed to make here some additional remarks on this subject also.

If moisture generally produced darkness of colour, all aquatic animals would be necessarily dark-coloured, which is notoriously not the case. Proteus anguineus, which lives in the caves of the Karst mountains about Adelsberg, is like other cave animals perfectly colourless. That the absence of colour is in this case due to the absence of light is shown by the fact that when the animal is kept in the light it becomes dark. And this fact alone proves the correctness of my view concerning the direct action of light, and shows that pammixis is not required in order to explain the colourlessness of cave-dwelling animals.

The view that the darker colour of Arion empiricorum on

the sea coast may be caused by the greater moisture of the atmosphere, notwithstanding the lowness of level, is supported by a remark of Leydig's [1] on Helix nemoralis. Leydig says:—

"The influence of light and warmth is very distinctly evident in the colouring of Helix nemoralis in this region. The fine citron yellow exhibited by the shell of this snail at Mainz and in the sunny vineyards of the Main valley is wanting on the banks of the lower Rhine.... On the other hand, it is interesting to notice how in the neighbourhood of Bonn and lower down the Rhine the red of this snail deepens into chocolate brown, and thus forms the beautiful variety above mentioned, which must attract the attention of every collector. It is perhaps too much to say that the moisture of the plain of the lower Rhine determines this change of colour, but we must keep in view the possibility that the moisture coming up the valley from the sea, which certainly here at Bonn has a distinct effect on the vegetation, has something to do with the matter.... And just as in a former publication I attributed to similar causes the formation of black varieties of our native reptiles, such as Vipera berus var. prester, Lacerta vivipara var. nigra, the black variety of Anguis fragilis, so I would maintain that there is a connection between the origin of the black varieties of Lacerta muralis which have meanwhile been made known by Eimer ... and be it noted all of them occurring on the small islands of the Mediterranean, and the action of the moist sea air."[2] Leydig goes on to mention that in the *Entomologische Zeitung* of Stettin, in 1877, a collector of insects stated concerning the Lepidoptera at Bilbao that they

[1] F. Leydig, *Ueber Verbreitung der Thiere im Rhöngebirge und Mainthal mit Hinblick auf Eifel und Rheinthal* in *Verhandlungen des naturhistorischen Vereins der preussischen Rheinlande und Westfalens*, 1881, p. 156 *et seq.*

[2] Cf. the subsequent section on the relation of adaptation to the origin of varieties among lizards. As long ago as 1874 (*Lacerta muralis cœrulea*) I considered moisture as having some effect in producing darkness of colour in these animals.

show a distinct tendency towards the darkening and blackening of the shades of colour, as in the north and on the Alps, and he also stated that the proximity of the sea—the moisture of the air therefore—seemed to cause this variation. Another lepidopterologist reports (*Ibid.* 1879) that a moist clayey soil seems in many species to produce a darker colouring.

I have noticed in my own studies of Lepidoptera that many species of the small islands of the Indian Archipelago, in comparison with their nearest allies of the Asiatic, and also of the European and American continents, are distinguished by remarkably dusky colours. The species of Papilio, Hermocrates, Nonius, and others, with the Australian Leosthenes, in fact form a well-marked dull-coloured group contrasting with the yellow Podalirius and its allies of the continent. The Antiphates forms of the Archipelago are also distinguished by their dark marking.

On the other hand, the Papilionidæ of tropical Africa have strikingly dull, dark shades of colour in comparison with their allies from regions where there is less heat, or less moist heat.

If we take a general survey of our field- and meadow-flowers in spring and summer we see that in spring, on the whole, the colours white and yellow predominate, while red and blue only appear later in the year. And, indeed, red precedes blue. This is in accordance with the physiological changes of the pigment, which show themselves more or less visibly in many red or blue flowers during their development, and also in some degree in the change of colour which occurs in foliage leaves in autumn.

What I wish to point out here is that a quite similar substitution of one predominating colour for another occurs also, on the whole, in butterflies with increasing warmth of climate. Taking the Papilionidæ again as an example, we find those of Europe, North Asia, and North America generally light yellow, or almost white; while in Africa, and to a great extent in India, and South America, in closely related forms, instead of light yellow, dark yellow, and greenish blue, then red and blue and finally black appear.

INFLUENCES OF LOCALITY ON THE VARIATION OF ANIMALS, AND THEREBY ON THE FORMATION OF SPECIES

Finally, on islands there are evidently some local external influences of unknown nature, but probably likewise connected with climate and nutrition, which control the evolution of species, *e.g.* in butterflies, especially affecting their size.

I need only remind students of the Lepidoptera of the Papilio-species of Java on one hand and of Celebes on the other. The original identity of the species on the two islands is obvious. But those of Celebes are now considerably larger than their blood-relations in Java, and exhibit at the same time slight differences of colour and marking more or less pronounced when compared with the latter. To almost every Java species there corresponds a larger, stronger species in Celebes. This I have proved not merely from the material already available in collections, but from numerous specimens kindly collected for me by Herr Seubert, forest-overseer in Java, and by Herr Bauer, merchant, in Celebes.

The larger, more luxuriant island possesses, therefore, in this case, the larger butterflies. Similarly the species of Papilionidæ which occur in the Antilles are smaller than those of the neighbouring mainland, although otherwise but slightly different from them.

The butterflies of Sardinia also are in general smaller and also more darkly coloured than those of the mainland. The same holds for the common fox in the Isle of Man.[1] It is well known that Sardinia possesses numerous peculiar varieties, or species of animals generally. Papilio Hospiton, which occurs

[1] A. R. Wallace, *Beiträge zur natürlichen Zuchtwahl*, German edition, by A. B. Mayer, 1870.

there and in Corsica, is nothing else than a smaller swallow-tail with much darker wings and without tails.

In all these cases and in numberless others, in addition to, and together with the difference in size, other peculiar characters are present, so that new species are established.

Wallace mentions in a chapter in his *Contributions to Natural Selection* entitled, " Variation specially influenced by Locality," other such cases, and among them also the striking instance of Java and Celebes to which I have referred. He remarks that almost the only case previously known of the direct influence of locality is recorded by Darwin in the *Origin of Species*, namely, that herbaceous groups of plants show a tendency on islands to become arborescent, while in the animal world no such facts are known. Then he continues : " In considering the nearly allied species, local forms, and varieties which occur in the Indian and Malayan regions, I find that larger or smaller districts or even single islands impress a special character on the majority of their Papilionidæ. Thus (1) the species of the Indian region (Java, Sumatra, and Borneo) are almost invariably smaller than the allied species which inhabit Celebes and the Moluccas; (2) the species of New Guinea and Australia are likewise, although to a less degree, smaller than the nearest species or varieties of the Moluccas ; (3) in the Moluccas the species of Amboina are the largest; (4) the species of Celebes equal or even surpass those of Amboina in size; (5) the species and varieties of Celebes possess a striking peculiarity in the form of the anterior wings, which is different from that in the allied species and varieties of all the surrounding islands ; (6) tailed species of India or of the Indian region become tailless as they extend eastwards through the Archipelago ; (7) in Amboina and Ceram the females of several species are darker coloured, while in the neighbouring islands they are more brilliant."

Specially striking are the peculiarities mentioned by Wallace in the form of the anterior wings in the species of Papilio of Celebes; the anterior wings are in general more elongated and sickle-shaped, the front margin is more strongly curved than in the species of other districts, and usually makes a sudden bend, an angle, near the base of the wing.

As this peculiar form of the wing in Celebes occurs not only in species of Papilio but also in Pieridæ and species of other families, Wallace believes its origin is due to an advantage in the struggle for existence; he supposes that these large butterflies must, in consequence of the sickle-like shape of the anterior wing and its curved front edge, be able to turn suddenly in flight with greater ease, and in this way baffle their pursuers. He thinks this the more probable because the only Papilio in Celebes which does not possess this peculiarity in the wings, P. Polyphontes of the Polydorus group, appears to be protected in other ways.

This is certainly possible, although considerations already noticed, showing that large wings actually protect butterflies from capture during flight, are opposed to it, and although Wallace is unable to name the pursuers which are baffled. For the very reason that the same peculiarity occurs in different but allied families, it should rather, however, be regarded as the consequence of a direction of evolution, a phyletic growth determined by peculiar conditions of climate or nutrition, and similarly with the disappearance of tails towards the East. The differences in size also of butterflies on islands probably have their causes in differences in the abundance of nutritive resources.

With regard to the tails, to Wallace's observations I am able to add that it is the higher species of the genus Papilio in which they disappear, so that we have to deal with their disappearance from the genus.

Moreover, the peculiarities of the colouring of butterflies

in particular islands, such as those Wallace mentions, and many others known to me, are certainly not to be considered as due to selection, but in part at least are to be recognised as stages of evolution in a definite direction, which I shall not discuss here in detail. Thus Wallace himself, joint-discoverer with Darwin of the principle of utility, says: "But this hypothetical explanation of the form and veining of the wings in the Celebes butterflies does not apply to other cases of local modification. Why the species of the western islands are smaller than those in the islands lying farther to the east, why those of Amboina surpass those of Dschilolo and New Guinea in size, why the tailed species of India begin to lose these appendages in the islands, and on the shores of the Pacific no longer show a trace of them, and why in three different cases the females of the Amboinese species are less gaily coloured than the corresponding females of the surrounding islands—these are questions we cannot yet attempt to answer.".

These are, therefore, characters whose utility in any case cannot be proved, in which the determining influence of selection cannot be recognised—are certainly acquired and inherited characters.

I can increase the number of these examples by many others taken from this same group—the butterflies; cases in which the variations determining the origin of species are so small, so obviously in constant progress towards definite directions of evolution, that their explanation by selection is in fact absolutely excluded.

IMPORTANCE OF THE STIMULATION OF THE NERVOUS SYSTEM IN RELATION TO ADAPTATION AND THE ORIGIN OF SPECIES

The reader will perhaps wonder that I did not assign a special *rôle* from the beginning to the action of the nervous system among the causes of the modification of colours in animals, since the influence of its action in this respect is well known. But it is clear that this is in most cases only an indirect influence determined by other causes, especially light and warmth, so far as it affects the question of the permanent modification of animal forms.

The nerves may influence the colouring of the skin in two ways: either by changing the supply of blood, and the deposition of pigment depending upon it—that is, indirectly, or directly by changing the shape of the pigment-cells. The first action depends more upon temperature, the second more upon light. Warmth increases the flow of blood to the skin, cold diminishes it—the former by dilating, the latter by contracting the blood-vessels; both, however, must act directly by expanding and contracting the pigment-cells, like illumination.

Finally, electric stimulation, or division of nerves, acts upon the colour of the skin. After removal of the cerebellum in a frog the skin becomes strikingly variegated. As excitement produces temporary paleness or redness in the human skin, and as irritable, malicious persons are always distinguished by a pale yellowish complexion, in consequence of constant excitement of the nervous system, so the colours of animals vary with excitement, sometimes to a striking degree. I may mention the familiar case of the chameleon, which when angry and after death turns yellow. The tree-frog also alters in colour in captivity, apparently through nervous excitement, as is usually said from discontent.

But in the latter case the direct influence of the colour of the surroundings must evidently also be taken into account, for the tree-frog, like many other animals, assimilates its colour to that of its temporary environment.[1]

It is well known that many other animals likewise change

---

[1] This process is usually described by the inappropriate expression: sympathetic colouring. I would substitute for this: stimulation-colouring.

their colours in accordance with their surroundings, and thereby become less conspicuous. Besides the chameleon, I will only mention the Sepias among the Cephalopods. This is a much more delicate action of light than that which merely changes the colour from light to dark, as occurs in so many cases—a sudden action instead of the gradual effect by which, for instance, darkness of tint is produced after some time in man; and physiologically different, for in the former case it is a question of change in the shape of the pigment-cells under the influence of the nervous system, in the latter of deposition of pigment.

Such sudden alteration of colour under the influence of light is strikingly shown in certain fishes, *e.g.* trout, grayling, minnows. In these the colour of individuals kept in the dark becomes lighter on sudden exposure to light in consequence of contraction of the pigment-cells.

We attach too exclusive an importance to "adaptation"—as I attempted to point out by examples previously given—when we think to explain by selection every similarity between the colouring of an animal and that of the ground on which it lives. For, as we have seen, animals may become similar in colour to their surroundings, actually adapted in colour, quite by chance; for instance, in consequence of the direct, necessary action of light, *i.e.* of the surrounding colours, and therefore without selection. Many really wonderful cases of adaptation, apparently due to selection, probably come under this category.

It is also to be noticed that even rapid changes of colour may also be due to a chemical action of light, and not invariably to nervous influences. It has been observed, for example, that the pupæ of butterflies during their development take on the colours of their surroundings. Pupæ have assumed the red colour of a cloth enveloping them, a colour to which they could scarcely be exposed under natural conditions, and to which in any case they could not have been adapted by selection without special conditions of life which ought to be explicitly pointed out. Yet, notwithstanding this, the phenomenon has been assumed without further question to be due to adaptation by selection.

This interpretation I would meet with the following proposition: The substance composing the envelope of the pupa possesses as a fact the property of being changed by light like a photographic plate, and the relation of this property to the outer world may be useful, but it does not necessarily owe its origin to selection.

That this substance is so constituted as to exhibit in action a process of colour-photography, the goal of so much human longing and striving, leads to another consideration.

Since the discovery of visual red in the retina of the eye, a substance which quickly bleaches under the action of light after death, and which is situated in the very cells of the retina on which light falls, we are brought near to the conception that sight, especially the perception of colours by the eye of the higher animals and of man is likewise a chemical process, a kind of photography.

A short step farther in the specialisation of nervous stimulation or nervous conductivity might well render comprehensible the wonderful fact above referred to, that the colours of the environment of an animal may be reflected in the colours of its skin. For it is self-evident that the path of action of the coloured light is principally through the eyes. Experiments prove this: after the eyes have been removed, the action of the colour of the environment on the colour of the skin ceases.

Thus, in many cases, chemical action and stimulation of the optic nerves may be closely connected in the process by which the colours of animals are affected by light.

Whether simple chemical action of the light upon the skin, or such action taking place indirectly through the eyes,

L

or simple general stimulation of the nerves, determines the change of colour, or whether, lastly, the gradual action of selection simultaneously plays a part in the matter, it will of course be often difficult to decide for any given case.

The following really astonishing case of "adaptation" of the colouring of an animal to the ground on which it lives forces upon me the idea that selection does not play a part in it, but that it is due to the direct external action of the colour-stimulus; the effect, however, probably not showing itself at once, but appearing more gradually than in many cases.

The grasshopper with red hinder-wings banded with black, which is so common with us (in Germany) in summer, Acridium germanicum (Œdipoda germanica), when it occurs on the reddish-brown Triassic clay of Tübingen, resembles this ground so closely with its wings folded, that it cannot be distinguished from it. A little above the clay on the hills of this neighbourhood there occurs a whitish sandstone, sometimes only for the breadth of a path or in somewhat larger surfaces, frequently surrounded by the former. On these small patches of lighter ground I find regularly only grasshoppers with quite light upper wings, so that they can scarcely be distinguished from the soil. And I have elsewhere observed the same remarkable adaptation. One of my friends, who is not usually accustomed to pay special attention to such animals, told me that he had been much surprised to notice that on the two banks of a brook on which the soil was of different colours, the grasshoppers were in each case exactly like the ground in colour. Without doubt these were Acridium germanicum or Acridium cœrulescens—the latter species appears to show the same adaptation.

Experiments must decide whether this is a case of stimulation-colouring. That such stimulation-colouring occurs in the tree-frog I have mentioned above, and it evidently occurs also in other Amphibia.

Years ago I was much struck to find in a pine-wood in the Tauber valley (at Bronnbach near Wertheim) that the numerous toads living there, without exception, had the yellowish red colour of the pine-needles covering the ground. The common frog also, Rana temporaria, changes colour according to the colour of the ground, not rapidly, but gradually.

It seems to me, we may safely assume that such stimulation-colouring has been in many cases the ultimate cause, the origin of the formation of a permanent and hereditary adapted colouring: for when any species or variety of an animal has lived throughout a long time on a particular ground, of which the colour has been constant, and has by nerve-stimulation taken on this colour from the beginning, this colour may at last fix itself in the animal unchangeably.

In this way a colour-adaptation, in a high degree unchangeable, may be produced without any selection, without the struggle for existence; and probably many colour-adaptations of animals of restricted habitat are to be explained in this way.

Among such cases, probably, belongs a remarkable colour-adaptation in frogs observed by Wiedersheim: the common frog in the Engadine has always the speckled colour of the granite on which it lives.

It goes without saying, that selection may favour such direct adaptation, and this has probably been the case with the wonderful degree of adaptation to the surroundings shown by the colour of wall-lizards, as I have proved by extensive researches,—colour-adaptations which, moreover, have doubtless been determined by the most various causes.[1]

My chief assistant, Dr. Fickert, has at my suggestion made the following experiment in order to test the action of

---

[1] Cf. the section on the bearing of adaptation upon the origin of varieties in lizards.

differently coloured surroundings on the common frog (Rana temporaria): Three frogs approximately similar in colour were placed in three glass vessels, of which the first stood on a black, the second on a green, and the third on a white surface, being surrounded up to a height of some five centimetres with the same colour. After about an hour and a half, the frog *a* on the black surface was the darkest, *b* on the white the lightest, while the frog *c* surrounded by green was intermediate in colour between the two. Hereupon the frog *a* was transferred to the glass on the white, frog *b* into the one on the black surface. After three-quarters of an hour they were again examined, and *a* was the lightest, *b* the darkest. Then *c* and *b* were interchanged, and in a quarter of an hour *c* was the darkest, while *b* was intermediate in colour between *c* and *a*. When, finally, *b* and *a* were interchanged, a change of colouring appeared immediately; *b* became light again, and *a* took the intermediate tint between *b* and *c*. The change of colour seemed to be due to the fact that when the colour became lighter, the pigment distributed over the skin was collected in particular places in small dark spots, and when it became darker, it was spread out more uniformly over the whole surface of the body.

When I repeated the same experiment with the same frogs on the following day I only observed a slight change of colour, although this was still noticeable. But one frog and a toad (Bufo vulgaris), which I had left during the following night in a quite dark vessel, had both become blackish brown in the morning. After I had placed them in the light on a light surface the frog immediately began to get lighter in colour, and continued to change gradually, so that after half an hour it was brownish yellow. In the toad the change of colour occurred much more slowly: only after some time could I see that it was distinctly lighter, and it was a whole hour before it became light brown.

Three other frogs changed colour in the course of about an

hour, as in Dr. Fickert's experiment—only the one surrounded with green was almost as light as that on the white surface, and the one entirely excluded from light, although it became distinctly darker in that time, even after four hours could only be called brown, not dark.

We have thus the remarkable fact that the change of colour in frogs begins at once, but only slowly reaches its full extent, and that it also takes place slowly in toads, which entirely agrees with my interpretation of the change in favour of adaptation without selection.

PARTICULAR FACTS WHICH PROVE THE INFLUENCE OF NUTRITION AND OTHER EXTERNAL CONDITIONS ON THE VARIATION AND FORMATION OF SPECIES IN LEPIDOPTERA

After I had in vain applied to various experts for trustworthy personal observations upon the effect of the nutrition of the caterpillars on the variation of Lepidoptera, and had concluded the preceding pages, a book was brought to my notice entitled *The Indo-Australian Lepidoptera, etc., with a Discussion of the Origin of Colours in Pupæ*, by Gabriel Koch (second edition, Berlin, 1873). This work contains some facts of much importance for my arguments, and its author, on the basis of these facts, expresses some views on the origin of varieties which completely harmonise with mine.

Koch states that his attempts to influence the colours of Lepidoptera by the feeding of the caterpillars date from the year 1832, when he succeeded in producing in Chelonia Hebe either fiery or dull red on the under wings, and in bringing out more strongly either the black marking or the white ground alternatively by feeding with different plants.[1] Other successful experiments were made on Chelonia Caja, and Nemeophila plantaginis.[2]

[1] Published in a paper on *Die Raupen und Schmetterlinge der Wetterau.*
[2] *Vid.* G. Koch : *Die Schmetterlinge des südwestlichen Deutschlands.*

Koch employs the facts he brings forward to support a theory set up by him concerning the origin of the colours, which physiologically is somewhat audacious. According to this theory, a yellowish slime overlies the imago's wings before its emergence. This slime consists of granules, which afterwards form the pigment particles of the imago. The caterpillars obtain the substance of this slime from the plants they feed on in the form of vegetable salts, acids, and tannin. Koch was, I believe, originally a simple artisan of Frankfurt (a tinman), and we cannot be severe upon him for explaining the production of colour as a simple laying on of pigment. On the other hand, it is highly gratifying to see facts employed by such a man scientifically, that is, so as to afford general conclusions, so that it must be said of him, that with simple unskilled judgment he hits the nail on the head.

Koch guards himself against the charge of not recognising the action of light on the formation of colours: he says this action is shown by the fact that diurnal Lepidoptera all have more vivid colours than the nocturnal. Warmth and light, however, according to him, do not act differently.

Koch's results are as follows:—

A change in marking is easily produced by a change in the plants fed upon (the "bear" and other species).

Nocturnal species which live exclusively on conifers have dull colours, usually gray, as for example our pine hawk-moth (Sphinx pinastri), or the pine-spinner (Gastropacha pini), and several foreign species. This is so invariably true that Koch was able to conclude from the colours of certain species from Sydney and Baltimore that the caterpillars lived upon coniferous plants, and when he suggested that they should be sought on such plants his conclusion was found to be correct.

It is known, Koch says, that when the caterpillars of our German "bears" (Chelonia s. Euprepia caja) are fed from their hatching to their metamorphosis with leaves of Lactuca sativa or Atropa belladonna, not one of the imagines produced resembles the original form; when the insects have been fed on lettuce the white ground-colour of the wings predominates; when fed on deadly nightshade the brown markings of the upper wings often coalesce and the white vanishes; in like manner the blue markings on the lower wings fuse together and displace the orange-yellow ground colour. To which I may add, that according to Koch's work, *The Lepidoptera of South-West Germany*, feeding with belladonna and poppies produces dark-coloured imagines. Koch made experiments with similar results on Chelonia plantaginis and Gastropacha pini. Finally, he refers to experiments with similar results on species of Militaea and Argynnis, as already known.

"Must not," concludes Koch, "similar processes occur equally, and even on a larger scale, in the natural life of the countless forms of the class in question? When a great number of individuals perish through an occasional scarcity of their proper food-plant, must not nevertheless considerable numbers survive by contenting themselves with other allied food materials, and so give rise to varieties whose origin we do not dream of, and which therefore we are led to regard as new species?"

Further, he brings forward instances to show that cold or warmth, moisture or dryness of climate, change the size and colouring of Lepidoptera.

Thus the chess-board butterfly (Arge Galathea) in middle Germany is quite different in appearance from its dark varieties (Procida and Leucomelas) which occur in the Tyrol and the southern parts of Europe; the speedwell butterfly (Melitæa Artemis) in middle Germany is always smaller and duller coloured than its highly coloured varieties (Desfontainesi and Beckeri) in Spain. Our buckthorn butterfly (Gonopteryx rhamni) acquires in Southern Italy and Portugal a large

orange-coloured spot on its anterior wings (var. Cleopatra). Further, not rarely a long period of dry or moist weather exercises a considerable influence on the size of the following generation. Immediately after a continuously dry summer butterflies are always smaller than after a moderately moist season. The second generation of Argynnis Selene, which takes flight in the height of summer, is always smaller than the first generation, which is produced in spring, and so on.—
"If, then, such important facts are observed in many species of our small continent, why should not similar effects be possible in extra-European species, which are spread over so large a territory where they are influenced by the most varied climates, differently composed soils, and to some extent quite different food plants? In a hemisphere where the conditions of temperature are so varied, why should not the tropical sun, under which varieties are generally acknowledged to occur, be also the chief cause of variation? Thus the origin of the immense Lepidoptera of the East Indies is to be ascribed to the moisture and heat of the climate (*e.g.* the Ornithoptera); and, on the other hand, the small size of the Australian forms is to be attributed to the dryness of the Australian climate.

"If Lepidoptera pass during the rainy season in their normal habitat into another region where the rainy season has not begun and where great heat and a dry atmosphere prevail, or if they pass from the monsoon region, which is neither deficient in water nor in rain, into the ever-dry and waterless continent of Australia, diminution of size in the very next generation is sure to take place. But as the climate in Australia is permanently dry, the diminution which occurs remains permanent in all subsequent generations in their new habitat, and the deteriorated condition of the species is established for ever."

Thus permanent local varieties are produced. There are therefore numerous Lepidoptera which in moist climates are larger, in Australia smaller. But there are also very large native species in Australia (*e.g.* the giant hawk-moths Brachyglossa triangularis and B. Australasiæ, etc.) Instances of species which become smaller in Australia are Chærocampa celerio, Sphinx convolvuli, Euchelia pulchra, and many others.

Koch further discusses the importance of the season of the year in which Lepidoptera appear, and mentions, after referring to Vanessa Levana, and Prorsa, that the second generation of Argynnis Selene, which takes flight in the height of summer, also has finer and blacker spots and markings than the one which comes out in spring.

Thus Koch also reaches the conclusion that the origin of permanent varieties is to be ascribed to external influences, and the most important result gained by his work is that he makes it probable that many forms at present considered as species are only similar permanent varieties. It is a remarkable proof of the hold which the idea of the permanence of species has upon systematists, that Koch never once refers to the question whether species might not arise or have arisen in this way, although he supplies instances thereof.

In the preceding I believe I have proved beyond a doubt that external influences—climate, light, warmth, moisture, and differences of food—modify organisms directly, even without the aid of selection, and that inasmuch as the modifications so caused are inherited, they will give rise and must have given rise to new species. I presuppose, however, that the inheritance of characters acquired through the action of external influences is an indisputable fact.

Characters Acquired by Use

It is a self-evident physiological fact that practice or use strengthens and improves the organs of the body, while disuse causes them to deteriorate.

The causes of this effect of practice lie firstly in the increased supply of blood which it produces, and secondly in the more delicate adjustment of nervous and muscular action.

That characters acquired through use or disuse are inherited, and must therefore aid in the formation of new species, can, I believe, be proved more easily than any other of the propositions I am maintaining. If I were to bring together all the facts which could be used as evidence on this point, I should never come to the end of them, for I should have to refer to all the facts of comparative anatomy and physiology. But I intend to show in particular that use and disuse by themselves must lead to the formation of new permanent characters, without the aid of selection, for even this I hold to be a physiological necessity.

Lamarck long ago explained the degeneration of the eyes in animals living in darkness by disuse.

But how are eyes evolved?

The pigment-spots which at the present day are the only representatives of these organs in many animals show evidently the first stage of their evolution.

The stimulus of light acting upon spots of the outer skin specially suited for seeing, *i.e.* for "feeling at a distance," has produced a deposition of pigment in these spots, because they, originally endowed only with the sense of touch, have constantly been directed towards the light in order to receive the more delicate stimulus of the ether waves. Pigment cannot have been the primitive organ of sight, nor can it at present, in any instance, be itself such an organ. But pigment is so far absolutely necessary for sight that it serves to separate the rays of light falling on the nerve-endings. Pigment is deposited around primitive tactile cells, and separates each one from its neighbours, or it is deposited in some of the touch-cells, leaving one free here and there. The touch-cells which remain unpigmented become sight-rods. I have described such superficial pigment-spots representing eyes in their simplest form, in Medusæ, *e.g.* in Aurelia aurita.[1] By their means, in my opinion, the simplest kind of sight is effected in this way, that of the rays coming from any object, one passes through each such sight-rod and is separated from neighbouring rays, and so the lights and shades of the object are communicated to the nervous system in a number of separate points, so that a representation of the object is obtained in the nervous system by a flat image composed of points. The lighter and darker points of the object are each represented in the nervous system, as when an object is seen through a number of minute holes in a card. This kind of sight by points can only be effected by these lowly developed eyes in such low forms as the Medusæ, not, as most zoologists, following Johannes Müller, assume, by the highly developed complex eyes of insects, in which images must certainly be formed. The origin of eyes capable of this kind of sight depends then on the deposition of pigment around touch-cells for the purpose of separating the different light-rays received from one another. We have already seen what an importance light has in the production of pigment. Without the stimulus of light, the pigment so essential to the formation of an eye could not be produced; without the continuation of the stimulus, that is, without constant use, the eye cannot continue to exist at all; but by constant use it is improved and perfected. The same stimulus which it is the function of the eye to receive, namely, light, created and still maintains the essential and fundamental element of the organ.

Exactly the same statement holds also for the other sensitive cells, for the cells of smell, taste, and hearing, as for those of touch and for sight-rods. They have all been ultimately evolved from indifferent epidermic cells, and the

---

[1] Th. Eimer, *Die Medusen*.

specific external stimuli, so closely allied to the stimulus of touch to which they respond, must have gradually produced their specific nature. Only by use, by constant exercise, can they be specifically maintained, and by use, at the same time, they are improved.

In these considerations, the truth of which is undeniable, lies the best proof of the great importance of use, the great importance of acquired and inherited characters, in the modification of organs.

Thus every organism owes its peculiar structure essentially to the use of its parts; nay more, owes its continued existence exclusively to this exercise and the unceasing action of external stimuli. Without any exercise all our organs would degenerate; we should perish as surely as if we were dead, if only for a moment external stimuli ceased their action upon us. Life is truly nothing else than the expression of the interaction between the organism and the stimuli of the external world, and natural gradual death is due to the fact that the organism becomes incapable of receiving stimuli.

Thus the statement that the eyes of animals living in darkness are atrophied through disuse may be more exactly rendered into this, that the cessation of the stimulus of light causes the pigment to fade, and must therefore cause deterioration of the eyes.

For the rest, in the preceding it was my intention only to point out that stimulation from without, *i.e.* use, according to my view, even without selection, must necessarily form organs, strengthen, and perfect them, and I did not think of denying the fact of the influence of selection, *e.g.* in the formation of eyes.

Among the examples which Lamarck brings forward in support of the modification of organs by use, and which are, it is true, in most cases, partially to be referred to selection, is one to which I have to devote further consideration. He says: "M. Tenon states in a communication to the Science Section, that in investigating the intestinal tube in several men who during a great part of their lives had been inveterate drunkards, he found it always extraordinarily shortened in comparison with the same organ in all men who had not been addicted to the habit.

"It is known that great drinkers, or those who have given way to drunkenness, take very little solid nourishment, that they scarcely eat at all, and that the liquids which they take in excess and constantly suffice to nourish them.

"As liquid nourishment, especially spirituous liquors, do not remain long in the stomach and intestines, the stomach and the rest of the intestinal canal in drunkards are no longer accustomed to be distended, just as in persons who lead a sedentary life, who constantly occupy themselves with intellectual work, and who are accustomed to take very little food. Gradually and in course of time their stomach is contracted and their intestine shortened.

"It is not a question here of narrowing and shortening caused by a wrinkling of the parts, which would allow of the usual extension as soon as the intestines, instead of being continually empty, were again filled, but it is a question of a real narrowing and shortening, so that these organs would burst before they were distended to the usual extent.

"Let a comparison be made between two men of the same age, of whom one is devoted to study and habits of mental work, which weaken his digestion, and has consequently become accustomed to eat very little, while the other takes plenty of exercise daily, goes out often, and eats well, and it will be found that the stomach of the first scarcely has any capacity, and that a very small quantity of food fills it, while that of the second has not only maintained its capacity, but even increased it.

"Here then is an organ which is considerably changed in

its dimensions and its capabilities during the individual life simply in consequence of a change of habit."

By this example the zoologist will be immediately reminded of the fact that many of our domestic animals have a larger intestine than the wild species from which they are derived. Other morphological differences between the wolf and the dog having been sought in vain, one zoologist has recently attempted to make this difference in the length of the intestine serve as such. But it is clear that this is only a difference due to physiological conditions; it is evidently caused by the difference in the food of the domestic and the wild animal. The wolf feeds upon flesh, the dog has to accustom himself to vegetable food also. The domestic cat, also evidently from the same causes, has a longer intestine than the wild cat. It is a universal fact that, among wild animals, the carnivorous have a much shorter intestine than the herbivorous. In the development of the frog the two types succeed one another, clearly in relation to the food. The tadpole, which feeds principally on Algæ, to some extent on Infusoria, has an intestine of surprising length; the adult frog, feeding only on small animals, has a very short intestine. It is very natural that a shorter gut suffices for the nutritious food of the flesh-feeders, while the less nutritious vegetable food, in order that the necessary nutriment may be got from it, must be taken in larger quantities and digested for a longer time. Therefore a longer intestine is useful to the vegetable-feeder.

But in what way has the gut been lengthened in the domestic animals which live partially on vegetable food?

That this has been brought about by selection cannot certainly be supposed, for although the longer intestine is an advantage to the nutrition of the dog and cat, yet the difference in this respect between the tame and the wild animal is not so great that the life of the animals in a domesticated condition could depend on it. There can, in my opinion, be no doubt that the elongation in fact depends on mechanical causes—that as the intestine in a drunkard is shortened during his life, so it has been mechanically lengthened in the vegetable-feeding animal by the greater quantity of food taken; and the new character has been inherited, so that now it is looked upon by one zoologist as a specific distinction.

It is a very remarkable fact that nearly all men use the right arm much more than the left. Since all races of men, even the most remote and most uncivilised, have this habit, we cannot suppose that it has simply been acquired by practice, and thus become hereditary. It is more probable that there is something in the structure of the body, *e.g.* in the direction and force of the course of the nutrient liquids, which gives the right side of the body its superiority, if the mere position of the heart towards the left is not the determining cause, on account of the greater disturbance which that organ would suffer if the left arm were chiefly used. Cases of left-handedness rather favour than contradict such a supposition, for in these cases parents and teachers have taken trouble enough to train children to the usual habit, but in vain.

It is to be supposed that this reverse habit is due to congenital reversion of nutritive relations, such as that produced in the simplest way in complete reversion of the *situs viscerum*. I do not know if it has been ascertained how far left-handed people, *e.g.* left-handed swordsmen among students, exhibit this inversion. It would be a praiseworthy task to investigate whether some recognisable anatomical conditions do not underlie their peculiarity of habit.

Notwithstanding this view of the connection between organic conditions and the priority of the right arm, I believe that the skill and strength of the latter, especially of the right hand, are partly to be attributed to use. Martins, the editor of Lamarck's *Philosophie Zoologique*, explains them exclusively in this way. He maintains that in consequence of greater exercise the right arm is thicker and heavier, and all its parts, bones, muscles, nerves, vessels are stronger than those of the opposite side. And he adds,[1] that the Dutch naturalist L. Harting has proved that the differences are already present in the new-born child which has not yet made use of its limbs, and that this gives rise, independently of example and instruction, to the innate tendency to employ the right arm in preference to the other.

The skill of the right hand as compared with the left, in fact, depends upon a number of congenital conditions, which can only be explained by the inheritance of acquired peculiarities of muscular action, that is, ultimately, of the musculature and nerve supply.

In this conclusion I am confirmed by another observation. I noticed that the Arabs in Egypt in a great number of actions make skilful use of the toes instead of the fingers. This is particularly striking in the turners, who in turning wood use the foot as well as the hand with much skill; and the same power was exhibited in the operations of weaving by the Dinka negroes recently travelling with Nubians in Germany. In the Egyptians the great toe plays an important part. Beggars sometimes pick up pieces of money by grasping them between the great and the second toes instead of grasping them with the hand. And the rapidity with which this is usually done gives an idea of the dexterity possessed by the toes. The skill, however, with which even young boys use their toes in similar ways again leads us to infer some congenital organic peculiarities.

Moreover, it cannot be doubted that the deficiency in the development of our toes and their want of dexterity are to be ascribed to degeneration, and as both must be hereditary, our

---

[1] *Op. cit. Biographische Einleitung*, p. 35.

toes afford an example of hereditary characters acquired through disuse, of which I shall treat in the following chapter under the title "Degenerated Organs," and, conversely, this deterioration itself shows that strength and skill are really inherited. This degeneration or deterioration must, however, among us, as we wear boots, be somewhat greater than in the barefooted Arabs, and therefore the latter must be from their birth more fitted to use their toes than we. At the same time it is known that our babies still make grasping movements with their toes.

Lamarck explains the origin of the elongated body in snakes by the exertions they make in order to force themselves through narrow holes. It is evident from the preceding that I should not hesitate in certain cases to allow that such efforts on the part of an animal to elongate any organ for a definite purpose might result in a hereditary modification. But Lamarck's views on the effects of use have been so much neglected, for the very reason that here, as in other cases, the examples brought forward by him are not convincing. It seems to me inconceivable that the elongated body of a snake could have been produced in such a way.

But the origin of the elongated bodies of creeping vertebrates is really a great enigma.

Not only snakes, but the Blindworm (Anguis fragilis) and other animals have acquired, along with the degeneration of the limbs and the development of a creeping habit, a worm-like form of body. If this could be considered as due to constantly repeated stretching out, as a rope or an india-rubber tube can be at last made longer by continual stretching, Lamarck's explanation could not be so flatly rejected. But in all these cases the elongation is connected with an increase in the number of vertebræ—in some a very great increase.

M

Pure Darwinism is contented to explain this elongation of body as adaptation produced by selection.

It is true that it permits of more rapid motion, and therefore gives an advantage in the struggle for existence to the animals which have acquired it. But Darwinism leaves us entirely without an answer to the question, Why, in what way, has the number of the vertebræ been increased?

Evidently the elongation of body is connected with the disappearance of the limbs. Snakes, like the blindworms, are descended beyond a doubt from lizard-like animals. But the long tail of the lizards shows that they themselves are descended from animals whose body was elongated, worm-like, composed of many similar segments (metameres), had many vertebræ. Their tail vertebræ are the more rudimentary the more posterior they are in position. Thus it may well be inferred, conversely, that the vertebral column as a whole has been shortened in lizards in proportion to the degree of development of the limbs—as also in other vertebrates, unless the tail has acquired a particular function, as *e.g.* in the kangaroo and the apes, etc. When the limbs began again to disappear, and the body again to move in a writhing, creeping manner, the original condition might again gradually return, if the shortening arose through disuse and degeneration, favoured by selection. Now selection contrariwise favours the elongation of the body, its reversion to the earlier condition; which, however, results directly from the fact that the original energy of growth attains again its unrestrained freedom, or may even proceed beyond the original limits.

The elongation of body in creeping vertebrates seems so extraordinary as to demand an explanation of its ultimate causes. But I can come to no other conclusion than that physiological growth unhindered or even favoured by actual conditions must play the chief part in the matter, while exercise (habit) and selection have only an auxiliary effect.

Use will be only so far essential that it determines the distribution of liquids through the whole body, so that its growth up to a certain limit is rendered possible; while in the lizard, for example, the tail is at a disadvantage.

Snakes have in some cases as many as 300 dorsal vertebræ, Saurians 15 to 100. The length of the neck of the Giraffe depends only on the elongation of the individual neck vertebræ, in birds partly on this and partly on an increased number of vertebræ. The Swan has 23 neck vertebræ, the other Lamellirostres and the Storks up to 17, the singing birds 10 to 14, the running birds (Cursores) 15 to 18, the Cormorant (Carbo) 18, Podiceps 19.

Since, therefore, very long necks are differently formed, even among allied forms, and since, on the other hand, short necks often have the same structure as long, effort cannot for this reason alone have determined the development of the former, and mere selection may likewise be rejected. The principal determining cause must rather have been a definite law of growth, which again must have been influenced by peculiar conditions in the life of the ancestors of each group.

Further, the view I have suggested, that the atrophy of the limbs physiologically, *i.e.* correlatively, contributes towards producing the reversion of the body to its original condition, that it may cause the recurrence of the original course of growth, is supported by the consideration that the development and maintenance of the limbs must withdraw a quantity of material from the rest of the body, and that this material is restored to the latter when the limbs degenerate. Since at the same time the previously rudimentary tail begins again to be used, several processes tend to the development of the snake-form.

I have to mention here still another case brought forward

by Lamarck as evidence of the effect of use in the formation of new characters, in order to employ it in my own argument.

Lamarck says of the Ruminantia, in reference to the origin of their horns: "In their ebullitions of wrath, which principally occur in the males, their emotions by their intensity direct the fluids more strongly to this part of the head (the forehead), and therefore there takes place in the one case a secretion of horny substance, in the other a formation of bone-tissue mixed with horn-tissue, by which solid processes are formed; hence the horns of the ox tribe and stag tribe."

Shortly before this he mentions that the ruminants fight together by pushing with the forehead, but it does not occur to him to seek in this fact the stimulus which causes the formation of horns. In my opinion the first beginnings of the formation of frontal processes in the Cavicornia, and of the antlers in the Cervidæ, are to be attributed to this stimulus. By the inheritance of a simple swelling on each side of the frontal bone, by the continued inheritance and increase of this swelling in consequence of constantly repeated stimulation, the frontal processes were produced in Cavicornia, and also the antlers among the Cervidæ. The latter, having a special form in each species, and regularly advancing in development and perfection each year up to a certain limit, afford one of the best examples in support of my conception of organic growth in the animal kingdom. For although selection plays some part in this development, it cannot possibly have had in view from the first the form peculiar to each species, and have brought it into existence unaided by other causes. That the ultimate cause of the development lies rather in regular processes of growth is shown by a fact concerning the antler of the roe-deer, which is of particular importance in relation to my whole view.

The antler of the roe-deer is known to be distinguished,

when it has peeled, by elevations on its surface, which are called "pearls." The antler very often has an extremely long pearl on the internal side of the main stem. In one of six points it is between the lowest branch and the base. This structure is obviously the commencement of a new branch in process of evolution, for it is occasionally as much as an inch long, and then has all the appearance of a new branch, and "may be counted as such in venery."[1]

In this we have therefore an instance of the origin of a new part of an organ by simple growth.

Only in this gradual way, by the inheritance of acquired modifications sometimes together with degeneration, numberless organs in the animal body can have been evolved. Consider the limbs of Ungulata and other animals: the foot of the horse, of the ostrich, the hind-foot of the kangaroo, in their partly degenerate, partly specialised structure, depend only upon inherited acquirements due to adaptation to the same physiological demand, *i.e.* diminution of the number of the toes, and increase in strength of the one or few remaining for the sake of greater firmness and more rapid progress on hard ground—in contrast, for example, with the five-toed foot of the elephant, which is adapted to prevent any sinking into the soil.

The treatment of comparative anatomy, especially of osteology, from this physiological point of view, which at present, to my regret, has almost entirely fallen into disuse, has its own special attractions, and affords at every point support to my views. In this respect so-called analogous organs are of the utmost importance—organs which, although occurring in very different animals not immediately allied by descent, have a very similar form, because they serve the same purpose; thus, for instance, the similarity in the formation of the limbs in the cases just mentioned, the development

---

[1] Cf. J. H. Blasius, *Naturgeschichte der Säugethiere Deutschlands*, 1857, s. 464.

of a sternal keel for the attachment of the pectoral muscles in flying and burrowing animals, in birds, bats, moles, etc.

It is frequently possible in such cases to form definite conclusions as to the future form of organs which at present are seen in process of modification. Thus my friend Wiedersheim in a very interesting paper [1] has recently drawn attention to some facts in human anatomy which indicate the future modification of the human body, assuming that existing causes continue to act. In man certain muscles are now increasing, namely, the adductors of the thumb and the musculi glutæi, in the face the mimetic musculature is on the increase, the pelvic girdle is increasing, the ilia are becoming wider apart—and all this in addition to the degeneration of organs through disuse, which in a still higher degree indicates the future structure of man.

It is clear that the progressive development of the intelligence has likewise gradually effected the increase in the size of the cranium in proportion to the facial part of the skull in man, by means of the increased size of the brain, so that the brain case now forms the principal part of the skull. But it appears that in the most highly developed races of man, provision is made for the farther increase of the brain-case: as Gratiolet has pointed out, in these races the union of the cranial sutures proceeds from behind forwards, so that a greater development of the frontal lobes, on which intelligence depends, is rendered possible; or rather, the vigorous growth of the frontal lobes has conversely given rise to this arrangement. In the lower races of man the union of the sutures proceeds from before backwards, as in the apes.[2]

I forbear here to enter more minutely into the importance of the inheritance at particular ages of life of characters

---

[1] R. Wiedersheim, *Der Bau des Menschen als Zeugniss für seine Vergangenheit*, Freiburg i.B. 1887.   [2] Cf. Wiedensheim, *op. cit.*

which certainly often depend upon acquirement, and which must contribute to the modification of the organic world, namely, the facts of the inheritance of characters of the parents by the children in such manner that they appear in the latter at the same age in which they occur in the former. This kind of inheritance has been called homochronous; it is obviously a necessary consequence of the law of correlation. Thus the characters which accompany sexual maturity can only appear when the sexual organs have reached maturity. But many characters belong here which have been acquired by use and habit, as the fact, to be discussed subsequently, and already mentioned by Darwin, that the peculiarities of handwriting when inherited by a son from his father only appear at a ripe age.

On account of the contrast in which it stands with this homochronous inheritance, I must here mention an observation which at the same time illustrates the influence of the condition of the organism as a whole, at the time of generation, upon the germ.

I observed that children of very old parents, especially of old fathers (male preponderance) have in their early years—as children—features of a remarkably old appearance.

I have repeatedly tested the correctness of this observation by appeal to impartial witnesses, and am able to state it with certainty.

A senile expression of face may certainly be called a character acquired through use.

In the following I will adduce a few additional instances of the inheritance of characters acquired by exercise, some due to habits of life, some to training, and some instances of mental characters, although I shall subsequently have to devote special consideration to the latter, above all, to instinct.

Is it not due essentially to better nourishment, to

the conditions of domesticated life in general, that our hens lay eggs and our cows give milk almost all the year round? How else are we to explain that our hens, ducks, and even geese have almost completely lost the power of flight? not to mention the modifications of structure, some of considerable importance, *e.g.* in the skeleton, which almost all our domesticated animals exhibit, when compared with the wild forms from which they are descended, and which cannot in any way benefit them.

Weismann points out as evidence against the inheritance of acquired characters, that articulate speech and the art of piano-playing are not inherited, but only the predispositions thereto. But each of these is obviously not an acquired "character" of the organism, but only an artistic skill, which has been learned. But the necessary tones of voice which constitute a gradually acquired character—indeed voice itself, and the acquired faculty of forming certain tones melodiously in a definite succession—are inherited.

In no less a degree faculties acquired through education, through training, by various domesticated animals, *e.g.* by our breeds of dogs, are transmitted by heredity—no one would doubt this who had ever observed the behaviour of a young untrained pointer in presence of game which he has never before seen, and compared it with the behaviour of a mongrel of the same age.

Here is an instance. Some years ago I was surprised to see a pointer in my possession which had never before seen a partridge, had never been taught to point at them, point at a covey of partridges perfectly correctly, standing motionless with head outstretched, fore-paw lifted up, and tail stiffly erected. Exactly the same thing happened a few days ago with a pointer now five months old which was sent to me by one of my cousins who is chief forester in the Taunus, who assured me that this dog had never yet had a partridge in front of his nose. There are thorough-bred pointers which require no training at all, but which have completely inherited the habits to which their ancestors were educated, so that only a slight restraint till they have become perfectly obedient is necessary to make them equal to the most perfectly trained dogs. How different it would be with a mongrel or a Pomeranian or a drover's dog.

Once I took home from the Black Forest a dog about two weeks old and brought him up. He grew up into what is called a *Wildbodenhund*, which in shape and size is between a badger-hound and a pointer—nearer the former than the latter—and which is used to drive game towards the sportsman by barking. My dog began as soon as he grew up, although he was never taken out shooting, to drive game on his own account in the neighbourhood of my house, which is in the country, and in spite of all punishment extended his operations farther and farther every day. At last he ran away in the morning, and only returned covered with perspiration and tired out in the evening. I had to give him away, but on account of his inveterate habit he was of no use anywhere, and he finally came to an end on the dissecting table of a university laboratory.

And even as I write this I can hear the barking of a neighbour's dog, which was sold to him when young as a badger-hound. But it is half a *Wildbodenhund*, and has the habit of running about madly through gardens and yards in the neighbourhood for some hours in the day, barking ceaselessly, with his head down and his nose towards the ground as though searching for game. In accordance with his descent he still retains in part the peculiarities of the *Wildbodenhund*, and, without making further use of them, shows them by running about in the attitude characteristic of those dogs and barking in the same way, not knowing why, but impelled by the mechanism of the brain-cells which he has inherited from his ancestors, and which were acquired by them. He acts like a wound-up watch, by "instinct."

From the point of view of evolution it is in vain to seek for any other explanation of instinct than that which regards it as inherited habit. Automatic actions, *i.e.* those which we perform involuntarily without being conscious of them, although they were originally consciously performed and consciously learned, enable us better to understand this inheritance of habit by our own experience. We act automatically in consequence of habit. If such habit is inherited, we speak of instinctive actions, of instinct. A good example of the origin of instinct is afforded by the above described behaviour of my young pointers. These cases show that not only in bodily but in mental life one generation is connected with another. Such inheritance of acquired mental characters is, in my view, of great importance in the social life of animals and the explanation thereof.[1] That all the ingenuity of animals, so far as it is not due directly to the exercise of intelligence and reason, *i.e.* of reflection at the time, can only be explained by such inheritance, is self-evident.

Without this inheritance of faculties developed by practice, neither instinct nor any higher mental evolution could be explained. Not only in their bodily characters, in their liability to particular diseases, in their bodily movements, but in their whole intellectual life, the nations and races of mankind, which must have descended from common ancestors, show quite peculiar characteristics, *e.g.* the Jews, which must have once been acquired by their forefathers. Among these characters those which are to be ascribed to imitation from generation to generation can easily be distinguished, but only for a short time, for imitation also is inherited, and becomes a permanent acquisition.

It is also an example of the inheritance of acquired char-

[1] Cf. my lecture, *Ueber den Begriff des thierischen Individuum.*

acters due to habit that the handwriting of the son often shows the same characteristics as that of the father, although there has been no imitation. I can instance myself in confirmation of this. As I left home at a very early age I had very little opportunity for imitation, and, morever, the similarity of my writing to my father's was not developed till later years, and, as I am able to affirm, entirely independently. The causes of this inheritance in any case lie essentially in the inheritance of peculiarities of the fingers, of their musculature, and in their consequently inherited peculiarities of position and motion, apart from the fact that certain peculiarities of mental character are expressed in the handwriting. But the expression of disposition in an acquired faculty is also inherited.

Without wishing to help to prove that character is indicated by handwriting to the ridiculous degree assumed by journals in need of circulation which trade upon human vanity, I would yet declare my agreement with the often-expressed opinion, that in general a vigorous large handwriting shows vigour and frankness, a small prim hand shows a reserved, subtle nature, fond of minute expedients and indirect ways.

I must return here once more to the question already touched upon, why most dogs carry their tails erect, while wolves and jackals, from which they are descended, do not. Only the dogs which are nearest to the original forms, as the jackal-dogs of the East and our sheep-dogs, which are not far removed from the wolf, carry their tail depressed like the wolf and jackal. The erection of the tail by the domestic dog is obviously an acquired and inherited character, and one in which selection is out of the question. It is difficult to discern the causes of this alteration. It seems to me, however, allowable to suppose that it has a psychical basis. It is known that the dog lets its tail droop when in fear of punishment, and generally when in a mentally depressed condition, or even draws it in tightly between his legs, while during the feelings of security, pride, or conquest he elevates it. Must not, then, the feeling of security, which the dog as a domesticated animal almost always possesses to a much greater degree than his wild ancestors, have gradually produced the habit of constantly carrying the tail erect? I know not of course how wolves and jackals carry the tail when specially well pleased—they may possibly then elevate it somewhat. Of the fox, however, apparently in contradiction to my interpretation, it is known that he raises his standard in flight; but this may be an expression of the consciousness the animal has of mental power, or it may be merely a consequence of a high degree of excitement, in which condition dogs too, *e.g.* when fighting, or in the case of the pointer when pointing, raise their tails.

I will add here another hypothesis similar to that of the origin of horns. It seems to me that the combs and wattles of our domestic fowls are to be ascribed to similar causes. The originals of these fowls in India (Gallus bankiva) have these appendages only slightly developed. It is known to every keeper of fowls how much they fight, especially in confined enclosures, pecking at each other's heads and combs, not to mention the similar habit of the cock in treading. In a free condition, the hens probably contrive to evade these attacks, and it seems to me allowable to ascribe the extreme development of the comb in our domestic fowls, although artificial selection has doubtless to be considered, to the constant external stimulus referred to, to find in this stimulus the causes of the greater development. The ultimate causes of the commencement of the development may of course in fowls, as with the horns of ruminants, lie in purely physiological internal phenomena of growth.

Inheritance of Injuries and Diseases

Weismann expresses himself concerning acquired characters as follows:[1] "Acquired characters are universally understood to be those which arise in consequence of the action of external forces upon the organism, in contrast to those which proceed from the constitution of the germ." To which it might be objected, according to my views, that the germ also may acquire characters and transmit them by inheritance, and that it is, moreover, always affected by the condition of the body. How far Weismann admits this will be discussed in the sequel. Characters acquired during individual life Weismann describes as transient, because, according to his view, they cannot be transmitted by inheritance, "for it is obviously a consequence of the theory of the continuity of the germ-plasm, that characters can only be inherited when their rudiments are contained in the germ-plasm, but that modifications which occur in the developed body in consequence of external influences must be limited to the organism in which they arise. The latter must therefore be the case with mutilations, and with the effects of exercise, or of the disuse of any part of the body.

"If then this is true, there is not only an end of all Lamarckism, *i.e.* of the view which derives the modification of species from the direct influence of the conditions of life, especially from the increased or diminished use of particular parts, but it becomes necessary to discover a new basis for one factor of selection, namely, variability. For variability has hitherto been attributed to the variable influences which act upon the organism without intermission. But if all the influences which might make bodies individually different have only a transient and not an inheritable effect, the individual variations which form the material for selection to work upon cannot arise in this way."

[1] *Biolog. Centralblatt, loc. cit.*

This exposition is in direct contradiction to my views.

But farther on Weismann makes some concession. He says, "But although I hold it improbable that individual variability can depend on a direct action of external influences upon the germ-cells and their contained germ-plasm, because —as follows from sundry facts—the molecular structure of the germ-plasm must be very difficult to change, yet it is by no means to be implied that this structure may not possibly be altered by influences of the same kind continuing for a very long time. Thus it seems to me the possibility is not to be rejected, that influences continued for a long time, that is, for generations, such as temperature, kind of nourishment, etc., which may affect the germ-cells as well as any other part of the organism, may produce a change in the constitution of the germ-plasm. But such influences would not then produce individual variations, but would necessarily modify in the same way all the individuals of a species living in a certain district. It is possible, though it cannot be proved, that many climatic varieties have arisen in this manner. Possibly other phenomena of variation must be referred to a variation in the structure of the germ-plasm produced directly by external influences. At present we cannot decide upon this; but this much may be maintained, that influences which are mostly of variable nature, tending now in one direction, now in another, can hardly produce a change in the structure of the germ-plasm, and this is the reason why the causes of inheritable individual differences must be sought elsewhere than in these varying influences."

... "No one has doubted," he says further, in reply to objections made by Virchow, "that there are a number of congenital deformities, birth-marks, and other individual peculiarities, which are inherited. But these are not acquired characters in the above sense. True, they must once have appeared for the first time, but we cannot say exactly from what causes; we only know that at least a great proportion of them proceed from the germ itself, and must therefore be due to alteration of the germinal substance.

"If Virchow could show that any single one of these hereditary deformities had its origin in the action of some external cause upon the already formed body (soma) of the individual, and not upon the germ-cell, then the inheritance of acquired characters would be proved. But this no one has yet succeeded in proving, often as it has been maintained." The inheritance of artificially produced diseases, according to Weismann, is no proof. "Unless I am much mistaken," he says, "the transmission of acquired epilepsy to the following generation, the occurrence of which is not to be denied, is not due to inheritance, but to infection of the germ, to the transference of living disease-producing organisms."

As Weismann himself in another place makes use of the expression that his view is a matter of conviction, that no proof of it can be brought forward, all that can be done is to bring against it, as we have done, the facts that contradict it. In particular, that view seems to me to have no prospect of being accepted so long as there is no peculiar pathology of the germ-plasm, and so long as the assumption that the physiology of the germ-cells is so exceptional in comparison with that of the other cells of the body, as Weismann's theory requires, is unsupported by any basis of proof.

Of course it is true that the consideration on account of which he proposed his theory seems to be in Weismann's favour, namely, that this theory would explain the portrait-like inheritance of characters, and reversion. But it is a question whether such an explanation is not possible on other grounds. "We are well aware," he remarks in one of his latest writings,[1] "that all bodily and mental peculiarities may

---

[1] *Ueber den Rückschritt der Natur, Berichte der naturforschenden Gesellschaft zu Freiburg i. B.* Bd. ii. Heft i. 1886.

be transmitted from the parents to the children ... but the ancestors possessed all these peculiarities by virtue of the properties of their germs."

What agencies, then, must be asked, in reply, have first introduced new peculiarities into the ancestral series? By what means have quite new characters arisen at all in organic beings? Sexual combination could not originate them: it could only work from the beginning with what already existed, was already present.

If I understand Weismann aright, the variability inherited by the germ-plasm from unicellular ancestors is supposed to have created these innovations. But such an assumption, it seems to me, is purely hypothetical. It stands entirely without proof, and finds its support only in the explanation of portrait-like inheritance and reversion which is hypothetically possible by its aid.

In opposition to this assumption I believe I can appeal to facts even with regard to the inheritance of injuries and diseases.

That injuries, when continued for an extremely long time, may be inherited is proved to my mind by atrophied (rudimentary) organs. The degeneration of these organs depends incontestably on disuse: in consequence of disuse the blood-supply is diminished, in consequence of the decrease of nutrition degeneration takes place. If we consider the course of gradual degeneration, *e.g.* of the tail as it must have taken place in the higher mammals, to have proceeded in this purely physiological manner from the tip towards the root, the process is much the same as if the tip of the tail had been in many successive generations amputated, and the shortening had been inherited, and then the shorter tail thus acquired had been farther shortened artificially, and so on. In any case in the degeneration of the tail an acquired character has been inherited by the offspring, a character which in the causes of its origin is closely similar to a perpetually repeated mutilation. Great periods of time, however, have been necessary in this case for the accomplishment of the final result.

I add here some cases of inherited injuries which seem to me to be authentic. A. Decandolle describes one such case with the assurance that it is perfectly true.[1] In the year 1797 a girl twenty-one years old was thrown from a carriage, and in consequence had a scar about five centimetres wide over the left ear and temple, which remained without hair. Married in 1799, she bore a son in 1800, in whom the hair was absent from the same area, and remained so. The son of this man, born in 1836, had no such defect, but it was present in his grandson born in 1866, and in 1884 in this last individual, when he was eighteen years old, the peculiarity was disappearing.

Dr. Meissen, of Falkenberg, records in the number of *Humboldt* for June 1887 the following case of inheritance of an injury in his own family: "When I was seven or eight years old I had the chicken-pox, and I recollect with complete distinctness that I scratched one of the pustules on the right temple, in consequence of which I had a small white scar at this spot. Exactly the same scar, which I had of course ceased to think of, on exactly the same spot, was present on my little son, now fifteen months old, when he came into the world. The resemblance is so perfect that it surprises everyone who sees the little mark."

My assistant, Dr. Vosseler, relates that his mother in her eighteenth year injured the ring-finger of her right hand by squeezing it between the door-latch and the door, so that it

---

[1] Alphonse Decandolle, *Histoire des sciences et des savants depuis deux siècles, précédée et suivie d'autres études sur des sujets scientifiques, en particulier sur l'hérédité et la sélection*, Genève, Bâle, H. Georg, 1885. Cf. also on the whole question: Lucas, *Traité philosophique et physiologique de l'hérédité*, etc., Paris, 1847; and E. Roth, *Die Thatsachen der Vererbung*, Berlin, 1885.

was bent towards the radial side between the last and the middle joint, and remained the rest of her life rigid in this position. Dr. Vosseler, who was born two years later, has had the same crookedness of the same finger from his birth, and a brother of his also. The peculiarity was more marked in his youth than it is now.

The following was related to me by my colleague Professor Dr. v. Säxinger: His father-in-law had a pair of long-tailed pointers, which had once produced a litter of long-tailed pups. In order to obtain short-tailed pups, he had the tails of both shortened. The bitch from that time produced repeatedly short-tailed pups only. As the most careful attention was paid to the parents, no error can be suspected in this case, which, moreover, appears to excite no surprise among dog-breeders.

Brown-Séquard,[1] as is well known, has shown that epilepsy is inherited by the offspring of guinea-pigs, in which it has been produced by division of the sciatic nerve, or of a portion of the spinal cord. In like manner, a peculiar alteration of the shape of the ear, or a partial closing of the eyelids, is inherited by the offspring of animals in which these changes were caused by dividing the sympathetic. Thirdly, exophthalmia was inherited by guinea-pigs in whose parents this protrusion of the eyes had occurred after an injury to the spinal cord, as also ekchymosis and dry gangrene, as well as other trophic disturbances in the ear, due in the parents to an injury to the *corpus restiforme*. Fifthly, loss of certain phalanges or of whole toes of the hind feet, which had occurred in the parents in consequence of division of the sciatic nerve. Sixthly, diseased condition of the sciatic nerve in the offspring of guinea-pigs in which this nerve was divided, and the occurrence of the phenomena which Brown-Séquard had described as characteristic of the increase and decrease of epilepsy. Further, Brown-Séquard possessed forty

---

[1] *Compt. rend.* tom. xciv. p. 697, Paris, 1882.

guinea-pigs in which one or both eyes showed more or less morbid change, and which were descended from three individuals in which one eye had become diseased in consequence of transverse section of the *corpus restiforme*. Lastly, twenty guinea-pigs exhibited muscular atrophy on the upper and lower sides of the thigh, in whose parents such atrophy had been caused by section of the sciatic nerve. Mention of Brown-Séquard's observation of the inheritance of the loss of phalanges of the hind foot in guinea-pigs caused by injury to the sciatic nerve, gives me an opportunity to describe an extremely remarkable case of the inheritance of congenital shortening or deformity of the fingers which I observed in a young man studying in Tübingen this summer.

Mr. —— was a law student twenty years of age, and meeting him by chance, I was much struck by the fact that the second and third fingers on both his hands were much shorter than usual, although his bodily structure was in other respects perfectly normal. He kindly placed himself at my disposal for the purpose of more exact examination.

The result of this examination and the peculiarities of inheritance in this case are so remarkable, that it would be difficult to find another case comparable with it. With regard to the inheritance, I will mention first that, according to the statement of Mr. ——, his mother and the youngest of his three brothers, who is fourteen years of age, have exactly the same defect, while the eldest, twenty-one years, and the third, sixteen years old, are without it. In the brother and the mother the shortening is not so great as in himself. It does not occur in other relatives. As to its causes nothing is known to him.

Exact investigation shows that in both hands the fourth finger is much the longest, on the right the middle finger, on the left the index finger is the shortest, so that leaving out the thumb the following is the order of the fingers in size:—

In the right hand: 4, 5, 2, 3.
„ left „ 4, 5, 3, 2.
In myself the order is:
right „ 3, 2, 4, 5.
left „ 3, 4, 2, 5.

In the latter case the fourth is scarcely longer than the second, in the former the second very little longer than the fourth.

The order of size is usually 3, 4, 2, 5.

Exact measurement gives for the fingers of Mr. ——:

	RIGHT HAND.	LEFT HAND.
2d finger	7·8 cm.	7·2 cm.
3d „	7·4 cm.	8·8 cm.
4th „	10·4 cm.	10·4 cm.
5th „	8·7 cm.	8·9 cm.

measuring always from the terminal knob of the metacarpal to the tip of the finger. The shortness of the second and third fingers in both hands is caused by a strongly marked shortening of the middle-finger joint. This joint is:—

	RIGHT HAND.	LEFT HAND.
In the 2d finger	1·2 cm.	1·0 cm.
„ 3d „	1·6 cm.	1·5 cm. long.

The joint is therefore only half to one-third the usual length.

The first joint of the third finger of the right hand is also shortened: it is only 3·5 cm. long; that of the same finger of the left hand is 5 cm.

The measurements of these joints is somewhat too great in proportion to the length given for the whole fingers, because they are taken from knuckle to knuckle with the finger bent.

Externally the shortened fingers give no impression of deformity, with the exception of the index finger of the left hand, the first joint of which is particularly slender, and which altogether is somewhat weak.

As the shortening of the fingers is very similar on both hands, it can scarcely be supposed that it has been derived from some injury incurred by an ancestor of Mr. —— during life. If the idea of self-mutilation were to suggest itself, it would be at once disposed of by the fact that the general shortening is due to the shortness of the middle finger-joint.

But comparison with Brown-Séquard's experiment suggests the reflection, whether the ultimate causes of the shortening or atrophy are not to be sought in some disease, or some injury of the spinal cord in an ancestor.

Another possibility is that the germ of some ancestor was in some way affected or injured, and such derangements of the germ are certainly very often the causes of congenital defects. In this case, since the shortening of the fingers occurs on both sides, the injury must have affected some of the embryonic cells which take part in the formation of the trophic centres or centres of innervation of the fingers in question.

On the other hand, how can the first origin of this merely local structural defect be explained by sexual combination? I confess I am unable to conceive such an explanation, especially as it is not merely a question of a character formed by the greater development of ancestral characters, but certainly to some extent of a pathological atrophy.

It is known that numerous other instances of the inheritance of injuries have been recorded, *e.g.* inheritance of the artificially shortened tail of a bull, of artificially produced hornlessness in cattle, many cases of inheritance in man of curvature in a finger caused by injury, inheritance of the absence of one eye which had been lost by the father during life by disease, etc. The opponents of the inheritance of acquired characters flatly assert that these statements are unauthenticated stories, anecdotes of no value as evidence. They attempt to deprive the experiments of Brown-Séquard, confirmed by others (Westphal and Obersteiner), on the inheritance of artificial epilepsy in guinea-pigs, of importance, by explaining [1] "that this transmission of acquired epilepsy to the following generation, the occurrence of which is not to be denied, depends not on heredity, but on inoculation of the germ, on the transference of living disease-producing organisms."

Since, however, this hereditary epilepsy was produced by partial or complete section of the spinal cord, or by division of the *nervus ischiadicus*, when epileptic attacks could only be produced some weeks after the operation by pinching the skin on the lateral parts of the head and neck (epileptogenous zone); since Westphal, who expressly states that he entered upon these experiments with distrust, was able to produce inheritable epilepsy by blows with a hammer on the head of the animals—that assumption seems to be entirely unfounded. But Ziegler, although he will not contest the possibility that a transmissible disease due to infection may have followed the operation, prefers to try to diminish the force and importance of the experiments of Brown-Séquard, Westphal, and Obersteiner, by asking if the guinea-pigs on which the operations were made may not have been already predisposed to disease; if the appearance of epilepsy may not have been due to general decrepitude, rather than to the inheritance of a disease experimentally produced.[2]

If epilepsy can actually be produced by blows upon the head, and then inherited, as Westphal states (it showed itself, according to that author, in the two young ones of a guinea-pig

---

[1] Weismann, *Biolog. Centralblatt*, 15th March 1886; Westphal, *Berlin. Klin. Wochenschrift*, 1871; Obersteiner, *Med. Jahrbucher*, 1875.

[2] E. Ziegler: *Können erworbene pathologische Eigenschaften vererbt werden, und wie entstehen erbliche Krankheiten und Missbildungen?* Jena, G. Fischer, 1886. Published separately, from the *Beiträgen zur pathol. Anatomie u. Physiologie* of Ziegler and Nauwerck, Bd. i.

made epileptic in this way), then the possibility of infection is excluded. The only question is whether in this experiment the animal was not already pregnant when the experiment was made. If it was, then the evidence is not conclusive. Apart from this, it is evident at first sight that there is more justification for Ziegler's objection than for the other. But why should all the guinea-pigs operated upon in Berlin, Vienna, and Paris have been decrepit, and why should guinea-pigs be generally decrepit? The fact that they are domesticated is no evidence for this assumption, for they are all domesticated, they exist only in captivity, their wild ancestors are not even known; for thousands of years they have been reared only under man's protection, for they were kept by the ancient Peruvians and Chilians, and they seem to thrive very well under this protection, as is proved especially by their almost incredibly rapid multiplication. Still Ziegler may be justified in demanding that the experiments should be repeated.

With regard to the other experiments of Brown-Séquard and Deutschmann [1] concerning acquired and inherited affections of the eye, Ziegler holds that these are to be explained as cases of infection and inflammation, and that therefore they afford no evidence of inheritance.

Some of Brown-Séquard's experiments, nevertheless, as the inheritance of an atrophy of the foot in the guinea-pig after section of a nerve, seem to me to afford unassailable evidence, unless indeed any one would assume that here too the cause lay in an infection conveyed to the germ. But with such purely arbitrary assumptions anything could be proved.

I repeat that a single certain case of the inheritance of acquired characters upsets the whole theory of inheritance being exclusively limited to the germinal cells; and it seems

[1] Deutschmann, *Ueber Vererbung von erworbenen Augen-affektionen bei Kaninchen* (Zehender's *Klin. Monatsblätter*, 1880).

to me that in the preceding instances enough of the inheritance of injuries has been given to prove this inheritance—putting aside my previous attempt to argue in favour of the same conclusion from the natural evolution of atrophied organs. The latter argument seems to me sufficient by itself to settle in the affirmative the question whether acquired characters and injuries are inherited.

Moreover, it is self-evident that the injuries of a body are not all hereditary to the same degree, and further, that injuries are inherited in different degrees in different animals.

The idea naturally suggests itself that injuries affecting parts far removed from the centre of circulation—parts less richly supplied with blood—are most likely to be inherited.

The question of the inheritance of injuries is further closely connected with that of the reproduction of parts removed from the body: the lower the grade of organisation, the less the degree to which division of labour is carried, the more easily will such losses be repaired, the less probably will injuries be inherited. Therefore it is to be supposed that such inheritance will occur much oftener in the higher animals than in the lower, scarcely at all in plants.

We have now to enter upon the inheritance of diseases of spontaneous origin. The negative standpoint in this question has been advocated by my friend Ziegler in the paper above cited, and in a lecture "On the Inheritance of acquired pathological Characters and the Origin of inheritable Diseases and Malformations."[1] He concludes that the former as well as the latter arise always from changes in the germ.

Ziegler holds that so long as it was assumed that fertilisation was a process in which a distribution and solution of the spermatic substance in the ovum took place, the trans-

---

[1] Separate edition, from the *Verhandlungen des Kongresses für innere Medicin in Wiesbaden*, 1886.

mission of acquired characters was conceivable; and Darwin, by his "pangenesis," and Haeckel, by his theory that reproduction and heredity depend on the transference of a definite form of motion (motion of the plastidula) from the parental organs to the organic molecules of the generative cells, offered an explanation of the phenomenon of the inheritance of acquired characters considered by both as proved. But Ziegler says that since newer researches have shown that fertilisation is a purely morphological process, the inheritance of acquired characters is excluded. This view is further supported by the argument that the sexual cells are not elements which could be derived from any cells of the organism whatever. The nuclei of an organ of the fully-developed organism could only possess by inheritance the property of producing, with the aid of the cell-protoplasm, tissue of a single kind. The structure of the sexual cells, on the other hand, must be so constituted, that from the offspring of two nuclei which coalesce, all the cells of the individual body, including new sexual cells, can arise.

I have already expressed my opinion upon the view which regards reproduction as a purely morphological process. I should hold it no less justifiable to regard life in general as a morphological process, for reproduction is a part of the life of the organic world.

That the sexual cells give rise to not only one, but to the several kinds of tissue-cells, is their most peculiar (specific) function in multicellular animals; but if multicellular animals have been derived from unicellular, this distinction cannot be a fundamental one, and must have been evolved in course of time—as an acquired and inherited property. Since the cause of the evolution of this distinction can only depend on the advantage of division of labour, following upon the advantage previously gained by colonial life, and division of labour consists in physiological relations between the cells of the organism and the outer world, the distinction must be essentially due to external influences—to which is likewise due the beginning of all histological evolution (differentiation) in the body, to which I shall again refer.

Ziegler believes that the causes of the origin of germ variations which lead to inheritable deformities and diseases may be of three kinds—(1) Union of sexual nuclei which are not adapted for copulation; (2) disturbance of the copulatory process itself; (3) injurious influences which affect the sexual nuclei or the fertilised ovum at a time when separation of the sexual cells from the body cells has not yet occurred. "If the embryo is injuriously affected at a later period, either a malformation or constitutional anomaly arises, which is not inherited, or only the sexual cells are injured, in which case the body-cells develop normally, and a disturbance shows itself only in the development of the next generation."

The union of sexual nuclei not adapted for copulation is, according to Ziegler, the most frequent and most important cause of hereditary local malformations as well as of hereditary morbid tendencies, or of a defect in any system of the whole organism. "When in a family whose members show no special talents a genius suddenly appears; the natural explanation is, that sexual nuclei unusually well-fitted for union have united in copulation, and that in consequence the nervous system has reached an unusually perfect organisation." In like manner arise tendencies to mental disease, etc. If the nuclei are altogether unadapted to one another, sterility results, as in the sexual nuclei of distinct species.

Thus what Ziegler is here discussing is the self-evident effect of crossing, the importance of which I fully acknowledge, and on which I have already expressed similar views; but it still requires to be explained in what way the sexual cells originally acquired the peculiarity of being adapted to the production of a genius.

With regard to disturbances of the process of copulation itself, it appears, says Ziegler, that, *e.g.* the simultaneous fertilisation of an ovum by two spermatozoa results in a double monster.

As instances of injurious conditions affecting the sexual nucleus or the fertilised egg, it is pointed out that substances taken up from without—poisons, for example—are brought by the blood to the sexual cells, and others produced in the body are conveyed to the sexual organs. Thus "it seems as if alcohol, for example, might have an injurious effect upon the sexual cells, not merely by destroying the health of the parents, but also directly.

"If the organism is reduced by suffering or weakened by age, it is probable that some deterioration of the sexual cells may be the consequence, so that their copulation produces badly-developed offspring."

Here one is reminded of my instance of the inheritance of the signs of old age. I will not go further into the comparison of the two cases, the special case just mentioned and the general case stated by Ziegler. But the latter by itself seems to me to grant by implication the inheritance of acquired characters, the influence of the condition of the whole body at a given time upon the properties of the germ-cells,—to grant that a reduced condition of body, *i.e.* a character acquired during life, is inherited by the offspring.

If, moreover, alcohol can act upon the germ through the blood, and any morbid change in any part of the body can act in like manner, and if such influence can be inherited, then the dependence of the condition of the germ upon that of the whole body, and the possibility of the inheritance of acquired characters, is again fully recognised, and the only difference between our two opposing views is one of words. The assumed importance of double fertilisation in the origin of double monstrosities seems to me to indicate plainly enough the importance of the effect of external influences upon the germ (and upon the offspring).[1] Indeed, according to the explanation I have given of the ultimate causes of sexual combination, fertilisation in general is to be regarded as an instance of external action upon the germ-plasm, although I maintain the original equivalence of the male and female portion of the germ. The view that fertilisation must be regarded as an instance of external action is expressed also by Virchow in his paper on "Descent and Pathology," and he adds : "In a more rigorous sense it can be described as an acquired modification of the egg-cell." He points out further, that in general very much of what is regarded as internal cause (*causa interna*), is really external influence (*causa externa*). "As a rule, a multicellular organism, when variation takes place, is not altered in all its parts; usually only a portion of the cells is the seat of change. Upon this portion, or to express it more accurately, on each of the cells affected, the remaining unaffected cells can exert an external influence, and conversely the cells originally unaffected can be influenced by those which are affected as by external agents. The conception of a *causa externa* applies therefore not merely to those influences which affect the organism from without, but also to those which arise from other cells or inner parts and affect the individual cells either on the surface or in the internal parts of the body. Only those causes which really arise from the mechanism of the cells themselves are true internal causes." Thus, then, correlation also is, in accordance with my views, regarded as a phenomenon which has its causes in external influences. Further : "When an infectious substance is generated in one part of the organism and acts upon another part, it is in relation to the latter just as truly

[1] Cf. on this subject the papers of G. Born : *Beiträge zur Bastardirung zwischen den einheimischen Anurenarten, Archiv. f. Physiologie*, Bd. xxxii. p. 477 ; and E. Pflüger u. Smith : *Untersuchungen über Bastardirung der anuren Batrachier u. die Prinzipien der Zeugung, Ibid.* p. 558, 559.

a *causa externa* as if it was generated outside the organism and introduced from without into it."

I myself, as might be expected from my general views, have never used the expression internal causes except in the sense of the action of the given constitution of the body, itself partly due to external influences, upon the direction of evolution.

The supporters of the doctrine of the continuity of the germ-plasm, since they deny the inheritance of characters acquired by the body during life, and admit on the other hand the heredity of those due to influences acting directly on the germ-cells, set up a completely artificial distinction between the nature and the properties of the germ-cells before and after segmentation. Such distinction is not only altogether hypothetical, but completely contradicts that uniformity which proves the morphological and physiological unity of the living world.

I have already attempted to demonstrate on other grounds the inconsistency of the supposition of the non-inheritance of acquired characters with that of the unity of the living world. It can also be shown by another consideration that the conception of a fundamental difference between the properties of the germ-cells and of the larva, or of the adult body, must be false, namely, the consideration just indicated, that the germ-cells contain in themselves the material and the properties for the formation of the germ-layers and of all the various cell materials of the adult body; and that the variety of the latter can only have its ultimate causes in external conditions—must therefore be acquired.

I will not here discuss further the inheritance of acquired morbid conditions, leaving the defence of my views in this point to those special authorities who are on my side, some of whom, like Virchow, have already expressed their support. In particular, with regard to evidence, I may refer to the paper of Roth already mentioned, a paper in other respects noteworthy, which contains numerous instances in favour of pathological inheritance, and in which the question of heredity in general is profoundly discussed.

I would here add a few words on the inheritance of but one kind of diseases, namely, mental diseases. That mental diseases are inherited to a considerable degree no one will deny.

But it is quite beyond doubt that among such hereditary mental diseases are some which can only have arisen from external influences on the nervous system, not through direct modification of the germ.

The restless chase of life especially which is so characteristic of our time leads in certain classes of the population to an over-excitement of the nervous system which often expresses itself in actual mental disease, and which—the over-excitement as well as the actual disease—is certainly inherited. Where disease commences cannot be determined, for by the nature of the case it is impossible to define the boundary between the two. Every over-excitement of the nervous system is morbid, even when it is confined within the limits of the socially permissible. The brain-workers especially, whose life is often a struggle with their nerves, know the truth of this, at least those who accept the conclusions of physiological inquiry.

But the question may be asked, Whether the cultured classes generally of the present day, using the term in a wide sense, are to be regarded as perfectly normal with respect to their nervous system? The same question may be asked of their children, even before the latter have begun to expose themselves to the abnormal conditions which have injuriously affected their parents and ancestors. Our peasants and their children, on the other hand, enjoy traditionally sound nerves.

As among single classes, so also among different nations

very different conditions in this respect are to be observed —conditions which form the most important peculiarities, constitute a part of their character. Many nations which have become exhausted by their own civilisation, or are becoming exhausted, prove this clearly enough, and there are examples obvious enough and even belonging to most recent times which I need not particularise. Such nations are sometimes conscious that they need to be reinvigorated by the introduction of healthy blood from another race. Crossing is here the only possible remedy remaining. But the disease can only be explained by the aid of the action of external influences upon the nervous system.

It is actually inconceivable that the system of organs, which not only is the instrument of the relations of the body to the outer world, but which of necessity owes its origin and its evolution to these relations, should not be capable of undergoing hereditary morbid changes in consequence of such relations when they are injurious.

It is intelligible that the direct action of external conditions, or certain states of the nervous system which are ultimately ascribable to injurious external influences, lead to "fixed ideas," to the tendency to self-accusation, to melancholy and suicide, and that these aberrations may be hereditary seems certain. But the origin of these aberrations cannot be comprehended unless they are acquired through relations to the outer world. It is therefore inconceivable that in germ-cells, whose ancestors were never exposed to corresponding influences, the conditions for the origin of these aberrations should suddenly occur. For mental diseases have their seat in the brain, and the earliest ancestors of man must have acquired the brain itself through relations to the outer world —and only these relations could produce those diseases in it. It is therefore certainly intelligible (even if it be not admitted that melancholia, for example, is a disease depending on definite changes in the brain, and to a certain degree definite and capable of hereditary transmission) that a morbid state of the brain which may have melancholia as a result (if external conditions occur which excite its outbreak) is acquired and transmitted by a human being provided with nerves and brain.

It is also intelligible that such a condition may be acquired through union with another sexual cell, provided that one or both of the uniting sexual cells possess the acquired and inherited tendency thereto.

It is, however, perfectly inconceivable, if the living world be regarded as a continuous whole, that such a condition should ever have come about without acquirement and inheritance.

Mental capacity is an acquirement, and mental diseases are diseases of relation.

As for the rest, it is certain that not only mere general morbid conditions of the nervous system are inherited, but also definite tendencies to mental disease, and likewise definite mental abilities. The terms by which we express both are of less importance, for, by the nature of the case, these are only too often more or less artificial and elastic (conventional).

It is obvious that the mechanism of the human brain has become by acquirement so delicate that a high degree of division of labour has taken place in it, which can only depend on the fact that definite groups of brain-cells each exercise and can only exercise definite functions. These groups form "centres" which have assumed not the simplest of such functions, elementary functions, but those of more compound character, often indeed those which respond to the demands of recently-developed civilisation.

I appeal, therefore, not to the numerous physiological experiments which, by the aid of the results of pathological anatomy, have long ago made known the localisation of elementary functions in the brain as a fact beyond dispute, but, for instance, to the fact that the faculty of speech, which was obviously acquired through the relations of men to one another, according to pathological evidence has its seat in a definite part of the human brain, in the left temporal lobe. Loss of the power of speech, aphasia, may temporarily occur as a consequence of excessive mental exertion. I have myself experienced this fact, and know that in this condition the power to speak is lost, but not the ability to produce sound. One is only unable to find and to utter words; one is perfectly conscious of this and is astonished that it is so, and makes vain endeavours to overcome it.

I consider as a similar condition of less degree another which I have often observed in myself during great mental fatigue, and which friends have told me they also have experienced, namely, that in writing one repeatedly omits letters from a word, or puts wrong letters instead of the right.

It is universally known that in great mental fatigue one often "forgets" the commonest names, and I have already made the observation on myself that this especially occurs when, as for example in systematic natural history, one has worked at names too much; the part of the brain which is concerned with such matters is then exhausted and fatigued.

The following highly peculiar case which seems to me to prove the fact that very complicated faculties, the acquirements of civilisation, are connected with definite parts of the brain, I was able to observe accurately within the last few months :—

A young man who was a connection of mine, and who was a student here in Tübingen, began to require an unusual amount of sleep. As the consequence of this was that he usually did not get up till near mid-day, it was at first attributed to causes which seemed very natural—to irregular habits of life. This suspicion became apparently a certainty when the student, notwithstanding the remonstrances, requests, and threats of his troubled parents, never wrote a letter home, not even for money. Arguments on my part succeeded in obtaining promises from him, but these were never fulfilled. In the meanwhile, discreet inquiries elucidated the fact that the student in other respects led a thoroughly regular life; that he generally went home before midnight, and went to bed; and in particular, that he exercised in all his other affairs a painstaking orderliness, as being the son of a merchant he had always been accustomed to do. His landlady stated that latterly, as always previously, when he went away for a day he gave up his key to her in a sealed packet.

On account of these facts I became convinced that the young man must be suffering from a local inflammation of the brain, or that a local inflammation of the cerebral membrane must be causing pressure on a certain portion of his brain, and that therefore a part of this organ which influenced the activity of the will towards the particular direction of letter-writing, and towards this alone, was for the time functionless; for the mental abilities of the subject and his power of will were, and are still, in other respects normal, apart from the effects of the great amount of sleep he required. His general force of will is, however, only moderately developed. I was confirmed in this conviction by the fact that at times symptoms appeared in him which might well be ascribed to temporary increase of a chronic local inflammation of the cerebral membrane, symptoms some of which had a similarity to those of cramp of the neck, of meningitis cerebrospinalis, although no fever ever appeared. After constant medical treatment the student now, after some months, is apparently well and healthy, only he cannot be induced to write a letter.

I brought forward this case because I followed it carefully with great interest. Insanity doctors doubtless know numerous others of a similar character.

It is, moreover, a fact generally known, that just such deficiencies in the power of will in respect of various but definite actions, in other words, of claims of human life, have helped to determine the character of individuals and their descendants, the character of whole families and peoples—that such deficiencies are hereditary.

And because these deficiencies, because also the nerve-cells of the brain in which they have their seat, owe their origin to relations to the outer world, therefore they must have been acquired.

The following, as an irrefragable proof of the inheritance of mental diseases as diseases acquired through relations to the outer world, I give in the words of an authority in psychiatry, Dr. von Krafft-Ebing.

It is *a priori* a probable supposition that the doctrine of inherited sin, which still finds place in the Christian religion, especially in the Catholic Church, is derived from a knowledge of the facts of heredity, just as the Mosaic and Mohammedan prohibition of the eating of swine-flesh probably has its ultimate cause in the experience of possible injurious effects of that food, although trichinæ as the cause of disease, leaving tapeworms out of the question, could not have been known.

But this view of the origin of "sin" is strongly supported by the following record of facts concerning heredity as one of the causes of insanity. Dr. von Krafft-Ebing says in the section on this subject in his *Textbook of Psychiatry* :[1]—

---

[1] R. v. Krafft-Ebing, *Lehrbuch der Psychiatrie*, 1879, Bd. i. p. 153, *seq.* Cf. Prichard, *Treatise of Insanity*, p. 157 ; Lucas, *Traité philosophique et physiologique de l'hérédité*, Paris, 1847 ; Morel, *Traité des Dégénérescences*, etc. Paris,

"By far the most important cause of insanity is the transmission of psychopathological tendencies, especially of cerebral infirmities, by means of procreation.

"The fact of the heredity of psychical defects and diseases was known even to Hippocrates. It is only the special action within the limits of this province of a general biological law which plays a stupendous part in the organic world, on which indeed the whole progress of the human race depends.

"After tuberculosis there is scarcely a class of diseases in which heredity works so powerfully as in psychical diseases, but on the percentage of cases in which its effects are evident authorities differ. The percentage of cases determined by heredity given by statisticians (Legrand du Saulle, *op. cit.* p. 4), varies from 4-90 per cent. It is obvious that a constant factor cannot produce effects in such a variable manner. The cause of the difference can only lie in the different method by which the statistical analysis was produced. Much depends on the question, from what classes of people the material of the statistics was derived. In aristocratic circles, in communities secluded from intermixture with surrounding peoples, in exclusive religious societies (Jews, sectarians, Quakers), where close interbreeding is practised, the percentage of hereditary cases is greater than in a floating population. But the point of view of the different statisticians has also been different. Many investigators have only recognised heredity when insanity could be proved in the parents (direct similar inheritance). But the definition of heredity cannot be so narrowly limited. Three essential facts are here to be considered.

1857; Idem, *Archiv. génér.* September, 1859; Hohnbaum, *Allgem. Zeitschr. f. Psych.* 5, p. 540; Morel, *Traité des Maladies mentales*, pp. 114, 258; *Ibid., De l'hérédité morbide progressive, Archiv. génér.* 1867; Voisin, *Gaz. des. Hôpit.* 1858, 16; Moreau, *L'Union médic.* 1852, 48; Jung, *Allg. Zeitschr. f. Pschy.* 21-23, *Ann. Méd. Psych.* November 1874; Legrand du Saulle, *Die erbliche Geistesstörung*, German by Stark, 1874; Ribot, *Die Erblichkeit*, German by Hotzen, 1876; Hagen, *Statist. Untersuch.* Erl. 1876.

"(a) Atavism.—The bodily and mental organisation and character can be transmitted from the first to the third generation, without any necessity that the second and intermediate one should exhibit the peculiarities of the first—thus the condition of the life and health of the grandparents are of interest for us.

"(b) Only in rare cases is the actual disease transmitted in procreation (congenital insanity, hereditary syphilis), as a rule only the disposition thereto. Actual disease only occurs when accessory injurious influences produce an effect based upon that disposition. . . .

"We must, therefore, consider also the state of health of the relatives (uncles, cousins, aunts), and, since here also the law of atavism holds good, the possible diseases of great-uncles and great-aunts.

"(c) Only exceptionally does the same disease develop in ascendant as in descendant in consequence of the transmission of morbid dispositions. On the contrary, there exists a remarkable variability in the forms of disease which may almost claim the value of a law (the law of polymorphism or transmutation).

"The transmutations are innumerable. The most varied neuroses and psychoses occur side by side in families with a hereditary taint, and one after another throughout generations, and teach us that from a biological, etiological standpoint they are only branches from one and the same pathological stem.

"The fact of the variability of the morbid conditions due to heredity necessitates a careful examination of the question, With what conditions and symptoms of morbid nervous organisation hereditary transmission, direct or modified, is connected?

"(a) In this respect there is no uncertainty about those cases in which psychoses present themselves in both the ascendant and the descendant (homogeneous inheritance). In many of these cases the psychosis has even in both generations the same form, and its outbreak depends on the same accessory causes, *e.g.* parturition (uniform inheritance).

"(β) The recurrence of suicide[1] throughout successive generations is an equivalent phenomenon belonging to the same class, *i.e.* the tendency to suicide, which is almost always a symptom of melancholia or of a neuropsychopathic constitution which is unable to endure arduous conditions of life. Particularly convincing are those cases of suicide in which ascendant and descendant destroy themselves under almost the same conditions of life and at a similar age. Genealogical tables actually exist which show that whole unhappy families have died out through suicide.[2]

"(γ) Equally certain are the hereditary effects of constitutional neuropathies, though they may only consist in a habitual migraine or in a hysteria or epilepsy.[3]

"The hereditary morbid factor may make itself felt in the descendants in a mere neuropathic constitution, in the production of neuroses, but also of psychoses, even to idiocy, as the worst form of hereditary degeneration.

"(δ) The hereditary influence of pathological character in tending to insanity is firmly established.

"Visionaries, perverse, eccentric characters, oddities, or hypochondriacs have not only extremely often ascendants

---

[1] Tigges, *Vierteljahrschr. f. Psychiatrie*, 1868, No. 3, 4, p. 334.

[2] Morel, *Traité des mal. ment.* p. 404; Ribot, p. 147; Lucas, ii. p. 780; *Ann. Méd. Psych.* May 1844, p. 389.

[3] Trousseau, *Med. Klin.* German by Culmann, 1867, p. 88; Moreau (*op. cit.*) found among 364 epileptics 62 epil., 17 hyster., 37 apoplect., 38 insane relatives, 195 times convulsions, consumption, scrofula, eclampsia, asthma, dipsomania, etc., among the parents or relatives; Martin, *Ann. Méd. Psych.* November 1878, shows that the children of epileptics die in large numbers in convulsions.

and collateral relatives who are diseased in mind or nerves, but also neuropathic, insane, even idiotic offspring.

"These problematic beings, who mostly from their childhood feel, think, and act differently from other people, are also themselves in constant danger of becoming insane, and are often candidates for a degenerate form of true insanity—namely, primary madness, to which also their offspring are more particularly liable.

"(ε) That criminal and vicious habits of life [1] stand in hereditary relation to insanity is proved by the frequency with which insanity and other neurotic degenerations occur in habitual criminals themselves and in their blood-relations, their ascendants and descendants. The contrast nevertheless remains between crime as a moral and insanity as an organic degeneration. The point of contact of the two lies simply in the fact that insanity may show itself under the clinical form of moral depravity (see moral insanity), and is frequently falsely taken for the latter. The passion for drink must also be included in the series of factors having a hereditary influence. In this case homogeneous inheritance seldom occurs, usually heterogeneous; the parents who have degenerated in consequence of excesses in alcohol give life to children who come into the world idiotic or hydrocephalous or with neuropathic convulsive constitutions, who soon perish of convulsions, while among those who survive, epilepsy, hysteria, mental diseases, and, in fact, the worst forms of psychical degeneration, develop from the morbid constitution of the nerve centres.

"Thus Marcé records the case of a drunkard who had

---

[1] Roller, *Allg. Zeitschr. f. Psych.* i. p. 616; Heinrich, *ibid.* 5, p. 538; Solbrig, *Verbrechen und Wahnsinn*, 1867; Legrand du Saulle, *Ann. d'Hyg.* October 1868; Despine, *Étude sur les Facultés Intellect. et Morales*, Paris, 1868; Laycock, *Journal of Mental Science*, October 1868; Brierre, *Les Fous Criminels de l'Angleterre*, German by Stark, 1870; Thomson, *Journ. of Mental Science*, October 1870.

sixteen children. Fifteen died young; the only survivor was epileptic. According to Darwin, the families of drunkards become extinct in the fourth generation. According to Marcé, the degeneration is as follows:—

"I. Generation: Moral depravity, excessive indulgence in alcohol.
II.   ,,   Drink mania, maniacal attacks, general paralysis.
III.   ,,   Hypochondria, melancholia, tædium vitæ, impulse to suicide.
IV.   ,,   Imbecility, idiocy, extinction of the family.[1]

"It is a wonderful fact, which is nevertheless established by cases brought forward by Flemming, Ruer, and Demeaux, that even children of parents usually temperate, when their generation has chanced to occur during an exceptional time of intoxication, have in a high degree a tendency to mental derangement and nervous diseases. This evil influence may even show itself from birth as congenital weakness of intellect or idiocy.

"Griesinger draws attention to the fact that genius [2] sometimes occurs alongside of hereditary idiocy. Moreau even went so far as to explain genius as a neurosis. That men of genius not seldom (Schopenhauer's grandmother and uncle were imbecile) have insane, psychically defective relations, and produce children who are weak-minded or even idiotic, is certain. It seems as if a higher internal organisation of the nervous elements common to both, in one case, under the

---

[1] Cf. the valuable work of Taguet, *Ueber die erblichen Folgen des Alkoholismus*, *Ann. Méd. Psych.* July 1877; Morel, *Traité des Dégénéresc.* p. 116; Jung, *Allg. Zeitschr. f. Psych.* 21, pp. 535, 626; Bär, *Alkoholismus* 1878, p. 360.

[2] Cf. Hagen, *Ueber Verwandtschaft des Genie mit dem Irresein, Allg. Zeitschr. f. Psych.* 33, Hefts 5, 6; Maudsley, translated by Böhm, p. 309; Moreau, *Psychologie Morbide*, 1859.

influence of peculiarly favourable conditions, reaches a higher development; in another, under unfavourable conditions, leads to psychical degeneration.

"Whether consanguinity in marriage[1] is to be regarded as a factor in hereditary degeneration must for the present remain undecided. The experiments of breeders of animals, who, it is true, breed only from perfect individuals, as well as the pedigree of the Ptolemies, go to disprove it. It is possible that it remains for a long time without effect, so long as the pairing individuals continue free from degenerative tendencies. If this is not the case, rapid degeneration soon occurs— albinism, deaf muteness, idiocy, sterility.

"Lastly, there can be no doubt that everything which debilitates the nervous system and the generative powers of the parents, be it immaturity or too advanced old age, previous debilitating diseases (typhus, syphilis), mercurial treatment, alcoholic or sexual excesses, overwork, etc., may give rise to neuropathic constitutions, and thereby indirectly to every possible nervous disease in the descendants.

"The importance of heredity in our province becomes particularly clear when the fate of families afflicted with psychical disease is followed through several generations.[2]

"A genealogical table, compiled from my own observations, will exhibit this:—

---

[1] Darwin, *Ehen Blutsverwandter*, German by v. d. Velde, 1876; Devay, *Du Danger des Mariages consanguins*, Paris, 1857; Baudin, *Ann. d'Hyg.*, second ser. xviii. p. 52; Mitchel, *ibid.* 1865; *Allg. Zeitschr. f. Psych.* 1850, p. 359. According to Beauregard (*Ann. d'Hyg.* 1862, p. 226), from 17 marriages between relations 95 children were born, of whom 24 were idiots, 1 deaf, 1 a dwarf, 37 entirely normal.

[2] Cf. the interesting tables of Bird, *Allg. Zeitschr. f. Psych.* 7, p. 227; Taguet, *Ann. Méd. Psych.* July 1877; Doutrebente, *ibid.* September, November 1869, (*Schmidt's Jahrb.* 145, p. 3).

First Generation.	Second Generation.	Third Generation.	Fourth Generation.	Fifth Generation.
Father mentally diseased. Mother normal.	A daughter, the only child, became mentally diseased.	1. Daughter, mentally diseased.	1. Daughter, history unknown.	?
			2. Daughter, mentally diseased.	None.
			3. Son, maniac.	None.
		2. Daughter, healthy.	7 healthy children.	?
		3. Daughter, mentally diseased.	1. Son, mentally diseased, suicide.	None.
			2. Daughter imbecile.	None.
			3. Daughter, periodically mad.	None.
		4. Daughter, healthy.	2 Sons, history unknown.	?
		5. Son, mentally diseased.	None.	
		6. Son, mentally diseased.	1. Son, healthy.	?
			2. Son, insane.	None.
			3. Daughter, healthy.	Daughter insane.
		7. Son, healthy.	3 healthy children.	?
		8. Son, healthy.	5 healthy children.	?

"Of these thirty-seven individuals, therefore, descended from ancestors mentally diseased, thirteen are insane and twenty-four normal, but the history of some is wanting, and others are still very young.

"A review of all the facts mentioned teaches us that insanity, regarded broadly, is a phenomenon of degeneration whose determining conditions are to be sought in congenital morbid dispositions transmitted to the germ in consequence of hereditary pathological states of the brain in the ascendants, or in injuries of the cerebral organisation of the individual incurred in the course of life.

"The morbid disposition produced by either of these factors, whether infirmity or actual disease, shows, according to the biological law of heredity, a strong tendency to be transmitted in some form or other to the descendants.

"Thus the words of the Holy Scripture, 'I will visit upon you the sins of your fathers unto the third and fourth generation,' have a profound significance, and the manner of life, the lot in life, and the selection of the ascendants in great measure determines the fate in life of coming generations. The conventional German expression *wohlgeboren* acquires from our point of view a sense full of importance.

"The kind of transformation which takes place in the course of hereditary transmission, the special form of the nervous or psychical infirmity, depends on individual and external, often accidental conditions. In this question science has not yet attained to the formulation of laws.

"It can only be said generally, that when two affected individuals unite for reproduction, or when unfavourable interfering conditions (drunkenness, debilitating influences, etc.), are added to the unfavourable constitution of the generating individual, the morbid affection of the descendants becomes continually more serious, and in the continued transmission of psychopathic degenerative tendencies a progressive aberration proceeds till the extreme forms are reached. Psychoses are in such conditions developed from neuropathies, at first still quite benignant and conforming to the type of psychoneuroses, afterwards continually more degenerate (cyclical, periodic, moral, impulsive insanity), till finally idiocy occurs. Then nature extinguishes the pathological family, which loses the physiological capacity for reproduction. Conversely, however, regeneration is possible up to a certain stage through crossing with healthy blood from an untainted, unaffected family, or through the effect of favourable conditions of life. The forms of disease then become continually milder, and if the crossing is continued the degenerative tendency may entirely disappear.

"The interesting question, answered by Morel in the affirmative, Whether hereditary insanity exists as a clinical form, must remain an open one.[1]

"According to my experience, hereditarily degenerative insanity is only a particular case of degenerative insanity in general.

"With regard to the above question, stress must be laid upon the difference between mere hereditary predispositions (latent tendencies) and hereditary disease, *i.e.* where the factor of heredity acts with a determining and injurious effect on the mental and bodily development and constitution of the individual.

"Insanity in cases of mere hereditary predisposition differs from cases which are not hereditary in no way except by its appearance at an earlier age, and in consequence of trivial accessory causes, its more sudden outbreak and more rapid recovery, as well as its more favourable prognosis.

"In the stages of transition to hereditarily degenerative insanity the forms become more serious and more organic, and certain symptoms of degeneration (stupor, impulsive acts, periodicity) become noticeable."

I have repeated the observations of Von Krafft-Ebing word for word, because they so often actually give verbal expression to my own views, and because, as the reader will find for himself, they supply in their details examples supporting my views than which none better could be imagined.

It might almost be supposed that I had based my remarks on these instances. But I only read Von Krafft-Ebing's work after my section on the inheritance of diseases was already written.

[1] Cf. Emminghaus, *Allg. Psychopath.* p. 322.

SECTION V

DISUSE OF ORGANS—DEGENERATION—PAMMIXIS

In the paper previously mentioned, "Retrogression in Nature," Weismann replies with greater detail and precision than on previous occasions to the objections which may be made—as they have been made by me—to his theory on account of the facts of the degeneration of organs in consequence of disuse.

Starting from the proposition that "the adaptation of living beings in all their parts depends on the process of natural selection," he infers that this adaptation must be maintained by the same means by which it was produced, and that it must again disappear as soon as this means, natural selection, fails to act.

In other words, he says: Through natural selection alone forms have come to be what they are. By the continuation of natural selection only are they maintained in their present state. If selection ceases, they of necessity retrograde. But selection with respect to a particular organ obviously ceases as soon as that organ is no longer necessary ("the reverse side of natural selection")—its cessation therefore produces the degeneration of organs.

It is, according to my view, self-evident that the cessation of natural selection[1] can as little cause the retrogression of

---

[1] Weismann employs the expression *Naturzüchtung* (natural breeding) only in the sense of *Auslese* (selection), herein following Darwinism in general. This use

an organ as natural selection can cause it to develop. Selection is, I must ever repeat, no physiological factor which could directly produce anything new, or whose cessation could annul anything existing. Organs are produced by external stimuli, or by use acting upon the material given in a given case, with the aid of general and of sexual selection. Weismann, while he refuses to recognise what I assume to be the principal influences as having any important effect, while he puts forward natural selection or its cessation as active agents, brings forward another factor to aid in the carrying out of the modification due to these agents, namely, sexual intermixture. According to him, the degeneration of the powers of organs and the organs themselves is not due to the diminution and cessation of its activity, with the aid of the cessation of natural selection and of sexual intermixture, but is due to the latter two causes only. He says those parts which are no longer useful are no further subject to selection —the individuals multiply sexually, therefore, equally whether those parts are developed or not. All individuals unite without reference to the character in question, and therefore, through pammixis, through general intermixture, the character must vanish.

Weismann's explanation of degeneration is certainly correct, so far as the proposition from which it is derived is correct—the proposition that *all* adaptation of forms depends on natural selection ; and, I must add, so far as all that exists in the forms of organic nature depends on adaptation.

I have, however, strenuously opposed these propositions,

of the term natural breeding may easily lead to the error of treating selection as an active force. It would be better always to use the term selection where the survival of the fittest in the struggle for existence is meant, since natural breeding, according to the view advocated by me, can occur without any selection. Directly acting external stimuli and the use or disuse of organs effect modifications of living forms, and this may be properly described as natural breeding.

and thus, to my mind at least, the general validity of Weismann's conclusion falls to the ground.

That this conclusion—putting aside the fact that it does not regard the ultimate causes of the degenerations—is fully justified with respect to a large number of characters is, for every one who accepts the principle of utility, a fact long established and indisputable. Its justice is proved most clearly and most simply, for instance, by the disappearance of the adapted colouring of the wild ancestors in our domesticated animals—*e.g.* rabbits—in consequence of domestication.

But the conclusion does not hold good for any indifferent characters, including those which depend on correlation, nor for those which are accidentally useful and which have arisen from and are maintained by external influences.

That indifferent structures may undergo degeneration appears from physiological considerations indisputable. In his short paper, "The Origin of New Species through the Decay and Disappearance of Old Characteristics," Oscar Schmidt has, as I have already remarked, pointed out a case in sponges which, in my view, bears upon this subject. It concerns the genus Caminus. O. Schmidt had previously said that the fine Caminus Vulcani of the Adriatic Sea probably belonged to the Tetractinellidæ, notwithstanding the absence in it of the four-rayed siliceous spicules characteristic of that order. Subsequently he obtained a Caminus (C. osculosus, Grube) which contained in no small numbers such spicules in process of degeneration.

In the specimen which Grube had before him, however, these spicules, as could be still shown, were very rare, and in some preparations entirely wanting. On re-investigating Caminus Vulcani, O. Schmidt found in this sponge also scattered remains of such degenerate four-rayed spicules, but none in Caminus apiarium. "And thus," says O. Schmidt, " in Caminus proof is afforded that by the disappearance of an important, formerly diagnostic, ordinal character a new form of generic value has been produced."

The causes of the disappearance of the four-rayed spicules in O. Schmidt's opinion remain in obscurity. But he also holds that their presence or absence in Caminus does not depend on selection, since other spicules, uniaxial and stellate, are present, and because the genus in no respect gives the impression of degeneration, rather seems "to stand at the acme of life." Schmidt classes these spicules among those characters which, as he justly says, are somewhat indefinitely called "morphological," that is, immaterial, indifferent.

Indeed, in Caminus the uniaxial spicules are beginning here and there to degenerate, but independently of the degeneration of the quadriaxial.

It may of course be objected that the quadriaxial spicules in Caminus were once useful, and that they are now disappearing in consequence of the cessation of selection, because other useful characters have appeared in this genus of sponges which make the presence of the quadriaxial spicules unnecessary. But proof of this is wanting.

And the extreme variation of the skeletal parts of sponges in one and the same species indicates in the clearest way that the form of these parts is a character whose variation, at all events within very wide limits, remains entirely unaffected by selection.

Oscar Schmidt points out further that numerous other cases in sponges have been described by Haeckel and himself, in which the organisms are beginning to change into new species by the disappearance of certain forms of their skeletal structures. And I am able to add that in the markings of animals—*e.g.* butterflies—characters everywhere degenerate whose present or former use cannot be discerned, which we must regard as non-essential.

Weismann supposes that even in those cases in which

adaptation is not demonstrated it is really present. But such an assumption belongs to the domain of faith.

We ought on the contrary to say: We know that definite stimuli must produce an effect on or in the organism—that they must give rise to definite changes of form, definite characters, whether these be useful to the organism or not.

When we maintain this we take our stand, not on mere assumptions, but on physiological facts. Normal physiology and pathology in like measure speak for us with the weight of all their fundamental truths.

Thus there is certainly a physiological basis for the belief that the above-described variations of the sponge-skeleton are simply to be ascribed to changes of external, *i.e.* of nutritive conditions, of the material composition of the body.

From this point of view I will permit myself to discuss the several instances which Weismann gives in his latest paper in support of his explanation of the degeneration of disused organs.

Weismann attributes the degeneration of the eyes in subterranean animals to the cessation of natural selection. Without any doubt, this cause is of very great importance. But with equal certainty physiological considerations lead to the conviction that continual absence of the stimulus of light by itself, without anything else, must gradually injure and finally destroy the capability of the eye to serve as the organ of sight. First of all, the profuse circulation of blood in the eye would be diminished, and thereby the nutrition of the whole be affected; the muscles, the accommodation apparatus, would become unfit for use; the retina would be altered. At the same time, the pigment of the eye, which is always connected with the action of light, would disappear, and thus already the eye as such would be rendered almost useless, brought morphologically nearly to the stage of an organ merely susceptible to light, from which it was evolved.

P

It is a physiological demand that constant, regular, not excessive use of any organ is absolutely necessary to its maintenance. Indeed, the stimulus to which it responds is undoubtedly the influence to which an organ owes its origin and its gradual evolution.

Let us dwell somewhat further on the eye.

Weismann says: "Short sight, it is true, can be acquired, but then it is not inherited, as I at least believe myself bound distinctly to suppose. In my opinion, we owe the extensive prevalence of shortsightedness, not merely to the excessive straining of the eyes and to the continual observation of near objects, but to pammixis, the cessation of natural selection in this direction; for we are as much subject to the effects of these causes as all other organisms. As it is pretty nearly indifferent in the present state of society—especially since spectacles were invented—whether the individual sees somewhat farther or a somewhat less distance, the organ of sight has come under the action of pammixis."

That Weismann's conclusion is possibly partially true also in this instance we readily admit. But proof is wanting of the principal assertion, that short sight is not simply hereditary. The question would have to be tested accurately by the oculists with respect to Weismann's statements, to be treated statistically. He himself does not speak very decidedly on the point.

He says further, moreover, in connection with this same subject, that the deterioration of bodily advantages in us, in comparison with our forefathers, is to be considered in relation to the high development of our intellect. But this intellect can surely only have been gradually acquired and transmitted by our forefathers. It depends upon experience, which certainly cannot be produced by variability of the germ-plasm. I shall have occasion subsequently to consider this question at greater length.

With regard to the eye Weismann himself admits that its action might be affected by want of exercise, "inasmuch as the chemical changes which take place in the retina when vision takes place must cease when the eye is no longer exposed to light." "But," he says, "how could the stamen of a flower be affected by the alternative, whether the pollen which it bears reaches the stigma of another flower or not. And yet we know that hermaphrodite flowers have in some cases reverted to the original separation of the sexes, by the atrophy of the stamens in one flower, of the pistil in the others. Whether this particular case is to be explained by the cessation of selection—whether active natural selection does not play a part in it—is another question. But let us follow it further. After the anthers in the evolution of the species have atrophied and entirely disappeared, their stalks persist, not rarely possessing considerable length and thickness. Gradually, but very gradually, these also degenerate, and we find them in many species still of considerable length, in others already quite short, in others again completely absent, and only occasionally appearing in a particular flower as a reminiscence of their former general presence. The filament of the anther is no longer used, but how could it be thereby directly affected and caused to atrophy? Its structure has remained the same, the sap circulates in it as before, and flows into it as much as into the neighbouring petals or the pistil. From our standpoint the matter is easily explained, for the mere filament of the stamen is perfectly immaterial to the continued existence of the species of plant in question; natural selection therefore withdraws its hand from the organ, and it gradually atrophies."

Here also, in my view, it must be claimed that the absence of external stimuli contributes to produce the effect —on the one hand, the absence of the stimulation of the female organs of the hermaphrodite by the process of fertilisation, on the other, the cessation of the stimulation caused by the action of insects in gathering the pollen.

Moreover, correlation plays a particularly important part in the sexual organs, at least in the animal kingdom, and it is well known how prone stamens are to degenerate in plants generally.

Weismann lays particular stress on the fact that winglessness as a character of the asexual workers among ants is inherited from the winged sexual individuals. "It is therefore impossible that the degeneration of the wings, possibly caused by disuse in the individual animal, should be transmitted to a succeeding generation."

The immediately obvious reply to this is mentioned by Weismann himself, when he says: "It might perhaps be maintained that the loss of the wings occurred previously to that of the power of reproduction, but such a supposition must on very definite grounds, which I must here omit to particularise, be rejected." But what if the degeneration of the wings and of the sexual organs proceeded simultaneously— correlatively? This appears to me probable, on account of the important correlative connections of the sexual organs. I shall subsequently have to speak further, in reference to bees, of the power of correlation, which shows itself especially in the modification of the sexual organs. What is there said will apply to the case of ants, which is not completely disposed of here.

The absence of hair in the larger marine mammals is attributed to the cessation of natural selection: "The hairy covering is no longer necessary in the water—the layer of fat supplies its place." Here again I cannot banish the idea that physiological causes are also concerned in this alteration. It is certainly striking that no animal which develops and always lives in the open water possesses a horny epidermis, and hairs are horny structures of the epidermis. Whales, to take these as an example, of course develop in the body of the mother, and the embryos possess an inherited light covering of hair, which is no longer present in the adults. That the action of water is unfavourable to the cornification of the epidermis necessary for the formation of hair—be it even merely in the succession of the hairs during the later independent life—and that the hair, therefore, and also because it could be dispensed with and was not an advantage to the motion of the animal, has been gradually lost, is not surprising. Cornification depends essentially on the loss of water. Only exceptionally, for particular purposes, is horn developed here and there in exclusively aquatic animals (*e.g.* the horny teeth of the lampreys, whalebone of whales).

On the other hand, we find cuticular structures—hard secretions formed by the epidermis—widely distributed in aquatic animals, and forming their egg-cases. Such cuticular structures often form a case for the body of the animal, into which it can withdraw itself. This secretion consists of other material than the cornified epidermis of the vertebrates.

The parts of the body which are included in such cuticular sheaths often degenerate. Similarly, the hinder part of the body of the hermit crab, which is inserted into a whelk shell, degenerates, and the hard cuticle of the crab which is protected by the shell has become soft and tender. Weismann ascribes this alteration entirely to the absence of natural selection. He says: " The use of the shell depends simply on its entirely passive presence. Whether the animal is by it protected against thrusts or bites, or whether it is threatened by no such dangers, is quite indifferent to the shell itself and its proper development; it loses and gains nothing thereby, and least of all does its perfection depend on its being as often as possible battered with thrusts and bites. It cannot possibly be caused to atrophy by the fact that in the whelk shell it is out of the reach of such blows. When, therefore, the cuticular armour atrophies exactly in proportion as the body is covered by the protecting mollusc-shell, this can only be explained by the consideration that on the parts of the body covered by the whelk shell the armour is superfluous and of no consequence; and that natural selection, therefore, could no longer be concerned in its maintenance."

In this instance I can only make the objection which I feel obliged to make in all the rest: By selection alone nothing whatever can be produced, and the cessation of selection is not the only cause of the disappearance of anything which has been evolved.

It is scarcely necessary to mention what hinders the perfect development of the cuticular armour (by which is meant that of the posterior end of the hermit crab ensconced in its protecting gasteropod shell). The armour, according to my view, was produced in consequence of external stimuli acting continually on the epidermis of the animal: the cuticular structure was excreted as the effect of these stimuli. The stimuli were the first cause of the commencement of its formation. It served the animal as a protection. Selection thereupon picked out those animals which were most efficiently armoured. After the crabs had concealed themselves in gasteropod shells, the external stimuli could no longer act upon their skin, and at the same time the selection ceased, and so ensued the degeneration of the hard cuticle.

By the co-operation of the degeneration caused by disuse, and not by pammixis alone, as Weismann insists, according to my view, all cases are explained in which the mouth parts have atrophied or even the intestine degenerated in animals which have altogether ceased to feed. The latter applies to the males of the Rotifera, the former to many nocturnal Lepidoptera, and to the Ephemera. Such animals either have a large store of nourishment in the body, or they live like the Ephemera only a short time.

Simple considerations of a general kind are sufficient to show that the mere cessation of selection—sexual union without selection—pammixis—cannot possibly be the exclusive or even the principal cause of degeneration. It is a fact that most organs reach *complete* degeneration very slowly. But their functional powers they lose very soon, and this can only depend on the fact that they *begin* to degenerate very soon after they cease to be used, and to degenerate in the most essential parts, in the co-ordination of these parts which is the condition of the performance of the function. That the mere morphological framework remains so long, and for such a long time constantly reappears, is due to the persistence with which characters acquired by the ancestors are repeated in the descendants,—the morphological being inherited long after the functional has been abandoned. The latter vanishes first, and with it the essential existence of the organ. When it has completely vanished, the organ can scarcely ever again be recalled to the same activity—if the same function becomes again necessary to the organism, it must help to develop a new organ. But I do not intend to exhaust this question here, and will only point out that the loss of the function means the loss of the most essential part of the organ, for the form without contents is dead, worthless. Thus for us degeneration is a certain result as soon as the function has completely departed; a revival of it is impossible. The atrophy of the organ must then inevitably follow, step by step; however distant may be its completion, its occurrence is certain.

If the degeneration of functions, *e.g.* of instincts, took place merely in consequence of the cessation of selection, of pammixis, this degeneration, especially that of mental characters, would proceed infinitely more slowly than it actually does. For it would require sexual union of all the individuals of a species repeated through many generations, while selection remained inactive. Pammixis, all-mingling, can only mean that all attain to sexual intercourse irrespective of selection; not only the individuals which in this or that direction are best fitted for the maintenance and improvement of the species, unite sexually, but in these directions there is no longer any best or better—in these directions no further selection is made: thus all individuals, whether well or ill provided in these particular respects, succeed in reproducing themselves, and so the characters in question are gradually lost. What length of time, I repeat, would be required for functions to degenerate, if they could only degenerate in this way, and how rapidly they do actually degenerate. Scarcely a few thousand years were required for the most powerful, most dominant nations to deteriorate, to degenerate mentally, and even bodily, to the very limits of capacity for existence. Two generations are sufficient in a community for the citizens who, from wealthy benefices and foundations, have become accustomed to an idle life, to degenerate, not by pammixis, but because they no longer exert their powers. In rapid alternation, after a small number of generations, the capacity and powers of the burgess families rise and fall; dignified or parasitic life continued through a few generations leads to completely unhealthy mental decay.

Not cessation of selection is here the principal influence, but primarily the neglect of mental and bodily exercise, whose effects rapidly and powerfully transmit themselves.

But I think I need not go further into the evidence that the degeneration of organs is a process which is to be explained as the inheritance of acquired characters, especially when what has been previously said on the subject and the countless facts which comparative anatomy and physiology contribute to this evidence are borne in mind. And the great importance which must be ascribed to degeneration in the formation of species is shown by the most elementary knowledge of animal forms.

PAMMIXIS

Pammixis, it is self-evident, only affects the question of the modification of species in cases of forms which are confined and isolated within a comparatively narrow territory, if by pammixis is meant the cessation of selection and the consequent levelling sexual intercourse of all forms without distinction—that is, in English, in-and-in-breeding. In small political districts, especially in those which are in mountainous regions, such as are found among Germans, especially in Switzerland, the effect of in-breeding must with respect to human beings be particularly evident, as far as the production of similarity of characters is concerned. But even here the result which first appears is obviously only that one or another character which has arisen from the action of some external conditions on a few individuals becomes prevalent. Thus in the Canton of Appenzell I noticed an extremely remarkable variety of man among the male population. This variety is distinguished by perfectly curly, luxuriant, reddish-brown hair on the large skull. Selection, either general or sexual, cannot by any means be assumed here. Rather the origin of the peculiarity is due either directly to the general action of external conditions, or a few ancestors have acquired it as a result of such conditions, or have introduced it, and impressed it on the whole population. But it follows from my previous arguments that such general distribution, unless favoured by external conditions, is very very gradual.

Nägeli found that when he planted numerous species of Hieracium from different regions all together in the botanic garden, new species with peculiar characters were produced by their crossing. This instance of the "formation of new species by crossing" shows that new forms may actually arise by sexual mixture, and it is well known that crossing, infusion of new blood, generally conduces to the vigour of the race. But in unrestricted nature crossing cannot evidently have such an important influence in the modification of forms. I mean, that if the importance of crossing were so great, then the differences between the members of a nation in different districts would necessarily be much greater than it actually is. For it is known that mixture among much the greater part of a nation (the population of a country) takes place only within quite narrow limits. In mountainous regions interbreeding is, as I have said, naturally much closer than in the level country. In the former, in secluded valleys, it is true, strikingly peculiar types occur, since in them remnants of an older period maintain themselves with very little intermixture; I will mention only the dark small (in many respects also degenerate) men in the valleys of the Baden Black Forest, who are probably remnants of the Celts who retired to those corners before the German invasion and there remained. But even in level regions, considering the obstacles to intercourse interposed by rivers, etc., distinct types of people would be produced by the natural limitation of intercourse to a greater extent than is actually the case. The natural conditions of intercourse are such that thousands of such varieties would have been formed in a relatively small area, each with a central point where the peculiar character was most pronounced. But simple calculation shows that perfect mingling, even within a small area, requires a very long time. In order that 500 people of each sex should each pair once together, 250,000 unions are necessary. What movements of population, not to mention time, would be required before fifty or sixty millions of Germans were mingled together? But since, as a matter of fact, mingling occurs only within each group, if mingling alone determined the evolution of forms, as many varieties would necessarily arise as there were groups within which mingling took place. But this is by no means the case. I have already discussed the long-continued persistence of the contrast between dark and fair individuals, and another similar case may be here adduced. In the open hill-country of Old Württemberg a brachycephalic (Sarmatic or Turanian?[1]) race of men, most distinctly characterised by small size, black straight hair, scanty growth of beard, prominent cheek-bones, low forehead, and strikingly slit-like eyes, has persisted among the well-developed Germans. They are possibly survivors of Hunnish intruders. One who has an eye for such things can always recognise these people among the other strata of the population which are of comparatively mixed race.

The figures indicated above acquire a much greater importance when it is remembered that if my assumptions are true, in many cases several unions are required to produce actual blending. For we reckon thirty years to a generation. We should expect to see the greatest effects of mutual blending among the Jews, who do not number more than 600,000 heads in the German Empire. They are still but little injured by close breeding! But they seem to me, on the whole, through the influence of our climatic and other conditions, to show some approximation to the rest of the population, even in their external appearance, in their bodily structure. In fact, there are many German Jews who in personal appearance look like Germans.

It will perhaps be objected that with regard to the importance of sexual blending I consider only comparatively short periods of time, while for the physiological modification of forms I claim extremely long periods. I make this claim by

---

[1] Cf. H. v. Holder, *Zusammenstellung der in Württemberg vorkommenden Schädelformen und deren Maasse*, Stuttgart, 1876.

no means for all cases. But I must insist that since the action of sexual mixing assumed by Weismann presupposes differences among the mingling individuals, the periods of time necessary for that action presuppose the time required for the evolution of these differences.

It is thus clear, that modification by sexual mixture requires not only the time which is necessary for the accomplishment of the mixture, but in addition, the time the external conditions have needed in order to produce the differences which such action starts from.

I by no means deny that sexual mixture has any influence at all in the production of new forms; but I believe that it cannot possibly be the only determining cause, and therefore I contend against it in detail. If it were all-powerful, great differences in the production of species would exist among the various groups of the animal kingdom, for those animals which possess but slight powers of locomotion would necessarily produce more species than the others, because among them sexual mixture is more limited. Moreover, some animals are more stationary than others, even when they possess similar powers of locomotion. What I previously said of the individuals of a nation must apply still more to the innumerable animals which are confined to the narrowest limits of habitat, and which keep to such limits more than is usually supposed. Since in these cases, *e.g.* in reptiles,[1] slight local barriers set limits to migration for a long time, if the view of the all-powerful effect of sexual mixture were correct, thousands of species and varieties would have arisen where only one exists.

Sexual mixture is, in fine, only one of the agencies which may promote the formation of species.

[1] Cf. my paper on the "Variation of the Wall-Lizard," p. 261 *et seq.*

SECTION VI

MENTAL FACULTIES AS ACQUIRED AND INHERITED CHARACTERS

THOSE who refuse to admit the inheritance of acquired characters must also deny that mental faculties have been produced and perfected by experience and the inheritance of this experience.

And yet the mental faculties *can* only have arisen in consequence of the mutual relationship between organisms and the external world.

THE FUNCTION OF THE BRAIN

The brain, which is the organ of spontaneous action, is nothing else than an apparatus for the storing up of faculties and experiences, which have been either acquired and transmitted by the ancestors or acquired during the individual life of its present possessor.

I say experiences intentionally, and will endeavour to justify the expression.

It is the function of the brain that, having the experiences of every time at its disposal, it shall enable the body to make use of any external demands upon it—stimuli—according to its requirements, not to respond to them directly in merely reflex action. This is rendered possible by the accumulation of experience, and by the acquired and hereditary faculty of the brain to bring experiences into their proper relations, to use them in relation to and in conjunction with new external influences at a given moment for the best advantage of the whole organism.

Reflex Action

The starting-point of the evolution of all mental faculties must be sought in reflex action. The lower animals are in the scale of mental evolution, so much the more do they respond to the outer world by reflex action—the lowest perhaps act only in this way: every stimulus which reaches the animal calls forth directly and at once a movement or an action. The more highly a central nervous system, a brain, is developed, so much the more is the sphere of reflex action restricted, so much the more can experiences and faculties be accumulated and transmitted, so much the less will the animal react directly to each stimulus, so much the more will it draw conclusions from that experience—act with reflection.

Thus the more highly the brain is developed, so much the more firmly stands the organism when confronted with the manifold claims of the outer world; and conversely, the more manifold the claims to which its conditions of life expose it, the more perfect must be the faculties of its brain.

When our brain is not healthy, or when our general state of health is imperfect—if we are merely, for instance, suffering from a disordered stomach—the brain no longer works properly, the regular relation between the experiences stored up in it ceases, reflex action again results from stimulation which would otherwise not call forth a direct response. The person is irritable, and behaves unsuitably—after the manner of unreasoning animals—he acts involuntarily.[1]

[1] It follows from the above that no fundamental distinction exists between voluntary and involuntary action. In this sense I expressed myself as early as 1873 :—

"By 'will' I understand the release, as the effect of some stimulus, of a portion of the powers which are accumulated in the brain cells, existing in a condition of tension,

Intelligence, Reason, Habit (Automatic Actions), Instinct

I conceive the distinction between intelligence and reason in the following way : Intelligent actions are those which only have in view the momentary and immediate personal interests; reasonable action is that which also considers in relation to its experience and faculties the general interests —fellow-men—and the future, knowing that by so doing personal interests are doubly protected—or which depends on general conclusions. These definitions by no means imply that there are no animals which act according to reason. Indeed, it can be proved by examples, to those who reflect without prejudice, that animals reason; and that in animals actions founded on reason become automatic and are inherited I have already pointed out in my Freiburg address.

I describe as automatic actions those which, originally performed consciously and voluntarily, in consequence of frequent practice, come to be performed unconsciously and involuntarily. Herein I differ from those who use the term automatic as synonymous with reflex. Instead of the expression automatic actions, that of habitual actions may be used.

and connected with the material of those cells. The powers are partially inherited, partially acquired or modified by adaptation. This adaptation is effected either by external stimuli, in an empirical way by means of the senses, or by internal stimuli proceeding from the condition of the brain, or body, itself at the moment. Since the condition of tension is altered by stimuli, cumulative effects of stimuli increase the tension to its limits, and a final stimulus is at last capable of effecting the release. Will is thus the result of a number of factors, which are partly hereditary and material, partly directly or indirectly derived from the outer world. Involuntary and voluntary action do not differ essentially, but only in so far that the latter presupposes an accumulation, a storing-up of impressions in a common organ (the brain), and the possibility of an interaction of these. The will can therefore never be free. The erroneous idea of its freedom depends in each particular case on the neglect of factors of which it is always the slave.

"By consciousness I understand the sensation of the condition, as affected by the outer world, of the brain at a given moment" (*Zoolog. Studien auf Capri,* i. Beroë *ovatus, ein Beitrag zur Anatomie der Rippenquallen,* Leipzig, Engelmann, 1873).

This effect of habit greatly facilitates our daily life, for it includes a number of actions which we perform from morning till night as it were involuntarily, although we have once had to take pains to learn them, from the actions we perform on rising in the morning to those of going to bed. Thus the brain is left at liberty to work in other directions, to acquire new faculties.

Such acquired automatic actions can be inherited. Instinct is inherited faculty, especially is inherited habit. Or to state it more accurately, instinct is the power of habitually acting without reflection so as to attain a given object—in such a way as intelligence or even reason might dictate—in response to internal stimuli depending on the condition of the body, and to external, or without the latter.

No other scientific explanation of instinct seems to me possible.

I divide instincts into perfect and imperfect. The former are those which are inherited in such a complete manner that they need no further stimulation, no kind of guidance, no practice. These are inherited automatisms. They include, for instance, the collection of honey by bees and other insects, the formation of the cocoon by caterpillars, numerous "building instincts," the impulse in young water-birds to seek the water and their power of swimming immediately, the love of parents for their children, brooding instincts, and so on.

Imperfect instincts require to be called into action by an external stimulus, or to be developed by guidance and practice. In these there is merely an inherited capacity for automatic action. To this class belong the building of suitable nests by birds, and for the most part the instincts of game-dogs. Here also belong many other faculties called forth by very short experience (compare the following).

Imperfect instinct passes, without any limit which can be distinctly defined, into faculty acquired during life.

The view I have given of the nature of mental faculties and of their origin, and particularly this view of the nature of instinct, will appear the more probable the more the reader assimilates and understands the consequences of the idea that the organic world is a connected whole, and that in particular the most nearly related forms are clearly to be regarded as differentiated by division of labour, as organs of a larger organisation. "Thus the individual," as I concluded in my address on this subject, "On the Notion of the Individual in the Animal Kingdom," "as the German term for individual justly implies (*einzelwesen*), is a constituent portion, not merely of its own species, but also of the totality of the animal world. It follows from this conception that the animal world, in connection with the rest of the universe, is a harmonious whole formed of correlated parts, in which no part deserves an absolute superiority over another. If the animal world be considered as such a whole, we reach the idea of our profound philosopher Oken, and regard the individuals as the organs of a co-ordinated whole."

It is certain that rigidly logical reasoning must necessarily admit the conception that individuals are organs, and species and genera, by virtue of their definite structure adapted to definite ends, are organs of a higher order, of the whole living world.

Thus—leaving aside anything more comprehensive—in any case, immediately related forms are connected together as members of a whole, within which they have, if we count millions of years as minutes, only "recently" differentiated themselves. We may, leaving aside species and genera, grasp the idea firmly merely with regard to the members of a widely branching family—and this we are surely entitled to do. When we do so, the supposition that mental characters are inherited with the bodily, with the brain cells also the acquired impressions which have produced and developed their infinitely sensitive and complex structure, cannot à *priori* have anything startling about it. That the supposition is actually true we possess evidence in abundance, not only in the facts of instinct, but in the whole mental evolution of animals and of mankind.

If the acquirement and inheritance of mental characters, which can only rest upon a corporeal basis, were not possible, neither man nor civilisation would exist.

The examples which Weismann adduces in support of the opposite view with regard to instinct seem to indicate, as is of course only consistent on his part, that he completely renounces the explanation of instinct as inherited faculties and habits. This would imply that he believes in the possibility of the evolution of all the mental faculties of animals without the aid of inherited experience; in other words, of the improvement of the nervous system by inherited acquirements.

Weismann appeals to cases " in which degenerations may affect only a single instinct, while the animal remains completely unaffected thereby in its general form and general functions."

He mentions that the instinct of flight possessed by the wild ancestors of our domesticated animals has in the latter more or less completely disappeared. But he says the case of guinea-pigs shows how slowly this instinct disappears under domestication; ever since the discovery of America, that is, for about four hundred years, have they been dependent on man, and they still start at every loud noise and try to flee. I have always referred to guinea-pigs as an instance of animals whose wild originals are entirely unknown, as unknown as those of the gold-fish, and which, like the latter, have probably been maintained from time immemorial and still exist only under the protection of man. If this be so, the peculiarity referred to by Weismann is still more in his favour. I know well this characteristic of guinea-pigs. Is it really the instinct of flight which has persisted from a period so remote? It is a fact that the little creatures are in other respects very confiding, and willingly permit children to play with them, without betraying fear of man. This peculiarity might therefore possibly be also explained either as due to continued experience with regard to danger, or to specially developed reflex action deeply rooted in their nervous system. The former supposition is made more probable by the fact that guinea-pigs are very quarrelsome among themselves, are apt to fight viciously with one another, and also to devour their young.

On the other hand, the rapidity with which animals can become tame and fearless, as soon as they are relieved of fears for their safety, is proved to my mind by the following facts. When I was staying in the Dutch island of Rottum, in West Friesland, the water-rail (Rallus aquaticus), which is usually so shy, ran about close to me in the ditches so fearlessly that I could almost have caught it with my hands. This island is let by the Dutch government to an egg-bailiff, whose duty consists in collecting birds' eggs, and therefore no bird is allowed to be hunted there; it is especially forbidden to shoot at them. On the roof of the bailiff's house, to which a ladder leads for the sake of the outlook towards the Dutch continent and our island of Borkum, a starling perched himself every morning close by me and twittered joyously and carelessly to the world around as though he did not see me. In the Delta of the Nile, and elsewhere in Egypt, I found in winter the same migratory birds, which when with us are so shy, extremely tame, because, on account of the Mohammedans' kindness to animals, they are not pursued by them—only by Europeans, especially by murderous Englishmen belonging to uncultivated, but unfortunately, out of England, the prevailing classes. In my garden every sparrow and every crow knows me from afar because I persecute these birds. Once, in the presence of a friend, I shot down a crow from the roof of my house, while the pigeons and starlings on the same roof, to the great astonishment of my friend, to whom I had predicted it, remained perfectly quiet. They had learned by frequent experience at what my gun was aimed, and knew that it did not threaten them.

I mention these instances in the conviction that Weismann himself knows many such, and that he has argued from but one peculiar case, which, in my opinion, is possibly not quite correctly interpreted, and, at any rate, is not by itself conclusive. When Weismann ascribes the disappearance of the instinct of fear to the cessation of natural selection, I must insist that my examples are entirely opposed to such a view. They rather confirm the well-known fact that timidity even in wild animals can be dispelled in a comparatively short time by the results of experience, and that even deep-seated, inherited timidity—the instinct of fear, if it must be called so—can vanish in consequence of experience. They show further, that this instinct of fear, because it can be dispelled by experience, must be founded upon inherited, acquired experience.

At the conclusion of his discussion of the instinct of fear Weismann says : " Among guinea-pigs, as among the various kinds of pheasants which are bred in the poultry-yard, the youngest animals are the wildest. The instinct of flight is therefore here still inherited almost without any diminution, and each individual has to be tamed anew. The tameness of the adult animal is here still an ' acquired ' tameness, *i.e.* a character acquired during the individual life ; it has not yet affected the germinal cells : or rather, it is not yet the result of such modification of the germinal cells as must gradually take place in consequence of general crossing, but it arises in exactly the same way as in a wild animal captured when young, a fox, wolf, finch, or rat, all of which can be tamed to a certain degree, *i.e.* can accustom themselves to the absence of enemies."

These sentences would agree perfectly with my own views, so far as they admit that acquired characters are taken up by the germ-plasm; but the words "such as must gradually take place in consequence of general crossing" bring into prominence the complete contradiction between our respective doctrines—by which I do not mean to say that I do not also allow some importance to pammixis in the degeneration of mental characters. However, in order to estimate the degree of inherited fear, *e.g.* in pheasants in presence of man, we must rear them ourselves, and not leave them to be reared by the hens; chickens reared by the mother-hen are shy from the first, those reared by man are tame directly after hatching (cf. the following).

According to my ideas, the evolution of instinct is as much due to the impress of experience as the taming of the individual animal; the loss of fear in the latter in consequence of experience gives the foundation for the development of instinct—when it is inherited we call it instinct. And the same holds true for the degeneration of instinct. I am unable in this case any more than in morphological relations to separate completely the individual and its descendants.

To Weismann's explanation of the degeneration of the food-seeking instinct, which he also discusses, I feel obliged to make not only the same objections as I have made to his explanation of the loss of the instinct of flight, but others in addition.

Weismann attempts to trace also to the cessation of natural selection, to pammixis, the fact that various animals have forgotten how to seek food, even in some cases how to feed; young birds (nestlings) are fed passively, and similarly certain species of ants, and certain individuals in ant communities. The red ant (Polyergus rufescens), as is well known, steals the pupæ of the gray ant (Formica fusca), and rears them up that they may feed its larvæ and itself; for it has itself entirely lost the capability of seeking food. Such degeneration of the instinct in question and of others extends to the workers, that is, to animals which produce no offspring. "The disappearance of such instincts cannot therefore possibly have arisen from the fact that the individual animal became accustomed no longer to seek its food for itself, and that this habit was transmitted in any degree whatever to its offspring."

The latter argument appears very cogent. It must, however, be pointed out that the fertile ants also are passively fed, and queen-bees too; and that, therefore, an inheritance of the degeneration of the instinct of food-seeking may very well have occurred. But even if this were not the case, the occurrence of inherited degeneration might, according to my views, very well be explained in another way; but for this I must refer to the case of the inheritance of the characters of the equally sexless worker-bees, which I shall discuss subsequently.

In my opinion, at all events, the absence of natural selection is neither necessary nor sufficient to explain the disappearance of the food-seeking instinct. Under ordinary circumstances, if an animal takes no nourishment, it starves, and pammixis is at an end. In the cases hitherto mentioned, the seeking of food has been replaced by the passive reception of food. There is nothing to prevent the assumption that the loss of the instinct of seeking food has been acquired by the individual as a result of the passive reception of food from the parents or other individuals, and been transmitted by inheritance. The Axolotls which I have kept in confinement for many years have forgotten how to seek food by their own action, because they are regularly fed with meat from the hand of the attendant. If this condition were to continue through generations, the absence of this spontaneous action would necessarily become hereditary, simply because, as a result of this indifference, a whole series of the animal's faculties must suffer decay—partly as a direct consequence of the disuse of organs, partly in consequence of the cessation of natural selection. But since the latter must be always of less importance in confinement, the action of the former cause, as further considerations will show, would chiefly determine the degeneration. Finally, the Axolotls, simply through disuse of their organs, would, like the workers of Polyergus rufescens, reach a condition in which they would be incapable of taking food in any other way than by being fed with the hand.

Particular Instances of Intelligence and Reason in Animals

I cannot by any means recognise as instinct all that in the preceding has been so described, especially what was referred to as the instinct of fear, or what was mentioned by me in discussing the nature of instinct. When starlings and pigeons remained undisturbed on the roof from which I brought down a crow with my gun, when certain birds recognise me in the garden as a friend, others as a foe, that is not instinct, but intelligence; for in comparatively short time the creatures drew from the facts certain conclusions respecting me, and behaved accordingly. Among all the facts I have mentioned, that which shows most clearly the exercise of intelligence is, that the same migratory birds which in Europe in consequence of their experience are most shy of man, in Africa, whither they go in winter, and where they are not pursued, are quite tame. It is certain that on the northern coasts also birds, ducks for instance, which are in many places protected there for the sake of their eggs, as in Rottum, distinguish between friendly and hostile shores. The eider-duck, for example, which, in places where it is protected, has voluntarily domesticated itself, and every year comes from a distance to take up its abode with man, doubtless avoids other places which are dangerous to it.

Those who seriously believe that mental powers are to be explained by the accidental variations of the germ-plasm, as those must who deny the heredity of acquired characters, will of course estimate the mental powers of animals according to the requirements of their belief, and may also describe all those powers without distinction under any indifferent name whatever. Instinct is such an indifferent name, when the idea of instinct is not connected with the inheritance of acquired characters. But all explanation, all comprehension of the gradual modification and perfection, and of the nature of mental faculties in general, is then excluded.

I recognise the instinct of fear, *i.e.* fear acquired through the experience of ancestors, and inherited. But the cases I have already described show how easily this "instinct" in many animals can be dispelled by experience, and give place to the knowledge that they can have confidence in man. And, on the other hand, my own observation shows how quickly their confidence is converted into distrust by experience. Countless other facts prove that fear and confidence, trust and distrust, in animals alternate in different cases according to the circumstances.

Some years ago a male chaffinch in my garden had become so tame with me that he flew after me everywhere in order to take the hemp seed and meal-worms which I offered him. Wherever I went or stood in the garden the finch appeared from the bushes, perched on the nearest branch or on the ground in front of me, and with his powerful chirp—"Pink, pink"—demanded his food. But if he had not noticed me, I had only to whistle in imitation of his chirp and he appeared. At last he used to come after me even into the house and follow me from room to room asking to be fed. Yet, in spite of his trustfulness, I could not induce him to feed out of my hand. It was evident that he constantly endeavoured to overcome the remnant of timidity which still survived in him, but he could not yet succeed. Still I strove to attain this end, and the visible progress made permitted the hope that I should shortly succeed, when an unfortunate accident suddenly altered the condition of affairs and put an abrupt end to his confidence. One day a sparrow on a tree in front of my window was piping indefatigably his monotonous shrill chirp, which pierces the ear the more irritatingly the more energetically it is uttered and the greater its well-known deficiency of cadence. As the fellow had repeatedly disturbed me at my work in this way, I resolved on his destruction, and, creeping within range, I fired at him with a small chamber-gun loaded with small shot. At the shot my beloved finch flew suddenly from the tree where he had been perched unnoticed by me. The shot must have passed around him. My sorrow for the accident was deep, for what was to be expected occurred—the finch afterwards carefully avoided me, and notwithstanding all enticements, I could only with difficulty induce him again to take from the ground at a great distance from me food which I had scattered. But after a short time he disappeared entirely from my garden with the family which he had established.

This happened two years ago. At that time, when I sat at a certain table in my garden, a red-tail often came with the finch, and like him picked up, though always with great caution, the meal-worms which I threw to him or laid on the table. After my misfortune over the finch, I took no further notice of the red-tail. A few days ago I sat again at the same table, and when I happened to throw away a match, the red-tail flew down, thinking I had thrown him a meal-worm as I used to do. And even when he found his mistake he remained close to me, flying from twig to twig, expecting I should throw him something for the young ones which he had to feed, and which were crying for food among the bushes.

Still further removed from instinct are the following instances of mental operations in birds.

The cunning of sparrows and their prudence with regard to danger is well known. One snowy winter recently, when the sparrows around the house were very hungry, I made an attempt to catch a number under a large sieve, the edge of which was supported on one side by a piece of wood, which was connected with a long string: the string was covered up in the snow, and passed through an opening in the door into the house, where my little son watched, ready to pull the string as soon as some sparrows went under the sieve. Corn was strewed about under the sieve and around it as bait. The sparrows gathered in dozens round the sieve, and picked up the corn up to its very edge to the last grain, then flew round and screamed at the sieve in hunger and rage, but not one was enticed under it.

Such facts are sufficiently familiar, yet the intelligence of animals astonishes any one when he sees it himself for the first time.

Recently a friend showed me how one can feed the tomtits in winter abundantly without having to fear that the impudent sparrows will steal the food. He has on his verandah boxes nailed to the wall and posts, in which he puts the food. The tits take the food out of these, but the sparrows are afraid the boxes might be dangerous to them, might contain something intended to destroy them, or that they might be caught in them, and therefore they leave the food untouched. As they know well that at other times they are pursued by the owner of the house and the tits not, the fact that the tits suffer no harm by taking the food in the boxes is not enough to remove their distrust. This high degree of prudence is evidence not for an instinctive fear in them, having its origin in variation of the germ-plasm, but of a high degree of capacity to reflect and draw conclusions—of a high degree of intelligence.

But the following observation on the mental capacities of sparrows surprised me most; it is a still more striking example of the memory of birds than the case of the red-tail previously described.

Every year I require a large number of birds for the zootomical studies of my students. As I am naturally averse to sacrificing useful birds, I turned my attention first of all to the sparrow, as the commonest among those of our birds which do far more harm than good, although our laws —I might say in an almost incredible way—still protect them. One day I read in an agricultural paper the advertisement of a trap with which twelve or more sparrows could be caught at a time. That was what I wanted. I sent for one of these traps. It consists of a wire-basket of cylindrical shape, about 18 in. high, and the same in diameter, the upper end forming a funnel-shaped tube projecting inwards towards the bottom of the basket. It is constructed, therefore, after the fashion of the glass ink-pots which may be upset without spilling the ink, and like certain mouse- and rat-traps or fish-traps. When the animals have passed through the funnel into the trap, on the bottom of which bait is strewn, they cannot find the way out again.

The result of the use of my trap was surprising: almost immediately quite a dozen sparrows were caught in it. These were brought away as carefully as possible, so that none were taken out in sight of their companions. The trap was again set, and this time nine sparrows were caught equally quickly. I was very pleased with the invention, for it seemed likely to put an end for the future to all my difficulties. But it was to be otherwise. I noticed already that all the sparrows caught were young birds, hatched the same spring, and therefore of little experience. Not a single old sparrow had entered the trap. And when I set it for the third time, not one sparrow went into it—it stood for week after week; the yard was full of sparrows, but I caught no more.

However, I looked forward confidently to the next year—then, I thought, young sparrows will get caught again; and about two dozen would have been enough material for my purpose. But I had reckoned without the—intelligence of the sparrows. When I got out the trap again next year, and had it set, not a sparrow went into it. But a curious spectacle was observed: apparently several sparrows had the desire and the intention to go into the trap, and these were obviously the young inexperienced birds which had been hatched since the trap was last set; but others, of course the older birds who had learnt the danger of the wire-basket from the loss of their families, kept them back by constant earnest warnings, for the males, as soon as one of the yellow beaks approached the cage, uttered their warning cry most loudly, the cry which they always make when danger is present, and which consists in a long shrill rattling "r-r-r-r-r."

But the most singular fact is this: It is now nine years since I set the trap for the first time. Since then I have repeated the attempt almost every year, and each time with the same result—not one sparrow more has ever entered the trap, not even last winter, when, in consequence of the deep long-lying snow, the birds were in great want. The knowledge or the tradition of the danger attached to the wire-basket has been maintained, even although I once omitted to set it for two years.

Every lover of living nature, and particularly every sportsman, knows the cunning of crows. They allow the harmless pedestrian to approach quite near to them, but from the sportsman with his gun they flee far away. I have repeatedly shown to friends the following example of their intelligence. My garden slopes up the side of a steep hill, covered with vines. Crows are fond of settling on the vine-props, resting without motion, and apparently careless and heedless of their surroundings. Formerly, when the trees of the garden were lower, they could look down on a table which stood there. On the table I placed my gun, and beside it a stick. As soon as I took up the gun, the crows flew away at once; if I took up the stick, they remained quiet where they were. They noticed everything which went on, recognised the gun, and rightly judged it was dangerous to them, even from the considerable distance at which they were perched.

The more one observes the higher animals in their natural free state, the more one is filled with wonder at their intelligence. Intelligence alone, and not instinct, will explain the fact that the cattle-heron (Ardea russata) in Egypt, when fleeing before the sportsman, shelters itself under the oxen and buffaloes, because it knows that it is there protected from his gun. For it is not to be supposed that this has become an automatic, a hereditary habit, simply because, as I have said, it is only slaughter-loving foreigners who occasionally shoot birds in that country.

A curious instance of fear in an animal came under my notice during my stay at Rottum. The egg-keeper had a young dog of large size, a kind of sheep-dog in breed. As no one on the island troubled himself about this animal, he attached himself to me uninvited for want of companionship, and followed me everywhere. One day I went down to the shore to bathe. I had chosen a place beneath a high sand-hill, where the sea deepened gradually, so that I could go out a long way without getting beyond my depth. The dog had followed me, and he sat down on the sand-hill and watched me undress. With growing curiosity he followed each of my movements as I took off one garment after another. At first this curiosity was an agreeable one: something like the pleasure of having a certain degree of comprehension of what was going on was expressed in the repeated nodding of his head, in the droll oblique turning of one side of it to the object of his attention with an upward look, which is a well-known attitude in young dogs. But when I proceeded further in my operations, and was about to take off my shirt, I saw signs of agitation and fear appear in the dog; once or twice he suddenly drooped his ears, hitherto erect, and turned his head as if to run away, then took courage to remain longer. These signs of anxiety gave place to the greatest terror when I had thrown off my shirt; the dog looked at me for a moment in horror, but when I went into the water and more and more of me was covered, so that perhaps I seemed to him to grow smaller, or to vanish, the prodigy was too much for him, he suddenly laid his ears back, drew his tail between his legs, and ran as fast as he could run inland from the place. When shortly after I appeared again in my clothes near the house, the dog looked at me from afar with the greatest distrust, and afterwards took care to keep out of my way. The whole occurrence gave me the impression that the dog was terrified at the appearance, to him previously unknown, of an almost unclothed and afterwards naked man, and that he did not give way to panic from the first, before I went into the water, because he had not yet drawn the conclusion that he had to do with a spectre.

This case proves how cautiously the idea of instinctive fear must be used, at least in the sense of fear of definite objects. The dog was afraid of something which he had never yet seen, because it appeared to him incomprehensible, mysterious, ghostly. Such a condition in an animal still further deserves notice, because the possibility of forming supersensuous conceptions is intimately connected with it, and because it may accordingly be suspected that animals are endowed with this power also. Fear of things incomprehensible is indeed the cause of such conceptions among men, and among many savage peoples is still at the present day clearly the cause of their belief in a higher being. It is stated [1] that dogs do not bark at naked men, which is probably to be attributed to fear of the unknown, like their fear of thunder. Also tamers of wild animals are said to go naked into the cages in order to train the animals. H. Spencer considers that even fetichism exists in dogs; that, judging from certain circumstances, they believe inanimate objects to be alive.[2]

G. J. Romanes [3] has made some interesting experiments, similar to mine, on dogs. I will quote the most important of them here. Romanes relates :—

"My dog was in the habit, like many others of his kind, of playing with bones, by tossing them in the air, so as to throw them a little distance from him, and so making them appear to be alive, whereby he procured himself the imaginary pleasure of killing them. One day I offered him for this purpose a bone to which I had fastened a long thin thread. After he had tossed it a little while in the air, I took the opportunity, when it had fallen a little distance from him, to draw it slowly away by means of the long invisible thread. The dog's whole behaviour immediately changed. The bone which he had before only treated as if he believed it alive, now became really alive in his eyes, and his astonishment was boundless. He approached the bone at first with great caution . . . ; but as the gradual movement did not cease, and he became quite certain that the motion could not be set down to that which he himself had communicated, his astonishment changed into

---

[1] Cf. *e.g.* Adolf v. Conring, *Marokko, das Land und die Leute*, Berlin, 1880. Mansur, when he steals at night into the tent to Fatme, who is sleeping beside her husband, goes "completely naked, because he knows that no dog barks at a naked man."

[2] Herbert Spencer, *Principles of Sociology*.

[3] G. J. Romanes, *Mental Evolution in Animals*.

terror, and he ran away to hide himself under some piece of furniture and look upon the incomprehensible spectacle of a bone which had come to life, from a distance. . . .

"One day I took him into a carpeted room, where I blew a soap bubble, and then made it glide along the floor by means of a current of air. He at once showed the greatest interest in the bubble, but seemed unable to decide whether the thing was alive or not. At first he was very cautious, and followed the bubble only at a certain distance; but when I encouraged him, he went nearer with his ears pointed and his tail depressed, evidently with great distrust, and retired immediately it began to move again. After some time, during which I had always kept at least one bubble on the carpet, he became more courageous, and the scientific spirit getting the upper hand over his sense of the mysterious, he became at last bold enough to go cautiously up to a soap-bubble and touch it with his paw. The bubble of course burst at once, and I never saw surprise so strongly expressed. When the burst bubble was replaced by another, I encouraged him to approach it for a long time in vain; at last, however, he went up to it again and cautiously stretched out his paw as before, of course with the same result. After this second attempt, nothing could induce him to make another, and when I continued to urge him he ran out of the room, to which no coaxing could bring him back."

Another mental quality which is to be considered, not as instinct, but as mental faculty of a higher kind, is curiosity, which, even among animals in which other mental powers are but slightly evolved, obviously plays a great part. When I was sketching at Rottum with my sketch-book in front of me, the cows which were grazing around came nearer and nearer, formed a circle round me, and stood motionless, stretching out their necks and staring at my paper to see what was going on. They came so near to me as to be a nuisance, and I had to drive them away with a stick. But they always repeated their attempts to discover the secret.

In my memoir on Lacerta muralis cœrulea, I described how the boys of the island of Capri take advantage of the curiosity of the lizards to catch them with ease, although otherwise their capture is so difficult. They take a long stiff withered stalk of grass and make a noose on its thinner end. Then they lie down at full length, and hold the grass-stalk in front of them with outstretched arm opposite the crevice in which a lizard has just hidden itself. The creature's curiosity is so excited that it comes nearer and nearer to the noose in order to examine it, and the boy is able to pass it over the lizard's head and catch it. In order to further stimulate the lizard's curiosity, the boys by spitting on the noose make a shiny bladder over it.

In a later publication[1] I attempted to explain the famous statue of the Apollo Sauroktonos by means of this method of catching lizards, the practice of which is widely spread in other parts of Italy also. This statue, as is well known, represents a youth not much beyond boyhood, who, in a watchful attitude, leaning with his left arm against the trunk of a tree, and holding in his right hand a portion of a thin staff, follows with his eyes a lizard crawling up the tree-trunk. Archæologists suppose that the boy is intending to transfix the lizard with an arrow, of which the stick represents a portion. But the lizard is crawling towards the boy. This fact and the whole attitude of the figure, as well as the way in which he holds the stick in his fingers, seem to me to indicate in the clearest way that in the Sauroktonos we have a boy who is snaring a lizard with a noose of grass. For that attitude is in every detail one of quiet waiting, almost of negligence, and the boy holds the piece of stick in his hands lightly and playfully, not firmly and securely as

---

[1] *Variiren der Mauereidechse*, Section "Sauroktonos."

one holds an arrow when about to kill something with it. Lastly, the expression of the face is peaceful, indicative rather of play than serious effort.

As I said, the meaning of the statue is only comprehensible on this explanation, and with it the work appears in all its life-like truth and harmony.[1]

Fig. 1.

A copy of the Sauroktonos, which was dug out in the year 1777 on the Palatine, exists in the Vatican; another smaller in bronze, found at S. Balbina, is in the Villa Albani in Rome; among others, one in Paris. The two first I know from my own observation. In the most known and most beautiful, that in the Vatican, the two arms from the shoulders are restorations; in that of the Villa Albani the right hand is said to have been restored or repaired; also in the Paris specimen the right forearm with the hand is said to be new, as well as the fingers of the left.

[1] Some archæologists have, it is true, expressed a view more similar to mine, namely, that the boy is intending to tickle the lizard with his stick. The view that he wishes to kill it with his arrow is derived, so far as I know, from the statement of Plinius (*Hist. Nat.* xxxiv. 70): "Fecit (ex aere Praxiteles, to whom he ascribes the statue) puberem Apollinem subrepenti lacertæ cominus sagitta insidiantem quem sauroctonon vocant." According to this author, Apollo wishes to foretell the future from the convulsions of the dying lizard. An epigram of Martial referring to our statue says:—

"Sauroctonus Corinthius (*i.e.* of Corinthian bronze).
Ad te reptanti, puer insidiose, lacertæ
Parce, cupit digitis illa perire tuis."

But in all the way the arm and hand are held, as well as the attitude of the whole body, point to the light handling of a grass-stalk.

I know not whether the method of catching lizards in a noose was also practised in Greece, or is so now. The close relations between Italy and Greece, even if this were not the case, would have sufficed to supply Praxiteles with the subject of his statue.

Such practices can often be traced back to very ancient times, and are inherited and maintained in subsequent ages with very great persistence. A fresco in the Etruscan Museum in the Vatican affords me another example of this. It represents a boy holding a string attached to the legs of a bird and allowing the bird to flutter, an act which is still at the present day one of the commonest of the cruelties to which animals are daily subjected in Italy. This amusement therefore has been practised by thoughtless children, at least since the time of the Etruscan people, a time reaching back into the darkness of an unknown antiquity.

To these reflections, made in my essay upon the variation of the wall-lizard, I would add, in reference to the subject here under consideration, that by the remark that such practices are inherited by posterity, I meant not inherited as an instinct, but handed down. But any one who is acquainted with the facts of Italian cruelty towards animals will agree with me that the contempt and mercilessness with which animals are treated there, in the land which contains the seat of the head of Catholic Christendom, does in fact almost convey the impression of an hereditary habit, as also does the compassion towards animals among the Mohammedans, which affords such a striking contrast.

In the essay alluded to I have also insisted that lizards are very stationary, *i.e.* that every lizard confines its movements to a definite restricted area, and that it has the most exact knowledge of its retreats within that area, and understands how to reach them quickly and hide in them when it is pursued. But if it is driven from its home it runs about helplessly and aimlessly, and can be captured with ease. This fact led me in that paper to describe an observation on a similar certainty, acquired during life, in the knowledge of its surroundings, even in a shore-crab : " Animals also which are credited with mental powers still more feeble than those of the lizard seem to have a similar knowledge of locality, and to find their way to their accustomed retreats with as much certainty. I witnessed years ago in Capri a comical proof of this in a shore-crab (Carcinus mænas). In a large pool hollowed out in the rocks, shut off all round from the sea, and only reached by the waves in rough weather, stood a fisherman trying to catch a crab of that kind ; he pursued it with his two hands joined together so as to form a scoop in which to lift it out. The crab swam in a straight line towards the opposite side of the pool, which was some metres away. Softly and cautiously the fisherman followed it, evidently pleased to see the crab going towards the rock, between which and his hands he expected to catch it. But just as his hands were about to seize the creature, it slipped beneath them into a hole in the rock, and the discomfiture of the fisherman was received with laughter by the numerous bystanders who had watched his attempts with keen interest. It must be inferred that the crab had swam across the whole pool in certain knowledge of the position of its hole ; and the jeers bestowed on the fisherman showed that the spectators attributed this conscious intention to the animal beforehand, and regarded the fisherman as the victim."

I know well enough that many of the examples I give here of actions in animals only to be explained by direct experience, that is, by intelligence and not by instinct, are far surpassed by others which are known; but I have preferred, as everywhere else in this book, to use new examples which have come under my own observation.

Experiments and Observations on Instinct in Newly-hatched Chickens

Last year I had some chickens hatched in an incubating apparatus. They had never seen a mother, and were never taught to seek food. I placed some millet seed in front of them in a vessel; they took no notice of it. Then I took some of the seed in my hand and let it fall on the hard wooden floor so that the grains rebounded. The chickens then at once pecked at it, and in a short time fed without assistance. But the thing that surprised me most was this: a fly flew close by the eyes of a chicken which had only left the egg-shell about half an hour before, and the little creature pecked at it as if it had been long accustomed to catch flies. Similar cases have indeed been frequently observed.

It may be objected that this was nothing but simple reflex action—and of course it is reflex action. But the wonderful fact is that this action was so perfectly adapted to the external requirements; the chick snapped at the fly in order to catch it; it made with accuracy all the movements suited to this purpose. If it had caught the fly, it would have bitten it up and eaten it as it ate the millet. It had evidently brought into the world, in its brain, a mechanism which included an inherited mental image of the flight of a fly, an inheritance of the effect of the stimulation which flies have exerted upon its ancestors during their life for ages past. A proof of this is that the same chicken responded quite differently to other similar stimuli acting on its eyes, *e.g.* to a rapid movement of the fingers. In short, the chicken's behaviour towards the fly was a consequence of innate instinct, like that of itself and brethren towards the millet seed.

In the present year I have again made similar experiments on chickens, and with more care. I took two chicks away from the hen into my own care immediately after they were hatched, and wrote down exactly their behaviour under the experiments I made on them. I will distinguish the two in the following according to their colour as the brown and the white. There is no need to insist that when I began to study them they knew nothing of the world, that they had no previous experiences whatever, but I am compelled to point out the impossibility of doubt with regard to this on account of the high degree in which they exhibited the faculties acquired by their ancestors. I was extremely surprised by their actions, and confess I should have considered them incredible if I had not observed them myself. They are *alone* sufficient to demonstrate the inheritance of acquired characters as an incontrovertible fact, and to show what a tremendous importance belongs to this inheritance, especially where mental acquirements are concerned, and what importance belongs to instinct in the life of animals.

I proceed now to the description of my observations, only remarking that the white chicken was from the first somewhat behind the brown in the exercise of its faculties.

After they were hatched, I kept the little creatures warm during the night in a basket with some wadding near the stove, and on the second day of their life I placed them on a board, on which millet and crushed buckwheat had been placed.

To my very great surprise the brown one pecked at once at the millet grains, without having its attention in any way directed to them; and indeed at the first attempt it touched a grain with as much certainty as if it had pecked at millet for ever so long. But it did not succeed in taking up the grain in its beak. A second attempt immediately afterwards likewise failed, but at the third, which succeeded without a pause, the grain was grasped and swallowed, and now the little creature went on feeding as though it had done so for years! The three first pecks at the millet were made in rapid succession, evidently and incontestably with the intention of eating the grain. There must therefore be for the chicken in the form and general appearance of the grain a stimulus which awakes innate conceptions, and causes them to be employed. It was probably not chance that the chicken pecked first at the millet and not at the crushed buckwheat; the natural roundness and brightness of the millet grains may be supposed to have been the cause. Soon afterwards the chick pecked at the buckwheat and ate some, but like its sister it always preferred the millet. Therefore the question, however hazardous it may seem at first sight, forces itself upon us, whether the chicken did not have an innate idea that millet was a more suitable food for it than buckwheat? The question becomes, however, perfectly natural when I state that the chicken, when I put sand before it shortly afterwards instead of food, pecked at it no more than at the board on which it stood. Unfortunately I omitted at the beginning to set before the chicks sand and other kinds of food besides those mentioned, at the same time.

The white chick also pecked at the millet grains, but missed them repeatedly.

Thus after the brown chick had eaten abundantly, and the white had made only unsuccessful attempts to do so, I put them both back into their basket.

Three hours afterwards I placed them again on the board, having placed upon it, besides the food, a large drop of water.

The brown chick immediately began to feed again; at first it took no notice of the drop of water, but when the latter was made to tremble by shaking the board, it immediately drank from it with perfect success in the well-known fashion of adult fowls.

Now for the first time it dropped its excrement. A little later passing by this it pecked at it, but spat away what it took up with disgust. Considering its previous behaviour one might wonder that the chick had not an innate aversion to its excrement. But it must be remarked that the first excrement has not the same nature as the subsequent, but is surrounded with a gelatinous covering, and consists chiefly of the remains of the embryonic nutriment.

Both chicks were at once attracted by a fly which crawled on the board, and pecked at it repeatedly, without being able to catch it. It is very curious that after their first attempt at catching flies failed, both afterwards took no notice of them. Indeed they acted upon every first experience in a really wonderful way.

The white chick now also ate millet.

I laid before them boiled egg, finely minced. They pecked at it at once and swallowed it, both yolk and albumen in equal quantities, although one at first preferred the yolk.

I offered them some earth-worms of moderate size. When those began to crawl the chicks looked at them with all the signs of fear, and withdrew from them.[1] On the other hand, they began to peck determinedly at one another's toes. They pecked immediately at some chopped salad put before them, then at first left it, but afterwards pecked at it again and ate of it.

On the third day the white one went by chance near a drop of water again placed on the board, and drank from it at once, without having watched how the brown one drank

---

[1] I may remark here that even the old fowls in my enclosed fowl-run are always uneasy at first when earth-worms are thrown to them; they show unmistakable signs of shrinking and even of fear, and it is often some time before any of them overcome their repugnance and devour the worms, while others never touch them at all.

before. Afterwards both drank regularly from such drops of water.

They looked at a bee with curiosity, and at flies also, but pecked at neither.

The white one fluttered its wings for the first time, as young fowls are accustomed to do, in the manner which indicates a feeling of contentment, and later it scratched repeatedly on the wooden surface in just the same way as the old hens when they scratch in sand or manure in search of small animals or other food. To see them scratching on the smooth board, from which there was nothing to scratch out, was exceedingly comical; and as the action was afterwards repeated without any result in the same way and under the same conditions, it afforded a perfect example of strongly inherited habit. Clean white sand was untouched by either chick.

In the forenoon of the fourth day water was for the first time put before the chicks in an earthenware saucer, while previously it had only been placed on the wooden tray in large drops. Spite of the unfamiliarity of the vessel, the brown one drank at once when it chanced to go near the vessel.

In the afternoon the chicks were placed for the first time in the open air on the gravel in the garden, in the sun. They looked about with surprise, stretching their necks, and appeared afraid. They would not take their ordinary food which lay before them on the wooden tray, but very soon began to peck at the gravel, and to *seek* food. One pecked at the side of a box, each began to take up a small stone in the beak again and again as if to swallow it, until it was lost, when the chick began on another. It seemed that they now required small stones and sand in the gizzard, and wished to swallow them. They also tried the most various objects with their beaks, such as bits of wood, etc., throwing away what was useless. Thus after a few moments of surprise they felt themselves as it were in surroundings long familiar to them, in their natural environment, all parts of which they knew how to use. They tried to escape from my control in every direction, and in all respects behaved as if guided by old experience of their own.

A wood-louse which ran by was pecked at, but afterwards avoided. Flies no longer attracted attention. The brown chick, after it had at last eaten its fill of the food before it, went evidently with intention to the drinking vessel it had used in the morning, and the white one followed the example. Both fluttered their wings repeatedly in the fulness of their comfort and contentment.

Fifth and sixth day. They peck—on the board in a room—vigorously at small pieces of egg-shells which have been put before them, and eagerly try to swallow them. Sand, which was put on the board as well as their food, they leave untouched. Chopped chives they try, but do not eat, but finely minced meat put before them they eat at once with enjoyment.

Thus they knew from the first how to choose the food that suited them either without trial or after a single trial. They acted at once on every experience, even against their own interest, for they have now become perfectly indifferent to small insects, even ants, after their first attempts to catch them failed.

The white chick scratches repeatedly with a sort of dancing motion on the smooth wood, the brown has not yet performed this action. The latter made the movement first on the eighth day.

Both chickens from the first showed alarm at unexpected noises, such as a loud cough. On the seventh or eighth day when I made such a noise while they were before me on the table, and they had just run away from me in fear, I imitated the voice of a hen calling to her fellows. At this both chicks turned suddenly round and ran, as though old memories were wakened in them, straight towards me as though to seek protection. But as I could not supply the place of the soft warm wings of the mother-hen, they suddenly stood still, and when I afterwards began to imitate the voice of the hen again, they took no notice. The cry "Cluck, cluck," with which the hen calls her chickens, although I uttered it as naturally as possible, had no effect. But the behaviour I have described as caused by the imitation of the hen's chattering was so striking that it cannot possibly be ascribed to chance, especially as both chickens exhibited it at the same time; and it ought not to excite surprise, that when the benefit instinctively expected was not received the behaviour was not repeated, considering the decision with which the young chickens act at once upon every experience. Still such experiments ought to be repeated on other chickens. It is also a reasonable question, whether the sitting hen may not chatter sometimes on her nest, and whether the chicks may not thus, even in the egg, hear something of the fowl's language. Experiments therefore should also be made with chicks hatched by an artificial incubator.

Moreover, the little creatures were from the first not afraid of me, but they did not willingly allow me to take hold of them, and the less so as time went on, because they found by experience that I sometimes thereby deprived them of liberty and put them back into their basket. As it had been stated by Herr Spalding that chickens were thrown into the greatest terror by a hawk, although they had never seen one before, I made the experiment of trying to terrify them with a sparrow-hawk stuffed with the wings extended. They took not the least notice of the object, which of course does not exclude the possibility of their behaving differently in presence of the living bird.

Up to the eighth day, in consequence of the abundance of food given to them, they had lived together without jealousy or strife. On this day I again gave them some earth-worms.

After I had repeatedly thrown these in front of them the brown chick began to worry one. Immediately the white one came to seize it away from him. And so the chicks pursued one another and quarrelled over every worm which one of them had seized.

On the twelfth day I took my two charges to the hen by whom they had been hatched, and who was in a dove-cot with her other chicks. My chicks immediately retreated from the old hen, in evident terror, into a corner, while she looked on them with a hostile eye. In fact, the hen only waited till I had apparently gone away, and then fell upon her unknown children with savage pecks. I had to take them away from their mother. The two orphans soon ran about in the yard alone, searching for food, showing no inclination to go through the wide railing to the old hens in the fowl-run. They learned to look after themselves, but kept always near the house, never going far from it. One day the white one disappeared, having probably fallen a victim to a cat. The brown one, left alone, used to come regularly to us at meals, which we took in the garden, to pick up the crumbs which were thrown to him. He would have nothing to do with his fellows in the fowl-run, or with the old hen which was confined there with his brothers and sisters and clucking after them, and at night he repaired to some unknown hiding-place. This behaviour of his towards his own race in comparison with his behaviour towards mankind is extremely remarkable, because it was the consequence of experience which therefore, as with respect to the catching of flies, had conquered and displaced inborn instinct. For the bird is now—it is just eight weeks old—so tame, that when it wants food it suffers itself to be taken up and stroked, while its brothers and sisters who were brought up by the hen are extremely shy. It takes food preferably from the hand, and usually perches for this purpose on the back of the garden-seat between us or on our shoulders. But from glittering substances like spoons and forks he will not take anything. He also runs after us in the garden, but between meal-times he will be his own master, and avoids the hand which tries to catch him. Many times when we took a meal in the house and not in the garden, I have seen him running about round the empty table at dinner-time obviously expecting us: his stomach is his clock. Very comical is his behaviour when he is satisfied. He sits then with his crop full on the back of the garden-seat among us, ruffles his feathers, and shuts his eyes, so that my children when they saw this for the first time thought he was going to die. But he is only struggling with sleep; he now and then cleans his feathers and skin, then puts his head in the usual fashion under his left wing, and begins his mid-day nap.

Meantime he has given up his indifference to small animals, and especially towards flies; at these he snaps everywhere, and catches them with great skill. He has never gone outside my yard and garden, although the open fence offers no hindrance to his doing so.

He appears, after eight weeks of life, like a "self-made man," as a "character" who knows his objects, although he has scarcely reached half the size of an adult fowl. But what he is, and what he can do, he is and can through the employment of the abundant powers which he has inherited from his ancestors, and which were acquired by these.

How is it, on the other hand, conceivable that the variability of the germ-plasm, inherited from unicellular organisms, in combination with continued sexual mixture, could lead to such innate faculties as our chicken has shown?

I must confess that the assumption of a variability in the germ-plasm inherited from unicellular ancestors, and continually increased, with the effects ascribed to it by its advocates, seems to me, in face of these facts, to make no less a demand on the understanding than miracles.

The difference between the chickens I studied this year and those last year in their first behaviour with regard to food is possibly due to the fact that food was first set before the former some time after hatching, so that they pecked at the millet as soon as they saw it from hunger, while the latter being fed immediately after they were hatched, had to be attracted by the falling of the grains. Only a repetition of the experiments can decide this question. But the fact that the young birds, without ever having seen seed or water before, began to eat and drink, is sufficient alone to excite wonder at so high a degree of inheritance of acquired powers. It is true that this fact is very general, for we must not forget that the young of countless other animals take food at once in the same way, for example, every newly-hatched caterpillar; but closer relations of course between the caterpillar and its food-plant exist and are inherited than in the case of chickens.

Experiments carried out by Mr. Douglas Spalding[1] on newly-hatched chicks have afforded results similar to mine.

That writer placed a cap over the heads of chickens which he took out of the egg before their eyes had been capable of sight. When after one to three days he withdrew the cap, the chicks seemed almost without exception stunned by the light, and remained for some minutes motionless.

---

[1] Douglas Spalding, *Macmillan's Magazine*, 1873. Cf. also W. Preyer: *Die Seele des Kindes*, Leipzig, 1882, p. 82. The chickens preferred the yolk of egg to the albumen from the first. The observation communicated to Romanes by Allen Thomson, according to which chickens hatched on to a carpet did not scratch as long as they were kept on it, but began at once to do so when some gravel was strewn over it, is explained by the latter as due to the absence in the one case, the action in the other, of the accustomed stimulus, the experience of which is inherited (Romanes, *Mental Evolution in Animals*, p. 174).

"When one of my little pupils was twelve days old, while it was running about near me, it uttered the peculiar cry with which fowls denote the approach of danger. I looked up, and saw a hawk which was circling at a great height above us. Equally striking was the effect of the voice of the hawk heard for the first time. A young turkey which I took possession of when he began to chirp in the still unbroken egg-shell was, on the morning of the tenth day of his life, just engaged in taking his breakfast out of my hand when a young hawk in his box close by us uttered a clear 'cheep, cheep'; the poor turkey shot like an arrow to the other side of the room, and stood there motionless and paralysed with terror till the hawk gave out a second cry, when he ran out of the open door to the farthest end of the lane, and there remained for ten minutes crouched in a corner. Several times more in the course of the day he heard that disturbing sound, and each time with the same symptoms of fear.

"Frequently I saw hens lift their wings when they were only a few hours old, that is, as soon as they could hold their heads upright, even when they were prevented from using their eyes. The art of scratching for food, which one might think must above everything else be acquired by imitation (for a hen with her chickens spends half her time in scratching before them) forms nevertheless an undoubted instance of instinct. Without any opportunity for imitation, chickens which were kept entirely isolated began to scratch at the age of two to six weeks. As a rule the character of the ground invited them to this; I have often seen first attempts at it, which looked like a kind of nervous dance, on a smooth table.

"As an example of unacquired skill, I can mention that when I put four one-day-old ducks into the open air for the first time, one of them immediately snapped at a fly and caught it in its flight. But still more interesting in my opinion is the deliberate artfulness with which the turkey already mentioned, not yet a day and a half old, hunted flies. He aimed carefully with his beak at flies and other small insects without actually pecking at them, and while he did this his head trembled like a hand which one tries to hold motionless by an effort. I observed and recorded this when I did not yet understand the meaning of it; for not till afterwards did I discover that it is an invariable habit of the turkey when he sees a fly sitting on any object to creep slowly and with cautious steps to the unsuspicious insect, and to stretch out his head very carefully and surely to the distance of about an inch from his prey, which he then seizes with a sudden stroke."

THE INSTINCT OF THE CUCKOO IN LAYING HER EGGS IN OTHER BIRDS' NESTS

This instinct has been much discussed on account of its peculiarity, especially since Darwin gave an explanation for it. Darwin says in the *Origin of Species*: "It is supposed by some naturalists that the more immediate cause of the instinct of the cuckoo is that she lays her eggs not daily, but at intervals of two or three days, so that if she were to make her own nest and sit on her own eggs, those first laid would have to be left for some time unincubated, or there would be eggs and young birds of different ages in the same nest. This explanation is not sufficient, for the American cuckoo makes her own nest, lays her eggs in it, and has eggs and young successively hatched, all at the same time. She also lays her eggs occasionally in other birds' nests." Darwin, as is known, then supposes that the ancient progenitor of our European cuckoo had the habits of the American cuckoo, and that she occasionally laid an egg in another bird's nest. "If the old bird profited by this occasional habit through being enabled to migrate earlier, or through any other cause; or if the young were made more vigorous by advantage being taken of the mistaken instinct of another species than when reared by their own mother, encumbered as she could hardly fail to be by having eggs and young of different ages at the same time ; then the old birds or the fostered young would gain an advantage. And analogy would lead us to believe that the young thus reared would be apt to follow by inheritance the occasional and aberrant habit of their mother, and in their turn would be apt to lay their eggs in other birds' nests, and thus be more successful in rearing their young. By a continued process of this nature I believe that the strange instinct of our cuckoo has been developed. It has also recently been ascertained on sufficient evidence, by Adolf Müller, that the cuckoo occasionally lays her eggs on the bare ground, sits on them, and feeds her young. This rare event is probably a case of reversion to the long lost aboriginal instinct of nidification."

Thus here again is conspicuous the importance ascribed to chance, on which Darwin bases his whole theory; and for this very reason I devote particular consideration to the instinct of the cuckoo. It seems to me *a priori* impossible that instincts can be explained by mere accidents, and it seems to me especially impossible that the young bird derived from an egg which was accidentally laid in the nest of another species should "by inheritance" be more likely "to follow the occasional and aberrant habit of its mother"—of the mother which it had never seen. If there is any greater inclination in the young bird, it can only be, in my opinion, the inclination acquired from the experience which it passed through in the nest of its foster-parents during its own youth;[1] through the impressions it received of its foster-

---

[1] Cf. Walter, Brehm's *Thierleben*, second edition, vol. iv. p. 217.

parents and of the nature of their nest. And it is in fact all the more probable that such experience is the principal factor in the case in question, and that it has at last become instinctive by inheritance, because we know that the experience of youth also in ourselves, and in the animals of which we know anything on this point, is retained with extraordinary tenacity. Thus, for example, a single punishment is as a rule sufficient to make a young pointer indifferent to hares. Accordingly, the young cuckoo is influenced, not by the inheritance of the occasional habit of its mother, but by the experiences of its education, and by the deficiencies of its education. The cuckoo can never learn to build nests, and must completely lose the nidificating instinct. But, apart from these considerations, it is a question if the occasional laying of eggs in other birds' nests was really the ultimate cause of the cuckoo's instinct. I deny this, and believe that the original progenitors of our cuckoo when they began to lay their eggs in other nests acted by reflection and with design. We have no certain knowledge that the cuckoo does not at the present day lay its eggs in other birds' nests after conscious reflection. We do not know, therefore, how far this action is due to intelligence. That it is not a perfect instinct is shown by the fact above mentioned, that our cuckoo at times sits on her own eggs and feeds her young; and it is especially noteworthy that it builds no nest for the purpose.

This is of course disputed. Brehm ascribes the statement to a confusion of the cuckoo with the goat-sucker, but Darwin himself regards it as satisfactorily proved. But if the fact is considered unproved, the other fact that the American cuckoo sometimes incubates its own eggs, sometimes lays them in other birds' nests, and *a priori* the assumption that the instinct of our cuckoo has been gradually developed, imply that there was a time when it still occasionally incubated its own eggs, as conversely the American cuckoo now only occasionally lays its eggs in other nests. Who can prove that reflection did not underlie this difference of action, determining it according to the different conditions? For the cuckoo certainly shows reflection in its choice of nests in which to lay its eggs, and in many other actions. But whether our cuckoo acts at the present day from pure instinct or still by reflection—even assuming the first—it is, I believe, necessary, in order to explain the instinct, to start from the supposition that it has arisen by the inheritance of habit originally intelligent. The conditions of life of the bird enable us to understand easily enough the ultimate causes of the habit.

The cuckoo lives the life of a vagabond. It wanders about restlessly; in the first place, it remains with us only a few months, from April to August; secondly, even in the region where it settles here it has no permanent station, it wanders now in one direction now in another. This restlessness is caused by its insatiable appetite and sexual desires. It is ever seeking food and mates. Its food consists principally of caterpillars, especially those which, like Gastropacha pini, occur only here and there in great abundance. The cuckoo must therefore move about in order to satisfy its need of food. When it has found a swarm of caterpillars it revels in excess, and its sexual requirements are increased. It enjoys on account of its loose life the worst of reputations; the bonds of marriage are unknown to it, especially to the lady-cuckoo. Brehm says of it, with regard to this: "Any one who doubts the intense lustfulness of the cuckoo needs only to visit its sleeping-places repeatedly. To-day are heard the voice of the female, the soft wooing of the male; to-morrow only the cry of the latter; the former is then blessing a neighbour or a distant mate."[1] Farther on: "Although he meets with no reluctance, desire seems to drive him out of his senses. He

---

[1] Brehm, *Illustr. Thierleben*, second edition, vol. iv. p. 215.

is literally mad as long as the pairing time continues, he screams unceasingly, so violently that his voice breaks down, constantly scours his neighbourhood, and sees in every other his rival in love, the most hateful of all adversaries."[1] But the female is the greater rover; the male remains within a more limited range. The former "roves throughout the whole summer, that is, as long as its egg-laying period lasts, irregularly through the ranges of various males, attaches herself to none, but abandons herself to all who please her, waits not to be sought, but starts of her own accord to seek adventures, and after her desire is satisfied pays no further attention to the mate who has just shown her favour. A female which I watched near Berlin, which was recognisable through having had a tail feather shot away, visited, so far as I could discover, the stations of no less than five males, but probably extended her excursions still farther. Every other female doubtless behaves in the same way, as other observations prove almost to certainty." "I have often," remarks Walter, "seen a male in company with a female when she extended her excursions to a more distant region, *e.g.* across a large lake, suddenly desert her and fly back, at first in a wide circle, and then in a straight line to his own district. If the female had already laid an egg in the latter she returned thither, although not till another day. Only in case she could not find another nest in the neighbourhood of the one first used, she remained longer away, and was not seen again for days." "On the other hand, other females now pass constantly through the same district, and thus every male, if not by every, at all events by some female, has his desires satisfied. These dissolute and vagabond habits in the female give, in my opinion, the most simple and most satisfactory explanation of certain facts connected with the deposition of the eggs, which have hitherto been mysterious."[2] These

[1] *Ibid.* p. 213.   [2] Brehm, second edition, vol. iv. pp. 211, 212.

habits in the female are doubly remarkable because the males are at least twice as numerous as the females. Brehm supposes that the females wander about principally to find nests in which to lay their eggs. It seems to me that it is rather to be explained by their insatiable sexual desires.

The female lays her eggs, not like other birds in rapid succession, but at considerable intervals. Most authorities say the interval is from six to eight days. In observed cases a female produced at least two eggs in a week, in others the second was not laid till six days after. There appears, therefore, to be great variation in this respect. And this agrees with my view of the cause of the peculiarity of the species which is related to the long duration of the egg-producing period. It was probably ultimately determined by the irregular habits of life, the irregular, but on the whole extraordinarily abundant consumption of food, and the varying demands of sexual excitement. Thus our domestic hens have been brought by continued high feeding to lay eggs throughout the greatest part of the year, and to receive the male throughout the whole of this time. The latter condition must tend to increase the number of eggs laid. In man and many domesticated animals even winter has no effect in diminishing generative activity, as it has in most of our wild animals. It is even possible, although not proved, that the cuckoo continues to lay eggs during the whole of its period of reproduction—during about two months—and so lays as many as twenty to twenty-four eggs (Brehm).

That the production of eggs is extended in the cuckoo over so long a period is not therefore the only and ultimate cause of its instinct. But because, in consequence of its mode of feeding and the sexual demands therewith connected, it forms no regular family ties, but is compelled to lead a vagabond life, therefore it does not lay its eggs in rapid succession like other birds, and cannot incubate like these. Its ancestors, therefore, not accidentally, but from reflection, laid their eggs in other birds' nests, and this habit has now perhaps become instinctive, but is also doubtless maintained by the effects of experience, in that the young cuckoos know the characteristics of their foster-parents' nest, but have never learnt to build a nest, and have inherited no instinct of nidification from their parents.

There are vagabonds in human society, in the lowest as well as the highest, who lead in all respects a cuckoo's life—our knowledge of these enables us better to understand the instinct of the cuckoo. Such people trouble themselves no more than the cuckoo about their offspring, but let others look after them, and even leave them at other people's doors, or at the door of the foundling hospital.

But it is extremely remarkable that the cuckoo has still so far retained the instinct of providing for its offspring in spite of its pleasure-seeking life, that it secures the maintenance of the species although the instinct of incubation has been entirely lost. Compare with this again the habits of our domestic fowls. In them the incubating instinct is obviously gradually being lost; it occurs only in a few individual hens. Most hens are indifferent to the fate of the eggs they lay, or even, like the domestic duck, simply let them drop anywhere. Domestic ducks, like domestic hens, have forgotten how to build a nest, like the cuckoo. But the original form of the domestic duck (Anas boschas) makes a nest in which she lays eight to sixteen eggs. The original of the domestic fowl (Gallus bankiva) in India also lays in a nest, only roughly prepared it is true, eight to twelve eggs, and, like the wild duck, lays them in rapid succession, so that they are hatched all together.

The difference in the persistence of the parental instinct in domestic fowls and ducks on one hand, and in the cuckoo on the other, I can only explain to myself by the fact that the eggs are taken away from the former, so that many of them have completely lost the habit of providing in any way for the maintenance of the species—lost it in consequence of experience acquired during life, and inherited; while every cuckoo must remain in a certain relation to her offspring. On the other hand, the cuckoo knows by inherited as well as by continually renewed memory that she has been reared by her foster-parents, whom she must recognise. She must come, therefore, to the reasonable conclusion that her offspring will be reared by birds like her foster-parents. It is to be remarked that there are in fact only a very limited number of birds to which the cuckoo entrusts its eggs. Our European cuckoo confines its attentions principally to the reed-warblers, the water-wagtails, the hedge-sparrows, and the pipits.

It will of course be asked why does the cuckoo take care to provide for her eggs? For she cannot know that she was produced from an egg. But this objection applies also to the care of all other birds for their eggs. Every bird must, from the first time it hatches its eggs, draw the conclusion that young will also be produced from the eggs which it lays afterwards, and this experience must have been inherited as instinct. Besides this, it is allowable to assume that physiological causes are present which give rise to the habit of incubation. At the period of reproduction an abundant flow of blood towards the sexual organs takes place, especially to the oviducts and ovaries, which in birds whose eggs are provided with a large yolk require abundant nutrition at this period. The blood thus required, after the eggs are laid, causes an increased sensation of warmth in the skin, which the bird perhaps, as other naturalists have supposed, originally tried to alleviate by sitting on the cooler eggs. This may have been the first inducement to sitting, which afterwards gradually became instinctive.

Moreover, it is undecided how far all birds, even before they begin to incubate themselves, obtain a knowledge of their origin from eggs by observing others incubating.

The cuckoo has no longer the impulse to incubate because in her, in consequence of the slower rate at which her eggs are produced, different physiological conditions occur than in other birds, and this is true likewise of domestic hens and ducks. The time must come, although it cannot be predicted, when the latter will cease to incubate altogether, if the present conditions continue. That the incubating instinct in birds must have developed gradually follows from the fact of their derivation from reptiles. It will be replied to this, that the Pythonidæ lie upon their eggs; but I am inclined to suppose that this is really a case of the protection of the eggs by their parents, which is observed in so many other animals,—at the same time I by no means reject the supposition that this protecting habit may have been among the conditions from which the habit of incubation took its rise in birds. When the gradual evolution of the high temperature of the blood, either connected or not with the desire to cool the skin, is added, the rest follows naturally.

The instinct of taking care of offspring is one of the most rooted, one of the most perfect of all which occur in animals, and it is surprising enough that it can be lost even under the influence of domestication, as in our hens and ducks. That the instinct has not entirely disappeared in the cuckoo I have attributed partly to the possibility of the persistence of a certain degree of relation between the old cuckoos and their young, without, however, laying much weight upon this reason. I felt, however, obliged to refer to that possibility, because the love of parents for their children and their care for them obviously depends ultimately on the advantages of society, or on the advantages which the members of a family afford one another, and particularly which the children afford to the parents in old age. To be alone in the world is in itself painful. Accordingly the instinct of caring for offspring is a social instinct, and it belongs, according to my practical view of "reason," to the class of reasoning instincts, since it connects the care for the future of the species with the care for the family. At the lower stages of its evolution the egoistical side of this instinct will still be strongly prominent, but as the relations are continually refined, and with the mental evolution of the organism become more manifold and more spiritual, the most ideal conduct results from it.

Darwin mentions that some species of Molothrus, a genus of American birds, allied to our starlings, have habits like these of the cuckoo. He believes that among the species of this genus a series can be arranged which shows the evolution of the instinct. According to Mr. Hudson, Molothrus badius sometimes builds a nest of her own, sometimes lays her eggs in another bird's after she has thrown out the nestlings of the latter, sometimes builds a nest for herself on the top of the one appropriated. She usually sits on her own eggs and rears her own young; but probably she occasionally leaves them to be reared by other birds, for young of this species have been observed clamouring to old birds of another species for food. Another species, Molothrus bonariensis, according to Darwin, has the parasitic instinct much more highly developed. As far as is known, this bird invariably lays its eggs in the nests of others; but it is remarkable that several together sometimes commence to build an irregular untidy nest of their own in singularly ill-adapted situations, as on the leaves of a large thistle. They never, however, so far as Mr. Hudson ascertained, completed a nest for themselves. They often lay so many eggs —from fifteen to twenty—in the same foster-nest, that few or none can possibly be hatched. They have, moreover, the extraordinary habit of pecking holes in the eggs, whether of their own species or of their foster-parents which they find in the appropriated nests. They drop also many eggs on the bare ground, which are thus wasted. But I should think that the habits of this species perhaps indicate rather a degeneration of the parasitic instinct than a higher development of it in comparison with the preceding species. The bird seems beginning to make less careful provision for its eggs. A third species, the M. pecoris, behaves exactly like our cuckoo. Now, it is an important fact in relation to my view of the evolution of the cuckoo's instinct that the M. pecoris also is polygamous, and this seems also true of M. badius. Mr. Hudson says, according to Darwin, they live sometimes quite promiscuously together, sometimes they pair.

Reasoning Instincts

The Bee as an Example of the Importance of acquired and inherited Characters in the Modification of Forms

I here adduce first a case which in many respects supports my views in the most striking way, especially with regard to the inheritance of mental characters, and with regard to the great part which is played by correlation in the alteration of the sexual organs.

We know that in bees the asexual condition of the workers is the result of insufficient nourishment in the larval state, for within the first eight days of their life the larvæ which are destined to become workers can by better feeding—by the feeding which the queen-larvæ receive—be reared into sexually-perfect queens, capable of reproduction.

This is, be it said in passing, one of the most beautiful, incontrovertible instances of the direct influence of external conditions on the development of structure.

Here also the workers possess a whole series of peculiar characters of the body which are obviously correlated with the repression of the sexual organs, for when the worker larvæ royally fed develop into queens, they develop all the characters of the queen.

But the most interesting point is that the worker bees, whose duty it is to feed the larvæ, only in case of necessity, only when no young queens exist, rear queens from worker larvæ. This almost compels the conclusion that they *consciously*, under ordinary conditions, rear only workers, an abundance of which is so necessary to the community. But even if they do this instinctively, how do they attain to their wisdom with regard to rearing queens and to many other matters? It can, it seems, be no more inherited from their parents, queen and drone, than the winglessness of worker ants from the winged parents—indeed, much less, for the wisdom of the worker bees is in many respects much greater than that of the parents. These mental faculties of the worker bee exist in germ in the worker larvæ, and if these do not receive the royal food those faculties develop; yet when the larvæ are fed with the food of larval queens, the peculiar mental characters of the queen, not those of the worker, are developed together with the sexual organs and other bodily characters. And all this is brought to pass by a little more nourishment!

But all this can only be explained with the aid of the inheritance of acquired characters, by a very large measure of this inheritance, and by correlation. Not in the least, on the other hand, by the continuity and variability of the germ-plasm, and by sexual mixture or pammixis.

The larvæ of bees must have inherited in their brains the possibility of a number of faculties which have been acquired by their ancestors. I say possibility intentionally, and not rudiments, in order to prevent misunderstanding. For we have here to deal not with rudiments, which would have to be developed into complete faculties by exercise; we have to deal rather with two sets of connected characters which develop into perfection without any need of further practice, according as the larva receives the extra food or not. These two groups of characters unite together to form a whole, just as from a number of pieces of glass in a kaleidoscope two symmetrical pictures appear, according to the way it is shaken. In consequence of the better nourishment, a different development of the brain takes place *pari passu* with the development of the other characters, so that a perfectly definite set of mental characters becomes simultaneously predominant.

As I have said in my Freiburg address, we must regard the different forms of bees, queen, drones, workers, as discontinuous organs of one whole, which have been evolved from a single indifferent ancestral form. And this differentiation can only be the result, not merely of external demands, but also of direct external influences—especially of experiences.

Only thus can we explain to ourselves the fact that the peculiarities of the workers, notwithstanding that they do not reproduce, are inherited—they are inherited not indeed through the workers themselves, but from the ancestors. The sexual union of drone and queen produces a germ which possesses all the characters which the original bee must have possessed at the time when as yet no separation into queen and drones and workers had occurred.

From the fertilised egg can proceed a worker, or with better nourishment, a queen, also a drone, but the latter not directly but still indirectly; the queen "buds," she lays eggs, and if these are not fertilised, drones are produced from them. We have, however, compared fertilisation with conjugation, and represented it as a kind of nutrition: if, then, the egg of the queen is from the first better nourished by the spermatozoon, it gives rise to a female or a worker larva.

The two sexes of plants and animals, according to the preceding considerations, appear in all cases as separate organs of an original form in which both were united, a view which is supported by other well-known facts. Accordingly, there is reason to maintain that the separation of the sexes depends partly on conditions of nourishment, and their peculiar characters (especially in mental attributes), partly on correlation. To this question I shall return, and then at the same time attempt to explain parthenogenesis by considerations closely connected with those I have just suggested.

The case under examination thus appears to me to have a peculiar importance in the treatment of wider questions, and therefore I will shortly summarise the results of my reflections upon it.

1. The facts that queen bees can be reared from worker larvæ by better nourishment, and that the former have bodily and mental characters quite different from those of the latter, show clearly,

    a. the influence of nutrition on the modification of forms,

    b. the importance of correlation in the same process— the importance of kaleidoscopic transformation.

2. This remarkable process, in particular the fact that the mental characters of the two kinds of bees, workers and queens, are so completely different, and in each case so distinctly form a complete whole, can only be explained on the assumption that both sets of characters were originally united in one form, and that their separation has resulted in consequence of the necessity of division of labour, through kaleidoscopic transformation, so that the two complementary forms appear as organs of that original form.

3. But the original form included the drones also. It is in consequence of richer nourishment of the egg that even worker larvæ are produced: on nourishment by the spermatozoon, that is, on fertilisation. If this nourishment is not

supplied, drones are produced. Thus drones, workers, and queens are to some extent to be regarded as organs of an original creature which united the characters of all three.

4. The effect of conditions of nutrition on the mental characters appears to me above all remarkable, considering the exclusive individuality which those characters present in the different kinds of bees, especially as they can only have been developed as a result of experience. Of course I do not maintain that the separation of the mental characters into three groups, as it exists now in its present perfection and exclusiveness, arose suddenly, any more than that the peculiarities of bodily structure in the three forms of bees so arose: correlation must have played a part at the very beginning of the differentiation. But these three forms of bees have each gradually advanced towards higher perfection in consequence of the action of external conditions (or of experience) and of selection, and the sudden transformation which now occurs, *e.g.* of a worker larva into a queen through better nourishment, forms an abbreviated repetition of the evolution which formerly proceeded gradually.

Moreover, I assume that for some time only males and females existed in the bee community, and no workers.

The conditions existing among the humble-bees enable us to understand the evolution of the forms of bees according to my view.

"In spring, when the all-vivifying sun has warmed the ground to a certain depth, a female humble-bee creeps forth from a hole dug by herself in the ground, usually in a position exposed to the sun, or from a rotten tree-trunk, or from a clump of moss, or from some other retreat in which she has passed her winter sleep." Thus Professor Eduard Hoffer commences his description of the humble-bees of Styria,[1] and he

[1] E. Hoffer, *Die Hummeln Steiermark's, Lebensgeschichte und Beschreibung derselben.* Graz, 1882.

goes on to describe the foundation of the humble-bee family as follows: At first the female humble-bee flies from flower to flower sipping honey, then she seeks a place in which to build her nest. When she has found a place, any suitable hole, she carries into it moss, grass, leaves, hairs of animals, and fine needles of the fir or pine, and builds a nest closed on all sides, and provided with only a single opening directed towards the rising sun, and usually concealed. Then she collects honey and pollen, makes a cell of wax, fills it with pollen soaked with honey, and lays a pair of eggs in it. From these larvæ soon emerge, which grow rapidly, and therefore require much food. The mother now toils energetically day and night for the welfare of her children, by day chiefly collecting, and feeding the larvæ, by night biting up and arranging the materials of the nest, coating it with a wax-like material, and warming her young. She allows herself little rest, except when the weather is bad. At last, in the beginning of May, or in some forms several weeks later, the first young humble-bees creep forth. These are workers, much smaller than the mother or queen—are in fact stunted queens. They fly forth at once to collect honey and pollen, which they bring into the nest. As long as there are but few of these workers the mother continues to fly out also to the fields and collect industriously, but afterwards she goes out less; she now remains much at home, laying eggs and tending them. At last she ceases entirely from going out, her wings, as a rule, becoming useless. Some of the workers tend their younger sisters, who are still in the cells, and feed them, work at the construction of the nest, keep it clean, and lick and warm the young bees when they creep forth. The workers raise a great humming if the nest is disturbed, and defend it by stinging the invader. Thus the queen lives with her handmaids several weeks, about three months, continually adding to their number. As a rule, about July young of a much larger size emerge, likewise resembling the queen, the so-called " small females " or large workers, *i.e.* females whose reproductive organs are developed, but who generally produce only drone-eggs, though under certain circumstances they can lay eggs which hatch into females or workers. These large workers and the small workers and the old female, all three kinds, now lay drone-eggs in large numbers, from which males are hatched. Not till the end of summer does the mother again lay queen-eggs. There exist now therefore in the family—(1) the old queen, who is frequently incapable of flight and destitute of hairs ; (2) numerous young queens ; (3) the ordinary or small workers ; (4) the large workers or small females ; (5) the drones or males. All the workers throughout the summer fly forth to collect about a quarter of an hour before sunrise, awakened by a peculiar humming, the voice of the trumpeter. The males also go to the fields, but only from ten in the morning to four in the afternoon, and only for themselves. They do no work in the nest—although Hoffer saw them occasionally assist, but only when the roof was taken away from a nest. On fine sunny days in July, August, September, and the first half of October, the young queens fly from the nest, settle on flower stems, broad leaves, fences or walls in the sun, and are there sought by the drones of their own or other nests, wooed, and then fertilised during flight. When the young queens are all thus in a condition which enables them to found new colonies in the following year, the whole family gradually disperses. The accepted males soon die, the rest fly about a great deal, return no more to their home even at night, but usually remain out on flowers, and gradually perish (a melancholy example of unsuccessful existence).

If one takes off the mossy covering from a nest at its most flourishing period he usually comes upon a wall of wax, and when he removes this he sees, not the regular honeycomb of bees and wasps, but a somewhat irregular mass, consisting of larger and smaller pupa-cells, shaped like a hazel-nut, and usually of a fine whitish-yellow colour, darker clumps of larva-cells, and smaller clumps of eggs of the size of a lentil-seed, or a pea, or even as large as a bean. Some of the thimble-like pupa-cells from which the bees have already emerged seem to have been converted by a coating of wax into receptacles for honey and pollen; but one also sees a large number of proper honey-cells which are made of wax, and of which, on a very favourable day, ten or more are constructed.

In the humble-bee family there are but two sexes, male and female, for the small and large workers are only females with sexual organs imperfectly developed in consequence of imperfect nourishment, especially of being insufficiently supplied with honey during their larval life.

The moral of these facts is the following: The humble-bee queen is not yet so exclusively a breeding mother as the bee queen. The latter is the mother, or reproductive individual, of the bee community, and nothing else; she is the female reproductive organ of the social organism; she collects not, she builds not, she does not even feed herself, but is fed by the workers: the humble-bee queen undertakes all these duties at the commencement of her public career. The same applies to the drones: the humble-bee drones feed themselves, and, although only exceptionally, take part in work; the bee drones on the contrary are fed by others, like the bee queen, and do no work. The humble-bee drones have almost lost all independence with regard to all other functions than the sexual, like the bee drones, but they have not travelled so far in this direction as the latter. The same also applies to the workers. The bee workers no longer lay any eggs. True, quite recently cases have been recorded in

T

our bee journals by bee-keepers in which workers copulated with drones. These cases have excited a degree of incredulity and of sensation for which there is no justification. For since the workers are merely imperfect females, it can cause no surprise if their sexual organs are in occasional instances more completely developed—perhaps through the larvæ being accidentally better nourished—or even if they show the instinct of copulation. It is, moreover, known that workers in hives that have no queen sometimes lay eggs, and that from these drones are hatched. These exceptional cases among worker bees afford a perfect transition to the conditions which normally occur among humble-bees. In these the workers still regularly lay drone-eggs, indeed, the earliest brood of workers, the small workers, can lay drone-eggs only, while the large workers or small females are said to be able, as remarked above, under certain circumstances to lay female-eggs. Thus the workers of the humble-bees are not yet, like those of the hive-bees, exclusively workers. But the influence of nutrition is here directly before our eyes. The first brood of workers, laid at a time when the queen has to do everything herself or with the aid of few workers, can of course not be so well fed as the second brood, produced when the whole of the first brood render assistance. This better nourished brood, according to Hoffer, have developed sexual organs: their more highly nourished condition shows itself also in their greater size, and they sometimes lay eggs from which females are hatched, which must therefore be fertilised. Lastly follow the young queens, whose larvæ, with the assistance of both broods of workers, can be fed best of all, and the eggs of which have likewise received by fertilisation an important addition of nutritive material.

The present conditions among bees must have had a similar origin, but in them the complete evolution of the community has taken place, while the humble-bees, and similarly the social wasps, only reach the status of a large family. The bees sacrifice the welfare of the individual with relentless determination to the advantage of the state, in that they kill their drones when they have become useless, while these among the humble-bees perish gradually. The humble-bee queen dies in her second year; but the bee queen in her second year, when she has reared offspring, is compelled in the interest of the state to leave the hive and found for herself a new state. But this state-organisation has, for the mother of the whole and for the workers, the advantage that their life is secured for a longer time, for the workers among the humble-bees do not survive the winter, while those of the bees do.

In these respects we have in wasps conditions similar to those of the humble-bees.

The foundation of this essential difference between humble-bees, or wasps, and the economy of the bees, obviously consists in the fact that the reasoning instinct, *i.e.* the forethought for the future—in this case primarily for food—is far greater in hive-bees than in humble-bees and wasps, which provide no winter store, and whose females have to pass the winter in seclusion.

The state-organisation of bees and ants has often been praised as the ideal of social and political arrangements. But it must be remembered that in these animal communities the individual is unconditionally responsible to the state for all its activity, as the state's servant, and if it does not fulfil this responsibility, or if it has served its purpose, it is expelled or destroyed. The queen-bee does not begin to provide for the reproduction of the hive until there are a sufficient number of workers to secure adequate maintenance for two hives, and the female humble-bee or wasp likewise begins her family by providing workers. This is according to reason; while the causes of the social problems in mankind depend in great part on unreasoning conduct with respect to the establishment of families, and for the rest on the burden of idlers in the state.

In the preceding the origin of the sexually imperfect workers in the communities of humble-bees and bees—and the same will hold for the social wasps and ants—has been ascribed to deficiency of nourishment. The question now arises: Is this development accidental, or can we recognise in its beginnings the evidence of reflection in the animals which form the family or state? Obviously the latter alternative must decisively be affirmed. The animals, when they founded families or states, must have seen the advantage of several members living together, the fruitfulness of co-operation. The bees, for instance, must also have seen that the gathering of stores for the winter was useful to them. In the explanation of this forethought chance is perhaps not to be excluded: it may be that in exceptionally favourable summers the animals gathered unusually abundant stores, so that at least a part of them, instead of perishing in autumn, survived the winter with the queen, and these were able then in the following spring to go to work again at once. Selection cannot here have had much influence, since the workers do not reproduce. In order to make these favourable conditions constant, insight and reflection on the part of the animals, and inheritance of these faculties, were necessary. Whereby it is to be borne in mind that the queen-bee originally collected and built along with the rest, and that she must have transmitted to her offspring the experience thus acquired. It might be objected to this, and pure Darwinism will make the objection, that the instinct of laying in a store might have arisen merely because the queens, who happened to be the most industrious collectors, survived those which were less industrious, and so at length a race arose which laid in a winter store quite mechanically without knowing wherefore. But if this were so, it would be impossible to understand why the instinct of storing in the queen has in reality not been improved, but has been lost. And this view is also improbable on account of the high mental endowments—not of an instinctive character—which are displayed by the hymenoptera.

The origin of the storing instinct might again be explained by the supposition that the winter, or in warmer regions the time unfavourable to collecting, has been gradually prolonged, and that only those communities of bees survived which were the most industrious. But the same objections as were made against the previous view apply to this, and in addition various others into which I will not here enter.

It seems to me, indeed, necessary to assume that a gradual prolongation of the period of the year unfavourable to collecting must have contributed to the development of the bee community in its present condition, and especially to the accumulation of larger stores; but it is equally necessary to suppose that reflection and foresight in the bees worked hand in hand with this cause, and have been increased by experience.[1] A proof that bees even now collect their stores in full consciousness that these are a necessity for them is afforded by the fact, known to every bee-master, that the bees, especially the Italian, angrily pursue the person who has taken away their honey. Even several days after the removal of the honey individual bees pursue any one in the neighbourhood of the hive and endeavour to sting him. According to statements which seem to me perfectly reliable, bee-keepers who allowed their bees no peace and continually took from them the last drop of their honey have been

[1] In Australia, where the bees find supplies almost all the year round, the instinct of collecting a store for future use is said in fact to disappear in a comparatively short time, so that it is necessary constantly to renew it by the importation of foreign queens.

suddenly attacked by them, apparently with a common purpose, and been very severely injured by them.

As reason in general cannot have arisen suddenly from particular causes, or in particular cases, but must rather have been gradually evolved under the influence of external conditions, so in bees the evolution of the existing reasoning instinct of collecting must have taken place quite gradually, and must have arisen at first from simple intelligent actions adapted to immediate needs. I do not intend to pursue this question in further detail, but will only point out, that the production of sexually imperfect workers in itself is a highly remarkable instance of conscious consideration by the individual of the advantage of the community. For the only possible explanation of the origin of the workers is that the original bee or humble-bee could not feed all her larvæ sufficiently, and that she obtained an advantage in her housekeeping by the workers thus produced. The insect must have recognised this advantage, and afterwards, accordingly, reared similar workers intentionally by insufficient feeding. Another explanation which might be given is, that certain individuals began to sacrifice themselves to the requirements of the community by neglecting to feed themselves to such an extent that their sexual organs no longer attained their full development.

Further Remarks on Reasoning Instincts and Intelligent Instincts in Animals

It is self-evident that all cases in which animals display forethought for the winter by storing provisions, or providing themselves with winter quarters in which to protect themselves, must be described as cases of reasoning instinct, as also the cases in which individual animals provide nourishment for the use of their offspring at a later time.

One of the most remarkable instincts of this kind is

that of the mason-wasp (Odynerus parietum), and allied predatory wasps. The mason-wasp makes a hole about 10 cm. deep in a bank of clay, moistens and softens the clay she has removed with saliva, and probably also with water, and builds with it a tube leading from the mouth of the hole and prolonging it. The tube at first stands out at right angles to the bank, then bends downwards. Probably this arrangement is chosen in order that the water may run off more easily, just as the cells of the paper-wasp (Polistes gallica), which are made of chewed wood, are turned downwards, so that the rain may be kept out as much as possible, while the bees, contrariwise, build their cells upright, so that the honey may remain in them. But the wasp casts out some granules of clay from the mouth of the tube. Taschenberg [1] supposes that this is the material with which she afterwards closes the tube. When the nest is ready, the wasp brings into it larvæ of beetles and other insects, which she has paralysed by stinging them in the ganglia which govern muscular action. This is one of the most marvellous instincts that exist: since the wasp operates on various larvæ with nervous systems of various forms, she must effect the paralysis in various ways, and even apart from this, she makes a physiological experiment which is far in advance of the knowledge of man. The wasp thus carries one motionless but living larva after another into her tube until it is full, and she rolls up the larvæ and packs them so skilfully that they take up as little room as possible. Finally, she lays her egg in the store of living food, and closes the opening with clay. Then she begins a new tube, and so lays one egg after another.

What a wonderful contrivance! What calculation on the part of the animal must have been necessary to discover it! The larvæ of the wasp require animal food. Dead food enclosed in the cell would soon putrefy, living active animals

---

[1] Brehm's *Thierleben*, second edition, vol. ix. p. 240.

would disturb the egg, and accordingly the wasp paralyses grubs and packs them like sacks of meal one after another in the cell. How did she arrive at this habit? At the beginning she probably killed larvæ by stinging them anywhere, and then placed them in the cell. The bad results of this showed themselves; the larvæ putrefied before they could serve as food for the larval wasps. In the meantime the mother-wasp discovered that those larvæ which she had stung in particular parts of the body were motionless but still alive, and then she concluded that larvæ stung in this particular way could be kept for a longer time unchanged as living motionless food. It may be suggested that the wasp only paralysed the larvæ in order to carry them more easily; but even if this were the case, she must, since she now invariably acts in this way, have drawn a conclusion by deductive reasoning.

In this case it is absolutely impossible that the animal has arrived at its habit otherwise than by reflection upon the facts of experience.

An accurate observer, Lichtenstein, states, with regard to the laying and incubation of the ostrich, that during the breeding-time one cock and three or four hens live together, and he continues: " All the hens lay their eggs in the same nest, which consists of nothing more than a round depression in the somewhat loose clayey soil, of such a size that one hen can cover it. Round the hole the birds scratch up the soil, so as to make a kind of wall, against which the outermost eggs are supported. Each egg in the nest stands on end, so that the largest number possible can be packed in it. As soon as there are ten to twelve eggs in the nest the birds begin to incubate, and this they do in turns, the hens relieving one another by day, the cock alone sitting at night, in order that he may repel the attacks of the jackal and the wild cat, who greedily watch for an opportunity to steal the eggs. In the meantime the hens still continue to lay eggs, not only until the nest is full, which is the case when it contains thirty eggs, but even afterwards. These last-laid eggs lie irregularly round about the nest, and seem to be intended by nature to satisfy the rapacity of the enemies above mentioned, to whom she prefers to surrender these fresh eggs than those already incubated. But these eggs have also a more important use, namely, to serve as the first food for the young ostriches, which when they are hatched are as big as an ordinary fowl, and whose tender stomachs are not at first able to digest the hard food of the adults. The old ones themselves break these eggs for them one after another, and in a short time with this nutritious food bring them on so far that they are able to seek their own food." The same observation has also been made by others, and Brehm doubts it on insufficient grounds.[1] That there are cases of young animals fed by the parents' eggs is proved by the black Alpine salamander (Salamandra atra). This animal brings forth living young without gills and with completely developed lungs, in complete contrast to all other tailed Amphibia, the young of which are born with gills, and this is well known to be due to the fact that the conditions of life of the Alpine salamander do not allow it to deposit its eggs in the water. The young, therefore, under the stress of the external conditions, develop, until the lungs are completely formed, in the oviducts of the mother. In these oviducts a remarkable struggle for existence takes place, so that only one larva develops in each, and this nourishes itself at the expense of the other eggs in the oviduct until it reaches its maturity. I mention this because Brehm regards it as something entirely novel that birds should be fed on their parents' eggs. Such a method of nutrition occurs only in this single instance among Amphibia.

Thus in the Alpine salamander most of the eggs which

[1] *Loc. cit.* vol. vi. pp. 198, 199.

would otherwise develop are devoured by two more favoured larvæ.

But the case of the ostrich could only be employed as an example of a reasoning instinct if it could be assumed that the hens have come to co-operate in incubating a large number of eggs with the *intention* of providing the first food for their young. As the conditions of the case are not known more minutely it is not possible to form an opinion upon it; but in any case it is intelligent instinct, or intelligent action, when the ostriches make use of these eggs to feed their young; and I chose this example to bring into view the distinction between the two kinds of mental action, as well as on account of its interest, because it is little known, and because ostriches are proverbially described as stupid animals.

But what was it that taught the beaver to dam back the flowing water by a regular weir built of tree trunks, sticks, and branches, in order that some of the burrows which lead to the face of the river-bank from the building which it so skilfully constructs might be always under water?

What taught the ants to carry on regular agricultural operations, to collect the seeds of certain grasses, let them germinate, then to bite off the germ—to malt, in fact, and to dry the seeds, and keep them for future requirements? What taught them to keep slaves, and make these feed them, until more than one species have forgotten how to feed themselves by their own exertions?

I will not enter upon a minute consideration of the wonderful facts presented to us by the social life of the Hymenoptera, and especially by the ants. These facts, which so strikingly illustrate instincts of reason, were long doubted, but have been completely established, particularly by Forel, who has fully confirmed the observations of Huber, and largely added to them, in his excellent work. I need not mention Huber's observations on bees, many of which every observant keeper of bees can personally confirm.[1]

Forel comes to the conclusion, from his own observations on ants, that they are capable of sacrificing themselves for the general good, the highest ideal of conduct to which we can require man himself to attain; and if such conduct has in animals become instinctive, the fact must cause us to marvel the more at their mental and social life, and human society must confess that it has scarcely advanced so far.

The Countess Nostiz, who died in Meran a few years ago, accompanied her first husband, the entomologist Helfer, in the journey to Asia described in the book *Helfer's Reisen*, which she edited from his diaries. This extremely talented woman, seldom at fault in the observation of nature, assured me that she witnessed the following incident with her own eyes: A train of ants in active migration arrived at the bank of a brook which they wished to cross. The vanguard examined at one or two places the part of the bank whence, as the result showed, the crossing was to be made; then a number of the ants, holding fast to some stalks of grass on the bank, went into the water; others mounted on those and held on to them, and others held on to these, and so on; so these ants drowned themselves and formed a bridge over which the body of the army then passed over to the other bank. It can only be concluded that these ants first carefully examined the breadth of the brook and the other conditions, and thereupon determined to choose this method of crossing. This may seem a rash conclusion, but only because people in general, and many naturalists, immensely underestimate the mental faculties of animals, and therefore the majority are always inclined to receive as fables the state-

---

[1] Forel, *Les Fourmis de la Suisse*, Bâle, Genève, Lyon, 1874; P. Huber, *Recherches sur les Mœurs des Fourmis indigènes*, Paris, 1810; F. Huber, *Nouvelles Observations sur les Abeilles*, second edition, Paris, Genève, 1814. Cf. also the works of Lubbock.

ments of those who have minutely studied the life of any class of animals. It is indeed the custom to set down the most wonderful mental powers in animals without distinction as instinct, because men think they can thereby reduce the importance of these faculties, and conceive them as something mechanical implanted in the animal's nature from the beginning. Men have an idea they must at any price preserve their fancied exceptional position in the animal world, even though it be at the cost of their delight in nature, and of their own intellectual progress. Moreover, those among them who suppose they serve the Creator by their belief, do not perceive that the case is just the opposite, since they reduce his power instead of magnifying it. But apart from this, such a view altogether excludes every explanation of instinct and of the mental faculties of animals.

On the other hand, when we explain instinct as inherited experience, its phenomena appear as the result of continued practice due to this experience. On this view, intelligence and reason, it is true, become mechanical, but in a way which does not diminish the importance of instinct, but which, on the contrary, brings into stronger light the acuteness of the original powers of reflection in animals, and their conscious direction to a definite object. Instincts of intelligence and of reason are in all cases a highly important factor in the evolution of mental faculties. By the establishment of these, and, in a less degree, of automatic actions, as already pointed out, powers are set free which can be applied to further mental effort; they evidently constitute a means for the progressive development of the mental life.

As in the evolution of the individual body an abbreviation takes place in course of time which affords time, material, and energy for further modification, so is it in mental evolution: instinct is evolved by a suitable abbreviation, simplification of the process of thought.

It follows from my view of instinct, as I shall show more fully hereafter, that we cannot always draw a dividing line between the voluntary exertion of intelligence or reason—between action depending on immediate reflection—and instinct. Those who take the lowest possible estimate of the mental life of animals will not admit the former at all, although both intelligence and reason are necessary antecedents of the latter. Some of the cases already mentioned show the difficulty of distinguishing between intelligence and reason on the one hand and instinct on the other, as well as the self-evident difficulty of distinguishing intelligence from reason. The ants who by the sacrifice of their own lives built a bridge across the stream for the benefit of the community possessed certainly reasoning instincts; but this particular case of self-sacrifice must have been due to the direct voluntary exercise of reason, because it must have been an exceptional case. Here, therefore, the application of the general reasoning instinct depended on conclusions drawn by reflections made at the moment.

Whether ants, when they carry their pupæ, according to the temperature, towards the centre or the surface of their nest, act on instinct or reflection, and whether they or the bees act by the former or the latter in numerous other such instances, we cannot from the nature of the case decide. But countless facts prove that here as in the case above considered, instinct and reflection supplement and alternate with one another, or else that reflection operates principally or exclusively.

A theological friend of mine who studies attentively the mental powers of his dog, an animal of no particular breed, resembling a jackal in size and form, tells me the following story about him : The dog, who is extremely active, pugnacious, and courageous, was persecuted for weeks by a butcher's dog of about the same size, apparently from envy because "Ami" at the daily visit to the butcher's shop received a scrap of sausage. The butcher's dog after every visit followed him down the street barking and yelping, and running close at his side. "Ami" was quite a match for his adversary, as the result of several fights showed, but it was evident that the strife at last became distasteful to him, and besides he saw that his master disapproved of it. He no longer allowed himself to be irritated, and then he gave up his visit to the butcher's shop. The provocation was, however, continued when the peaceful "Ami," accompanying his master in his usual walk, went past the butcher's shop. From the street in which this shop stands two side streets diverge some distance on one side of the shop, joining the main street again farther on. One day "Ami" left his master at the place where these streets diverge, ran through one of them, and awaited him at the place where they unite again. Afterwards he regularly made the same detour, choosing sometimes one sometimes the other of the side streets. He thus avoids meeting the brawler, not from fear, as his own character proves, for he has never shown any fear of his enemy, but because he wishes to avoid strife.

Dr. Fickert tells me he has observed two quite similar cases in another dog.

If men behaved as "Ami" did—which would be an exceptional occurrence—we should say they behaved reasonably. It seems in fact that the dog drew the general conclusion that constant strife is useless, and it is better to avoid it; but of course it cannot be decided whether it was so, or whether he simply concluded that his opponent might at last handle him severely. But whether the behaviour of the dog be called reasonable or intelligent, it cannot be called instinctive.

Dr. Fickert tells me also the following story about a dog which he possessed: This dog found by experience that when he tried to swim across the rapid stream of the Neckar he was carried down by the current, and therefore could not reach a particular place on the opposite bank which was suitable for landing. He therefore used to go along the shore some distance up the river before he jumped into the water, and so swimming straight forwards, and being carried down by the stream at the same time, he crossed in a slanting direction to the landing-place. This was intelligent reflection, not instinct.

Professor Leuckart observed a highly remarkable example of intelligence in ants. The insects were crawling up the trunk of a tree in order to reach some aphides. Leuckart painted a ring of tar half-way up the trunk to see what the insects would do. They could not go either up or down across the sticky obstacle. For a time they ran up and down restlessly, but at last those who were below the ring descended to the ground. After a time they came back each with a grain of earth between her jaws; one after the other placed her morsel of earth on the tar, so that gradually a bridge was formed across it, over which the insects passed in safety. That they should have recourse to an expedient so remarkable was not instinct but intelligence; but if they intended to rescue the ants on the upper side of the ring, then they acted by reason. However, it is known that ants in such cases know how to help themselves by dropping from the branches to the ground.

Father Gredler of Bozen records the following observation, made in his monastery:[1] "One of my colleagues had for some months placed food on his window-sill for the benefit of a community of ants (Formica aliena, Fœrst.), which held regular processions from the garden to the window of his room. When I described to him the experiments made long ago by Gleditsch, and recently by more modern students of ants, the amusing idea occurred to him to fasten an empty ink-bottle by a thread

[1] *Zoologischer Garten*, vol. xv. p. 434.

to the cross-bar of the window. In this vessel the ants' food, some crushed sugar, was placed, and in order that the ants might be aware of the new position of the food provided for them a number of individuals were placed in it. The busy little creatures took up their crumbs of sugar, soon found their way up the only means of communication, the suspending thread, along the cross-bar and down the window-frame, and reached their companions on the window-sill, to continue from there their usual journey down the high wall to the nest in the garden. It was not long before a regular traffic over the new road from the window-sill to the store of sugar was organised, and so a few days went by without offering anything novel. But one morning the traffic ceased at its old terminus, and the ants again obtained their groceries from a nearer source—namely, from the window-sill. Not a single individual made the journey thence to the vessel of sugar. Yet this had not been emptied. By no means; but a dozen individuals were working vigorously and unweariedly in the vessel above, simply carrying the grains of sugar to its edge and throwing them down to their comrades on the sill below."

P. Huber[1] made the following observation: He watched an ant building a covered way out of earth: first it erected a vertical wall on one side, then it raised a second vertical wall parallel to the first, in order to support a roof on the two, and so form the gallery. It had begun to extend the roof out horizontally from the upper edge of the second wall; apparently it was intended to rest upon the first wall on the other side, but this had been carried too high. Then a second ant came by, looked at the defective structure, examined it, drove the unskilful builder away, pulled down the walls, and built the whole anew.

P. Huber also records the following observation in humble-bees, which build their combs horizontally one over the other:

[1] P. Huber, *Recherches*, p. 47.

A very irregular piece of comb placed on a very smooth table vibrated so much and so continually that the humble-bees could not work on it. In order to prevent the vibration, two or three of them held the comb fast, by setting their fore feet on the table and the hind feet on the comb. By relieving one another they continued this for three days, until the necessary supporting columns of wax were finished.[1] Darwin, according to Romanes,[2] justly remarks in his posthumous manuscript, that such a case could scarcely have ever occurred in nature. Self-evidently the animals must have acted entirely from reflection upon the particular circumstances; they acted like the ants in the two last-mentioned cases, not instinctively, but intelligently.

The construction of the honeycomb by bees seems to be the work of pure instinct. But it is proved by experiments made by Huber that the insects vary the structure of the comb in accordance with external circumstances, and these likewise are experiments which have no counterpart in nature, so that the bees here again must have based their arrangements on the results of intelligent reflection. It is clear that the building of the comb requires very considerable instinctive dexterity, that it must with regard to its essential character be reckoned among constructive instincts; but reflection seems not to have been completely eliminated from it, it seems to be by no means a perfectly unmixed instinct. The view that the cells, as Herr Müllenhoff[3] has recently attempted to prove, arise quite mechanically through the mutual pressure of the bodies of the bees at work, whereby they must necessarily become hexagonal, is certainly unfounded. Any one who has once examined the edge of an unfinished comb will admit this. Towards the edge of such a comb the cells gradually diminish in height. The

---

[1] P. Huber Fils, *Observations sur plusieurs genres de bourdons, bombinatrices de Linné, Transact. Linn Soc.* vol. vi. pp. 214-299, London, 1801.

[2] Romanes, *loc. cit.* p. 225.

[3] Pflüger's *Archiv für die gesammte Physiologie,* vol. xxxii. pp. 589-618.

external wall of some of the outermost cells last commenced is still wanting. The floor of the cells here is as thin as elsewhere. Evidently the bees build up this wall in its proper shape, and do not produce it by pressure. Similarly the construction of the half-cells which the bees begin a comb with, those which attach the comb to its surface of suspension, is inexplicable on this view. For these half-cells resemble cells divided longitudinally, which alternate with entire cells. It is self-evident that only by this means can a foundation be made from which to proceed with the construction of six-sided cells according to the regular plan. Does the construction of this commencement of the comb, which consists of cells different from those of the rest of the comb, depend upon instinct? In such cases it has been usual to talk of variations, of various directions of instinct, and E. v. Hartmann has assumed this explanation for the present case.[1] It is indeed obvious that in this way instincts in various directions may arise in response to definite and persistent external requirements. But how can we explain by such definitely-directed instincts the fact that the bees, when in the course of a comb of worker-cells they wish to build the much wider drone-cells, make each succeeding row of cells somewhat wider till they have finally attained to the size of the drone-cells? The art of the performance depends of course in this case upon instinct, but such a gradual change, which has to be arranged according to the external conditions, seems to me distinctly to imply reflection also, like so many other facts in the mental life of bees which are usually lightly described as instinct.

That bees really exercise even a high degree of reflection in building their combs is shown most completely by the following observations and experiments of F. Huber: Once the bees

[1] Eduard v. Hartmann, *Das Unbewusste vom Standpunkt der Physiologie und Descendenztheorie*, second edition, Berlin, 1877, p. 189. Hartmann calls such instincts polymorphous, but he also admits the importance of reflection in apparently instinctive action, *op. cit.* p. 200.

had made on a wooden surface the beginnings of two combs, one to the right, the other to the left, in such a way that the latter should support an anterior, the former a posterior comb, and the two when finished should be separated by the usual distance between two combs in a hive. But the bees found that they had not allowed sufficient distance. What did they do in order to avoid losing the work already done? They joined the beginnings of the two combs into one. The curvature necessarily produced was in the continuation of the comb completely levelled, so that the lower part of the comb became as regular as one properly commenced.

In another case the bees had commenced a comb on the lower edge of a glass plate. As the work proceeded it was evident that the comb would become too heavy for the narrow surface of attachment. The bees therefore built the comb at first regularly upwards on both sides of the glass, attaching it to the latter by the surface; but farther up they contented themselves with merely carrying a layer of wax over both sides of the glass plate until they reached the wood in which the plate was fastened, and there obtained a secure hold for their structure.

The skill of the garden-spider in building her web no doubt depends on instinct, but only with regard to the main process: here also reflection is exercised on many points. In the mere choice of the place where the net is to be spread, the spider needs to take many things into consideration: direction of wind, sunlight, abundance of insects, and, above all, the assurance that the web will be safe from disturbance in the place selected, require a host of intelligent conclusions —the question of security from disturbance alone requires a number. And yet how correctly the spiders usually judge in this very respect. And when the locality for the net is chosen, the spider has to decide the points at which the frame of the net to which the spokes are fastened shall be attached, and so on. In each single case again the threads which connect the spokes to the frame must be adapted to the conditions of the latter; and lastly, it is known that the spider not only in this regular work, but still more in repairing her web, in doing which she fulfils the highest requirements of mechanical adaptation, gives undoubted evidence of complicated reflection.

Of course those who attribute to animals neither intelligence nor reason, but only instinct, will call that which I describe as evidence of reflection, variation of instinct, or endeavour to bring it within the conception of definitely directed instincts.

But these facts only become comprehensible by the explanation of instinct as inherited experience, *i.e.* as intelligent and reasonable actions which have become automatic. I fully admit that in the construction of a net or a honeycomb various directions of instinct may have been developed, but other actions of animals have not yet become purely instinctive, and others again are exercised quite freely and spontaneously. But even spontaneous action in animals is often only slightly removed from instinctive, as is shown by the facts I have described concerning the chick, facts which prove that these animals have become by inheritance capable of making use of experience immediately in a surprising way. Such an animal as this chick is extremely clever. I can now add to the account already given, that when I had fed it a single time, instead of on the table on which it was usually fed, in front of the door of my study, to which a bridge of considerable length leads from the garden, it came regularly to this bridge expecting food. I am certain that much which gives us the impression of pure instinct depends on such a rapid employment of experience by animals.

The mental activities of many animals are obviously very limited in extent, but we must always remember that we know in general much too little in detail of this mental life, and that from the nature of the case we are only able to recognise its manifestations to a very limited degree, that therefore our low estimate of these mental powers is based on very little information. This limited extent of the mental activities is compensated by definite instincts, whose operations fulfil the essential requirements of the animal, and which are so conspicuous as to attract our attention, while the rest of the animal's actions due to mental faculties escape notice. But if the explanation of instinct which I have given is recognised as correct, even animals which act chiefly from instinct must be described as "clever."

The more manifold the demands upon an organism from the outer world, in particular, the more they are subjected to change of circumstances, so much the less can fixed instincts be developed, so much the more will the brain have to be developed, and to be capable of immediate and free response to these external demands. Accordingly, in man there is but little instinct, while voluntary action is very highly developed; but this is also connected with the fact that his body is imperfectly adapted in many respects to external requirements. Our senses, for example, as they exist are only sufficient because their low degree of acuteness is compensated by the high development of the brain,—this alone, for instance, has enabled man to live under climatic conditions which, like those of even our temperate zone, would not supply his primary wants without special appliances.

The slow and long development of the mental and bodily powers of man is connected with this higher stage of mental evolution which he has reached. Since man, in consequence of the great complexity and the great variability of the conditions under which he lives, cannot inherit experience, therefore he must acquire his experience during his life, and thus he attains independence at a relatively late period of life. In this respect the anthropomorphous apes are nearest to man, and among different human races it is obvious that the rate of development is nearly inversely proportional to their standard of culture, and clearly both are under the influence of climate. The preceding considerations explain how certain constitutional predispositions in a single direction, such as musical talent, arithmetical talent, etc., can appear at so early an age (phenomenal children), and are so frequently hereditary. Such faculties approach most closely in character to the instincts of animals; they are due to the inheritance of the results of practice, although occasionally such talents may arise from favourable sexual mixture, from crossing. Just as in animals, men with such special faculties are by no means necessarily possessed of great general mental power; on the contrary, other mental faculties often diminish to a corresponding degree, simply because only the special powers can be employed or applied in the struggle for existence. Thus in many cases it is not the nations which stand highest in culture in which musical talent, for instance, has become quite general. I mention only the gipsies, Magyars, and Tschechs. The latter nation has, besides the services rendered by its wandering musical companies, done another great service to mankind, it has contributed the invention of the dance known as the polka.[1] For the Tschechs are also born dancers. The resemblance to instinct in such inherited talents, in the musical talent of such nations, shows itself especially in the fact that it is not a case of the higher music, of faculty for lofty composition, but merely of the inheritance of the faculty of performing the ordinary music.

Reflex Action and Instinct

It is customary to connect the idea of instinct with nervous

[1] It was invented by a peasant girl, who danced the step for her own pleasure. Cf. R. Andree, *Tschechische Gänge*, 1872, p. 272, asserted on the authority of A. Waldau.

action, and I have done so too, but with a significance not hitherto usual, since I have expressly included in instinct reasoning action inherited through habit. The extent of the idea of instinct is also usually limited within the range of the action of the nervous system by the exclusion of reflex action from it. Only actions or inclinations, *i.e.* faculties for action, are called instincts which take place as if they were due to reflection without actually being so.

But there are evidently two kinds of reflex action : one purely involuntary, in the development of which conscious experience, a brain, has had no influence in its earliest phyletic origin. To this kind belongs, for instance, the peristaltic movement of the intestine under the stimulus of food taken into it. Also the rhythmical motion of the heart. To it also belong numerous motions of the parts of lower organisms or of the whole of such organisms excited directly by stimulation. In another group of reflex actions, those which were originally voluntary, it is evident that they were once, at an earlier period, under the control of the brain, of experience ; they are really automatic, in all their characters are now removed from all relation to the action of the brain, in such a way that such relation must be deduced from special reflection. Here belong some of those reflexes which take place involuntarily, but which can even now be performed by the will, *e.g.* the execution of purposeful co-ordinated motions by particular groups of muscles on stimulation of the appropriate nerves. The mechanism at work in this case can only have arisen in consequence of practice originally influenced by the brain, that is, in consequence of experience.

The reader will be here at once reminded of the well-known experiment with acetic acid upon frogs, which is usually quoted as evidence of the very great importance of reflex action ; when acetic acid is placed upon the skin of a decapitated frog on one side, the nearest limb of that side makes the appropriate motions to wipe away the acid. If this limb be cut off, the frog tries to remove the acid with the limb of the other side. I cannot admit this to be reflex action; it is evidently voluntary action set up by nerve-cells of the spinal cord which possess the power of volition.[1] In Annelids and Arthropods voluntary action is much more clearly exhibited by the posterior part of the ventral nerve-cord, which corresponds to the spinal cord of vertebrates; and other lower vertebrates, *e.g.* tortoises, likewise exhibit indications of voluntary action in the decapitated condition.

On the other hand, the contraction of the fingers, which follows upon stimulation of the nerves of the arm, and which is generally subject to the will, is an instance of reflex action which was originally dependent on the brain. Possibly the involuntary winking of the eyelids is also a member of this class, of reflexes which are fundamentally automatic, but not the respiratory movements, although these can to a certain extent be influenced by the will. A reflex action of this second class is not necessarily adapted to an end, it may even be very inappropriate; in any case, it can only by accident imply intelligent or reasonable motives. But the actions which I call automatic are always purposeful, intelligent, or reasonable.

I have still a lively recollection of the circumstances in which I was first in danger from a shell in the war of 1870-71. We were at the beginning of the campaign before Strassburg in one of the small forts on the right bank of the Rhine at Kehl, and I was standing with some comrades in a spacious room in the fort at a table, occupied at the moment in following on a map according to the latest information the advance of our troops upon Paris, when a shell came through the

---

[1] Compare also Goltz, *Beiträge zur Lehre von den Funktionen der Nervencentra*, Berlin, 1869.

barred window and burst over our heads. I remember perfectly that immediately upon the explosion, which filled the room with gunpowder smoke and the dust of the plaster of the shattered walls and ceiling, I threw my body and arms about in rapid motions, at which I afterwards laughed heartily myself, for these motions were as purposeless as possible—instead of contracting my body within a smaller space, I greatly increased its surface by these reflex motions. Very quickly, however, I learnt in future on similar occasions to act appropriately. Later on, while we lay before the fortress on the left bank of the Rhine on the island of Sporen, when a shell from the Strassburg citadel fell in our neighbourhood, I no longer beat about with my arms, but crouched together into the smallest space without further reflection—by force of habit, automatically. If I had acted thus appropriately on the occasion of the first shell, as by habitual action inherited from my forefathers, I should have acted from instinct. When on a subsequent occasion, it was in the fight at Chateauneuf, a chassepot ball flew so close past my ear that the pressure of the air produced the impression that the ball had struck me, I actually clapped my hand to my ear, evidently in consequence of the inherited habit of grasping any part of the body suddenly threatened, so as to protect it—that is, I acted by instinct.

Moreover, it is well known that in war men soon become so accustomed to shot and shell that they expose themselves with more and more indifference, almost with heedlessness, while the automatic motions of self-protection become rarer and are only exhibited in the most dangerous circumstances. When the first large projectile, a shrapnel-shell, from Strassburg burst over the open space in front of the church of the village of Kehl, and scattered its balls around, I was much astonished to see our colonel continue his walk across the place with folded arms, without looking round, although somewhat later, after some practice and experience, such behaviour seemed to me perfectly natural.

Our soldiers did not even consider it worth while to rise from their beds when a shell came through a loophole into the narrow corridor where they slept, and there burst with a frightful crash, after the bombardment of our little fortification already mentioned had been continued for several nights. But when, after the surrender of Strassburg, it altogether ceased, we slept badly at first on account of the unaccustomed silence, like a miller whose mill has stopped.

These examples show in the clearest way that reflex and automatic action cannot be sharply separated; how easily the latter is acquired by us during life; and further, that it passes directly into instinct, as my view supposes.

But since it is impossible, as will be shown more particularly, to establish a point in the animal scale beyond which a brain can be demonstrated, for voluntary action certainly takes place even where no defined brain is present, and since in lower animals it is very difficult to distinguish voluntary from involuntary action, and nervous action from the results of ordinary physico-chemical reaction, therefore it is difficult to say whether this or that action in the lower animals is to be referred to such reaction, or to reflex action, or to habit, or to instinct.

Thus, according to the preceding considerations, automatic action may be described as habitual voluntary action, instinct as inherited habitual voluntary action, or the capacity for such action.

According to the preceding, instinct cannot be distinguished from some kinds of reflex action, from the reflex action which was originally voluntary, by its mode of origin; and the reflex action originally voluntary, again, cannot be definitely separated from pure reflex action. Thus there is some temptation to describe reflex action also as instinct, and to speak of a reflex instinct, and an instinct of intelligence and of reason. In fact, this course has been adopted by E. v. Hartmann in the criticism[1] which he published, at first anonymously, of his own *Philosophy of the Unconscious.* In the chapter on the " Instincts of the subordinate central organs of the nervous system," he says : " When an isolated and washed frog's heart goes on beating for hours, the cause of this can only be sought in the predisposition of the cardiac ganglia to rhythmical action, which excites the muscular fibres of the heart to contractions in the same rhythm. Such a predisposition in the ganglia, of which the typical active manifestations have as much of the character of spontaneity as the instinctive expression of will in any animal can ever have, must be called instinct as undoubtedly as its functions must be called will, since the unconscious purposefulness of its results is not to be called in question. . . .

" What is true of the movements of the heart of course holds good for the movements of the stomach and intestines, and for the tone of the viscera, the blood-vessels, and muscles in relation to the sympathetic nervous system, as well as for the respiratory movements in relation to the medulla oblongata ; it likewise holds in relation to the cerebellum for those spontaneous movements and actions which birds and mammals execute after the extirpation of the cerebral hemispheres. . . . Here we arrive at once at the chapter of reflex movements, and, in fact, instinct and reflex action cannot be separated, for in instinct also some external motive of action must always be present, and the action follows upon this motive necessarily, therefore, in a reflex manner." And farther on : " As we have seen that all forms of bodily dexterity are acquired, inherited, and when inherited improved by practice, so we must

[1] *Das Unbewusste vom Standpunkt der Physiologie und Descendenztheorie,* Berlin, second edition, 1877, p. 206, *et seq.*

assume the same to be true of all those faculties which, whether they have their seat in the cerebrum or in the lower nerve centres, have in an eminent degree a reflex character, and therefore are described in the narrower sense as reflex movements."

In fact, it seems to be an unavoidable and self-evident consequence of my view also that all reflex action is a property acquired by practice, that it is inherited dexterity produced by exercise. But if reflex action is included under the term instinct, one must logically go so far, as E. v. Hartmann does, as to speak of an instinct of the lower nerve centres, *e.g.* of the sympathetic ganglion cells which govern the movements of the heart and the intestine. Thus V. Hartmann explains " the purposefulness of the reflex instincts of the lower nerve centres" partly as the "outflow or as a *caput mortuum* of former conscious purposeful action of the cerebrum"—as though all purposeful action must have been originally influenced by the latter. Nay, further, he goes so far as to maintain "the essential similarity and continuity of transition between instinct and natural recuperative power," and to express the opinion that it is impossible to establish a different principle of explanation for the two phenomena. "Accordingly, the way in which the type of the whole worm is contained in the ganglion of the worm-ring in process of regeneration is to be described as a molecular ganglionic predisposition established by inheritance." Further, "We can consequently have no scruples in supposing the existence in the ganglia which govern the vegetative sexual functions of predispositions for the regulation of the development of eggs and spermatozoa—the most important and most delicate products in the whole of organic life—predispositions of the same nature as those for the regeneration of lost parts of the body, or for the construction of the comb of the bees, of the web of the spider, or the shell of the Nautilus."

Therewith V. Hartmann evidently partly returns to the mysticism of the *Philosophy of the Unconscious*, which he is endeavouring in his self-criticism to refute. On the other hand, he comes to consequences according to which all development, all growth, would have to be described as instinct.

This example shows that we must limit the idea of instinct, and I believe that this is best done in the way I have adopted, namely, by applying the term only to those inherited habits which are so adapted to a purpose that they appear to be due to intelligence or reason.

V. Hartmann's error of attributing will to the ganglia of the sympathetic nerve system is evidently due to a complete misunderstanding of the functions of the nervous system. Such ganglia are nothing more than the collecting foci of the nerve paths, and at the same time apparatus for the reinforcement of the stimuli (accumulators).[1] That their action also depends on acquired and inherited properties, after the preceding arguments, does not require to be insisted upon; but it has never even been under the influence of the will, still less are they capable of originating spontaneous voluntary action.

In accordance with his view, as above sketched out, V. Hartmann defines instinct as "purposeful action without consciousness of the purpose."

Impulse and Instinct

Although we may rightly talk of the instinct of seeking food in higher animals, it is impossible among the lower to establish a limit above which the taking of food is influenced by a choice due to nervous action. In the same way such a limit cannot be established for the commencement of the sexual instinct; indeed, in my opinion, the commencement of sexual activity depends ultimately on processes of nutrition.

[1] Cf. the next section.

Since all nervous action must have been gradually evolved through relations of the organism to the outer world, the difficulties referred to lie in the nature of the case, and can therefore only be considered as in favour of the justice of the view I advocate. But these difficulties constantly make themselves felt to an uncomfortable degree in the attempt to distinguish between impulse and instinct. Ought we really to describe the sexual impulse as an instinct? Certainly it has as much right to be so called as the migratory impulse, which appears in spring and autumn, even in birds which have been reared and kept all their lives in cages. The migratory impulse is evidently excited by hereditary changes in the circulation which occur at those times, by local blood pressure, which again influences the nervous system, while the latter, on its side, is excited by this change in the condition of the body to the inherited automatic production of a number of acts which are connected with this change. In principle the same is true of the sexual impulse.

And yet impulse is not instinct, as has often been hastily assumed. Impulse is merely the urgent desire to get rid of an unsatisfactory condition of the body, to remove an unpleasant feeling. Instinct finds the appropriate means of effecting this: thus *e.g.* not the sexual impulse, but the purposeful endeavour to satisfy the impulse, is an instinct. But the two things, the impulse, and the endeavour to satisfy it, cannot be separated: thus, to use the case of the migratory impulse again, the physiological condition of the body which produces it, and the expression of the impulse itself, cannot be separated. Similarly, hunger and thirst are so far instincts that they cannot be separated from the instinct of seeking food. Physiology describes both as general feelings: they show themselves as needs felt by the nervous system, caused by an internal condition of the body acting as a stimulus—needs which must so far be called instinctive, that they include so much of inherited experience as enables them to find directly the means for their own satisfaction.

Thus we get back to the difficulty of finding the limit of instinct towards the lower end of the animal scale. The need of nourishment, in other words, the faculty of assimilation, is a fundamental property of protoplasm; upon it depends, in animals provided with nerves, the feeling of the need of taking nourishment, and through inherited experience the instinctive practice of satisfying the need. The question has often been argued, whether a new-born child seizes the mother's nipple itself or must be directed to it. Without doubt the former is the case: the child feels the need of nourishment, it expresses it at once by means of acquired and inherited faculties by crying, by means of the same faculties it makes sucking motions with its mouth, as we also do in the adult condition quite unconsciously, when we have a great desire for food and drink and think of them—motions which in many self-indulgent epicures have produced a quite characteristic form of mouth, with soft prominent lips, seen, *e.g.*, in so many clerical gentlemen who find the world pleasant. These movements are the same which the tongue makes in tasting, they consist principally in the pressing of the tongue against the hard palate. The child feels with its lips the warm soft breast of the mother, and then at once seizes the nipple and sucks, in accordance with its acquired and inherited faculties. Every one who has watched new-born animals when sucking, knows these facts, and knows that the mothers do not need to guide their young to suck.

But although the necessity of nutrition is a fundamental property of the protoplasm, the question arises, at what point does the inception of food cease to be a mere physico-chemical act and begin to depend on nervous action? Where does the feeling of hunger commence? This feeling must be at once connected with the impulse to satisfy it, although at first inherited experience cannot have taught how to satisfy it. Not until the inherited experience of proper method has been acquired does instinct exist—but here again the impulse itself, and the endeavour to satisfy it in accordance with inherited experience are difficult to separate.

I have considered instinct as a faculty which can be called into exercise by external or internal stimuli. These internal stimuli consist in the temporary condition of the body, and show themselves as impulses. The sexual impulse appears first at a definite period of life, as hunger at a definite time of day, and with them the instinct of satisfying them. If the sexual organs are removed, both impulse and instinct cease. Hunger we cannot remove.

Concluding Remarks on Instinct

My experiments on chickens show that the exercise of instincts is determined by definite inherited mental images of things, *e.g.* of the kind of food. But I consider it especially necessary again expressly to point out that it is also in a high degree the general acquired faculty of learning what actions are suitable, of making use of the slightest experience, which enables the animals so quickly to make themselves at home in the world, and not always, however it might appear so at first sight, perfect instinct, *i.e.* the inborn complete capacity for performing certain actions.

The faculty of rapid learning implies in the particular cases of its application a small remnant of independent reflection, which beautifully shows how instinct arises. When this reflection disappears in a particular case, we have pure instinct. If we do not regard its importance, we believe we have nothing but instinct before us. This explains how difficult it often is to decide how much in a particular action is to be ascribed to instinct.

The advantage of the faculty of learning rapidly lies in the abbreviation of the process, whereby time is saved.

In instinct this abbreviation is still more complete, has reached its highest stage—now the organism works intelligently and according to reason in the directions concerned with the same certainty and accuracy as the reflex mechanism from which intelligence and reason have been evolved, but which was only sufficient to meet the most common and most general demands of the outer world.

Darwin gives no special explanation of instinct, he only states his agreement with Peter Huber in the belief that with instinct is mingled a small amount of judgment or intelligence, even in animals which stand at a very low level in the scale of nature.[1] He says it is easy to show that quite distinct mental faculties are usually included under this name. In his posthumous essay on instinct, published by Romanes in his book on mental evolution in animals, Darwin still only gives examples of the faculty, and as in his previous works his object is exclusively to explain its evolution as a weapon in the struggle for existence by the principle of utility. Thus his posthumous manuscript concludes with the words:[2] "It may not perhaps be quite logical, but in any case it is in my view much more satisfactory, that I do not need to regard the young cuckoo who casts its foster-brethren from the nest, the slave-making ants, the larvæ of the Ichneumonidæ which consume the living body of their victims, the cat playing with the mouse, the fish-otter and cormorant with living fishes, as instances of instincts which have been specially bestowed on each animal by the Creator, but that I can consider them as partial expressions of one general law which leads to the progress of all organic beings—the law: Defend yourselves, vary, let the strong live, the weak die."

Romanes describes instinct as reflex action with which is

---

[1] *Origin of Species*, sixth edition, p. 287. [2] *Op. cit.* p. 437.

mingled an element of consciousness.[1] This definition of the term implies of course that he gives the name of instinct to much which in my opinion is not instinct. For further details I refer the reader to the book itself.

Irritability and Sensation—Will

In the preceding I have assumed that a brain is necessary for the exercise of voluntary action, but it will be seen from the following discussion, that I do not demand as the apparatus of voluntary action in multicellular animals a brain morphologically recognisable as an organ, since in the lower multicellular forms and in their larvæ the ectodermic or ectoblastic cells obviously are the instruments of voluntary action. Of course beyond a certain stage of evolution certain definite cells have more and more assumed the function of central government. Where the seat of voluntary action is to be sought in unicelullar forms, is a question which I shall consider in the next section. Suffice it to say at present, that it seems to me necessary even in these to assume a definite organ in the cell as nerve-centre, which stores up the impressions of the rest of the cell and makes use of them as if by will —that the whole cell-plasm together cannot be brain. And in like manner there must be paths in the cell which conduct the external stimuli to the central organ. These paths are probably protoplasmic threads, and on grounds to be subsequently set forth I believe I may claim the nucleus as the central organ. For such nervous substance not yet morphologically recognisable the term neuroplasm or nerve-plasma may be used. But if voluntary action is to be assumed in organisms which possess not even a nucleus, as in Monera,

---

[1] *Op. cit.* p. 169. Cf., on the other hand, Wundt, *Grundzüge der Psychophysik*, Leipzig, 1874, p. 809, *et seq.*; this author advocates a view of instinct which in essentials agrees with mine.

then a certain mass of the plasma might be imagined as the organ of will, since it is certain that the nucleus develops from the ordinary plasma, appears as a "precipitate" therein. However, we shall shortly see that much which appears to be voluntary action in such low organisms is certainly nothing of the kind.

E. v. Hartmann, Haeckel, and others, speak of the will of atoms: in order to surmount the difficulty of the limitation, in other words, of the origin of the voluntary action, they attribute will to all protoplasm and all matter. This assumption implies another, namely, that atoms are endowed with sensation, that sensation and capacity for stimulation cannot be separated (Zöllner). In this way the question of the first appearance of sensation is also got rid of. All organisms possess sensation, plants included, and in the unicellular the protoplasm is endowed with both sensation and will.[1] Nay, even the atoms of inorganic bodies have sensation. "On what," says Haeckel,[2] "finally rests the generally accepted chemical doctrine of the relations of affinity of bodies, unless on the unconscious assumption that the atoms which attract or repel one another are animated with definite inclinations, and that they, following these sensations or impulses, possess also the will and the power to move towards one another and from one another? . . . If the 'will' of man and the higher animals seems free in contrast with the 'fixed' will of the atoms, this is a deception caused by the contrast between the extremely complex voluntary motions of the former and the extremely simple voluntary motions of the latter. The atoms will the same thing everywhere and at all times, because their inclination towards the atom of every other element is . . . unchangeable and defined. . . . On the other hand, the

[1] Botanists, in fact, always speak of the sensation of plants, make no distinction between irritability and sensation.

[2] E. Haeckel, *Die Perigenesis der Plastidule oder die Wellenzeugung der Lebenstheilchen*, Berlin, 1876.

inclination and voluntary motion of the higher organisms seems free and independent, because in the ceaseless circulation of matter in them the atoms are constantly changing their relative positions and manner of combination, and hence the final resultant of the innumerable voluntary motions of the constituent atoms is highly composite and ceaselessly varying.

" While we thus from the mechanical standpoint of monism conceive all matter as animate, every material atom as provided with a constant and eternal atomic soul, we fear not to bring upon ourselves the charge of materialism. For our monistic standpoint is as far removed from one-sided materialism as from empty spiritualism. In this view only can we find a reconciliation of the crude atomic, and the empty dynamic theories of the universe, which have hitherto so violently contended against one another, and both of which in their one-sidedness are dualistic. As the mass of the atom is indestructible and unchangeable, so the atomic soul inseparably connected therewith is eternal and imperishable. Transitory and mortal are only the countless and ever-changing combinations of the atoms, the infinitely manifold modalities in which the atoms unite to form molecules, the molecules to form crystals and plastids, the plastids to form organisms. This monistic conception of atoms is alone in harmony with the great laws of the 'Conservation of Energy' and the 'Indestructibility of Matter' which the natural philosophy of the present day rightly regards as its irremovable foundations."

I would only ask what is materialism if this view of Haeckel's is not—on what grounds does he assert that it is not? To say that the atoms are "animated" is merely to give his doctrine an attractive appearance in the eyes of the opponents of materialism.

Although I am very much at one with Haeckel in giving

a mechanical conception of the idea of will, as follows from the explanation of will given on p. 223, I must strenuously protest against his application of the word. I connect the idea of will exclusively with nervous substance, or with the nerve-plasma which represents it: will is not a property of protoplasm, still less of matter in general, but a property of nervous tissue, in most cases of definite nerve-cells — the will is an acquired and inherited property. The nerve-cells of the brain alone are capable of storing up external influences, *i.e.* the action of stimuli, as experience, in such a way as is required for voluntary action in the sense of the "release of forces existing in a state of tension." The will depends upon movement, as does the whole of mental life, and life in general. Irritability and sensation likewise depend upon motion, but sensation also is a property of nervous substance, or of the nerve-plasma, is not possessed by plants any more than by inorganic bodies; but the latter are also destitute of irritability, which, however, is a property of protoplasm. Inorganic bodies possess only the property of combining together according to regular laws in consequence of attraction and repulsion. Irritability is a fundamental property of protoplasm; capacity for sensation, on the contrary, is a property acquired by it.

Haeckel's conception depends upon a quite arbitrary use of the terms. It arises simply from the desire to find unity in nature, even in the province of mental phenomena; for it appears indeed at first sight dualism to suppose that one class of organisms is endowed with sensation and will, another not. But this demand for unity has no justification —it forgets the fact of the modification of the properties of organic nature by the action of external stimuli, it confounds acquired powers with the fundamental properties of matter.

It is absolutely unnecessary and in contradiction to

important facts, to assume, as the botanists do, that irritability and sensation are equivalent (identical); the very fact that in animals a special nervous system, nerve-fibres and nerve-cells, have arisen, which are absent in plants, proves of itself that the motions which constitute the reception and conduction of stimuli in animals must be quite different to those in plants, for the stimuli are, as we shall see, the causes of the origin of the nervous system. Nervous substance is something peculiar to animals. It is true there are animals which possess the power of voluntary action, and therefore sensation—the latter is a necessary antecedent of the former—which yet possess no morphologically demonstrable nerves, *e.g.* the lowest multicellular and the unicellular forms. But here, as was previously argued, portions of the protoplasm must act, with regard to the stimuli they receive or conduct, as nervous substance—only these parts are not yet externally recognisable as nervous paths. For the single reason, then, that the substance which reacts to stimuli in animals, which in any case in the higher animals is the instrument of sensation and voluntary action, is quite peculiar to animals, it is allowable to infer that the two latter faculties belong to animals only. We might thus in a certain sense recur to the old aphorism of Linnæus: "Lapides crescunt, plantæ crescunt et vivunt, animalia crescunt, vivunt et sentiunt."

In fact, the division of the organic world, according to the evidence of the capacity for receiving "nerve-stimuli" and acting upon them, is still that which should be first established. The Linnæan proposition, interpreted as meaning that all animals, in contrast to plants, are capable of sensation and the exercise of will, would therefore certainly be more justified than the doctrine of the will and sensation of atoms. But I attribute to animals merely the special faculty of reacting to nerve stimuli. This does not imply that they all are capable of sensation and will. It is an evident fact, and one that corresponds with the whole evolution of the nervous system, that multicellular animals act more and more by reflex action the lower their position. Voluntary action must in its origin be traced back to reflex action. It is a question, indeed, whether the movements of the lower multicellular animals, like the sponges, are not exclusively reflex.[1] The same holds for the unicellular, in which paths in the protoplasm must supply the place of nerves.

But it matters not whether we explain the power of sensation as a property of all animals as distinguished from plants, or as a property which first appears in the animal kingdom—it is even on the first supposition not a fundamental property of protoplasm, but something which has been gradually acquired.

But how is it to be explained? If we regard all the effects of stimulation as motion, we are naturally led to explain sensation as a special quality of this motion, a quality indeed which resides in certain nuclei, in the higher animals in the nuclei of the ganglionic cells, an "excitation" the manifestation of which constitutes consciousness, the general sensation of the whole organism.

The fundamental property of protoplasm is (I repeat it), not sensation and voluntary action, but irritability (response to stimulation). From that property are evolved by acquisition and inheritance the irritability of nerves in animals, and the power of sensation and will.

I leave it, therefore, undecided whether the latter faculties belong to all animals. But in any case, a number of the phenomena of movement in lower organisms, like those of spermatozoa, which seem to be voluntary, are, according to my investigations, nothing but the result of definitely-directed

[1] Although the actions of the free-swimming larvæ of these animals might be properly ascribed to volition, yet the whole activity of the adult sponge might be reflex, as a consequence of degeneration of the nervous faculties, which occurs often to such an extraordinary extent in other animals.

movements of the protoplasm, which are themselves the result of the reaction, which has become regular, of the protoplasm to stimuli. Such, for instance, are obviously the movements of the cilia of ciliated cells, so far as these are not affected by nerves. The forward movement of the spermatozoa of the toad and of the salamander is caused by rhythmical streamings of the protoplasm succeeding one another like waves, and I have shown that this rhythmical motion occasionally passes into the amœboid,[1] so that we have some ground for regarding the latter also as involuntary, although it appears to be voluntary. In fact, I cannot accommodate myself to the idea that the amœboid cells which wander about like amœbæ in our blood-vessels and throughout our bodies are independent beings endowed with sensation and will.

Botanists have made more investigations than zoologists into the stimuli which cause locomotion in the lower organisms, or which accelerate or give a definite direction to their movements, movements which at first sight appear to be altogether voluntary. I will quote here an instance of the first case; another of no less importance is afforded by the experiments of Pfeffer, discussed on p. 334.

The movement of Myxomycetes is influenced by [2]:—

1. Moisture (Hydrotropism): In their young stages they wander from the parts of the substratum (*i.e.* of the deposit on which they are creeping), which are gradually drying up, towards those which continue moist longer; "it is even possible, by bringing moist bodies into the proximity of any ramifications, to cause the production of pseudopodia, which elevate themselves from the substratum, and soon come into

---

[1] Th. Eimer, *Unters. über d. Bau v. d. Bewegung der Samenfäden, op. cit.* Accordingly the ultimate cause of the motion of the spermatozoa of the Mammalia also must lie in amœboid movements in constant directions. With regard to ciliated cells compare the following.

[2] E. Stahl, *Zur Biologie der Myxomyceten, Botan. Zeitung,* 1884, No. 10-12. Abstract in *Sitzungsbericht der Jenaischen Gesellschaft für Medizin und Naturwissensch.* 1883, Sitzung vom 16th November.

contact with the moist object, so as to enable the whole mass of the plasmodium to migrate on to it." On the entrance of the plasmodia into the fructifying condition, positive hydrotropism gives place to negative; the myxomycete quits the moist substratum and creeps upwards on to the surface of dry objects.

2. Unequal distribution of warmth in the substratum, and

3. Unequal supplies of oxygen also cause locomotion in the Myxomycete.

4. Chemical substances soluble in water have a similar action. Contact of the plasmodia on one side with solutions of common salt, saltpetre, carbonate of potash, cause them to withdraw from the dangerous spot, while infusion of tan, or a dilute solution of sugar, produces a flow of the protoplasm and ultimately translocation of the whole plasmodial mass towards the source of nourishment. Some solutions have an attractive or repulsive effect according to their degree of concentration.

5. Finally, they withdraw from light (negative heliotropism).

With regard to the acceleration or definite direction of movement produced entirely by stimuli, compare the following:—

"The knowledge of the remarkably delicate reaction of the plasmodia under external influences enables us to comprehend how these tender structures, destitute of every kind of external protection, are able to carry on their existence. The plasmodia which are not yet ripe for reproduction are kept in the moist substratum by positive hydrotropism, which is assisted by negative heliotropism.

"But within the darkness and moisture of the substratum the plasmodia by no means remain in one place, because the differences in the chemical composition of the substratum cause continual migrations. The plasmodia have the faculty in a wonderful way of avoiding harmful substances and, traversing their substratum in all directions, of taking up the materials they require.

"When the internal changes have proceeded so far that the plasmodia approach the fructifying condition, they are brought by the negative hydrotropism which now sets in from the moist parts of the ground in the forest or wood to the surface, where they creep up various upright objects, often only forming rigid reproductive capsules at some height from the ground.

"When in autumn the substratum becomes gradually colder, a change which takes place from the surface downwards, the plasmodia migrate into deeper regions still having a higher temperature. When the cooling proceeds very gradually, which especially happens in large tan-heaps, the plasmodia may in their migration reach somewhat considerable depths, where they then change into sclerotia. To find the sclerotia of Æthalium in winter it is therefore not seldom necessary to search through the mass of tan to a depth of several feet. When the temperature again begins to rise, the sclerotia again germinate, and movement in the opposite direction takes place from the deeper and cooler parts to the upper portions already warmed."

In the locomotion of the Myxomycetes, then, we see extremely interesting cases of movements due to stimulation. Heliotropism, geotropism, hydrotropism, trophotropism, in general, are stimulus-movements, and ultimately all growth depends on stimulus-movement. It is the most primitive kind of protoplasmic movement. Stimuli in fixed directions and constantly repeated produced, but only secondarily, fixed paths of conduction, and responses of a quite definite kind (reflexes). Thus arose nerves and finally apparatus for storing up stimuli, arose sensation and will—as acquired and inherited faculties.

SECTION VII

ORGANIC GROWTH : THE MORPHOLOGICAL AND PHYSIOLOGICAL EVOLUTION OF THE LIVING WORLD AS THE RESULT OF FUNCTION

IN Section IV., under the heading "Characters acquired by Use," I have already indicated in a general way the influence of use in the modification of organic forms. I have now to show in detail that all organisation, and above all, the first development of organs, and further, all higher physiological evolution depends on use, is to be traced to the inheritance of acquired characters.

THE ORIGIN OF ORGANISATION IN UNICELLULAR ANIMALS —THE FUNDAMENTAL BIOLOGICAL LAW

Whoever assumes that complexity of organisation has arisen in consequence of the gradual evolution of the living world must also assume that all living beings, both plants and animals, have been derived from the simplest beginnings, from unorganised particles of protoplasm; in other words, from organisms which were destitute of organs.

The term organism was first employed by Aristotle, and was based upon the fact that each part of a living being is an instrument, ὄργανον, for the whole. In his time organisms without organs were yet unknown; they were first discovered by the aid of the microscope. They are represented by the simple minute morsels of protoplasm to which Haeckel has given the name of Monera (μονήρης simple). These are cells possessing neither nucleus nor investing membrane, which nevertheless lead an independent existence, and of some of which the whole life cycle is known. In these beings organs only appear at the moment they are needed, and at any part of the body, they are not developed beforehand, and have no definite shape. These organs are processes of the protoplasm of variable form, "false feet," or pseudopodia, by means of which the body crawls about or takes in food. When the cell returns to a condition of rest the pseudopodia flow back again, are withdrawn and disappear, while the body assumes a spherical shape. The conception of an organ as a part of the body constructed for a definite function is therefore not applicable to these pseudopodia, it applies only to the permanent instruments of the more complex beings which alone are in the full sense of the word organisms.

In the creatures next above these in the animal scale, in the Amœbæ, a permanent organ first appears, namely, the cell-nucleus. Besides this there are also present the so-called contractile vacuoles, which, however, in some cases appear now in one place, now in another, and then again disappear, and whose claim to be considered permanent organs is therefore doubtful. Locomotion and the inception of food are still effected by pseudopodia. In the next stage in the series of unicellular animals a membrane appears as a permanent organ, while cilia take the place of the pseudopodia, to which are necessarily added a mouth for taking in food, and in many cases an anus. The body is still in most cases capable of changing its form, and owes this power sometimes to the differentiation and modification of the outermost layer of protoplasm lying immediately beneath the membrane, this layer becoming striated or fibrillar. The fibrils are always arranged in the direction which enables them most easily to produce the usual contraction of the body, viz. perpendicularly to the plane of this contraction. In the stalk of the bell animalculæ (Vorticellidæ) the fibre to which the contraction of the stalk is due, according to the researches of Kühne, has exactly the same physiological properties as the muscular substance of multicellular animals.[1]

Thus these ciliated Infusoria are cells with a very delicate organisation, which must have been gradually developed in the course of ancestral evolution, and this certainly cannot have taken place as a result of the variation of their germ-cells, for the simple reason that they do not possess any. It must have taken place in consequence of acquirements due to use, and in consequence of the inheritance of such acquirements.

It is most clearly shown in the appearance of a specially differentiated layer of contractile substance beneath the surface of the body, and in the appearance in the stalk of the Vorticellidae of muscle substance equivalent to that of the higher animals, that activity or function calls forth organic or physiological differentiation.

This important proposition, which I might call the elementary fundamental law of all biological science, implies the most complete opposition to the denial of the inheritance of acquired characters, affirms the opposite.

Protoplasm has the property of being altered and transformed by the action of external stimuli.

Among external stimuli are to be understood both those acting directly and those which act indirectly by affecting the functions of the organism.

---

[1] "The stalk of the Vorticellidæ behaves exactly like frog's muscle: even when isolated from the rest of the animal it can be made to contract, and even be tetanised, by the stimulus of variations of an electric current," etc. Cf. W. Kühne, *Myologische Untersuchungen*, Leipzig, 1860, p. 216 or 213.

The evidence for this fundamental law lies in the following facts : (1) that numerous animals, although they possess no organs structurally adapted thereto, regularly perform actions which pre-suppose corresponding powers in the protoplasm, or even imply that the evolution of such organs has already begun ; (2) that various organs at first serve for many purposes, and that division of labour, *i.e.* the increase in the number of organs and their specific differentiation appears only by degrees in the series of organisms; (3) that the larvæ of even the most highly organised animals only possess powers corresponding to those described under 1 and 2,—that, therefore, in this matter also the biogenetic law completely holds good.

The two first of the above propositions are sufficiently proved by unicellular animals alone when their activities are compared with those of the multicellular.

The second proposition is further supported by the general description of the evolution of organisation given above ; the evidence required for the third can only be afforded by multicellular animals.

The striated layer in the Infusoria already mentioned is also evidence of the truth of the first proposition. The ciliated Infusoria which possess that layer exhibit the commencement of the evolution of a muscular layer, which commencement can only be ascribed to the constantly repeated contraction in a given direction of that part of the body. And in the stalk of the Vorticellidæ, by the continuance of the same activity, a muscular fibre has arisen, which behaves physiologically in just the same way as the muscles of multicellular animals.

The actions of the ciliated Infusoria in relation to the outer world are such that will must certainly be ascribed to them. This is proved by the simple observation of the character of their movements, apart from all else. Their movements are entirely spontaneous, and the variety of these is due to their power of moving all or different sets of cilia more slowly or more quickly, or keeping them at rest. The hypotrichous Infusoria,[1] *e.g.* Euplotes Charon, at one moment shoot through the water with their cilia in rapid motion, at another run about on the bottom on algæ or other objects, using their cilia as legs, and moving like Isopods. The behaviour of such Infusoria towards one another, according to some observers (Engelmann), looks as if they were running after one another in play.

Yet in these Infusoria, endowed with mental faculties, we can discover no nerves and no brain. It seems to me probable, from comparison with the relations of the cells of multicellular animals, that their nuclei, while at the same time connected with reproduction, or one of them, also acts as the central organ of the nervous system. This idea is supported by the fact recently demonstrated, that when an Infusorian is cut into two only one part grows again to a complete animal, namely the part which contains the nucleus, or at least a piece of it.[2] The same holds good for snails in regard to the brain. If from a garden-slug (Limax agrestis) the head with the brain is cut off, it no longer grows again into a complete animal, as it does after mutilations in which the brain has not been removed.[3] A. Gruber, on the contrary, believes that

[1] Those provided with cilia on the under surface only.
[2] A. Gruber, *Beit. zur. Kenntn. d. Physiol. u. Biologie der Protozoën. Berichte d. naturf. Ges. zu Freiburg i.B.* 1886 ; M. Nussbaum, *Sitzungsber. d. niederrhein. Ges. f. Nat. u. Heilkunde in Bonn*, 1884, p. 259, *seq.* Fr. Schmitz and J. v. Hanstein found the same was true of the division of plant-cells. Iu this connection, in my opinion, the general importance of the nucleus as the organ of life comes into consideration. Cf. the following.
[3] The reproduction of the head of snails was shown to occur by Spallanzani as early as the year 1764. The Rev. Schaeffer was astonished to observe that the snails in his garden whose heads he had cut off with the shears in order to destroy them again became perfect (Schæffer, *Versuche mit Schnecken*, 1768-70). According to Tarenne (*Cochliopérie, recueil d'expériences sur les Hélices terrestres*, 1808), the brain and the buccal nerve mass also grow again. But it seems certain that the retention of the œsophageal nerve-ring is indispensable to the continua-

at least in the higher protozoa, in the ciliata, the seat of nervous activity is to be sought principally in the outer layer of protoplasm, and that the will resides in every protoplasmic element, that nervous action is not limited to definite paths. He supports his view by the fact that the movements in the two parts of an Infusorian in process of division exactly correspond so long as they are connected by a strand of protoplasm, however thin (Stentor). The same thing is observed also after incomplete artificial division. This proves, he says, "that the nervous power of the cell is diffused." In the figure given in illustration a Stentor is represented diagramatically as divided transversely, that is, so that the nucleus also is transversely divided. Whether pieces without nucleus act voluntarily is not stated. But it would be necessary to determine this. My experiments in dividing Medusæ, to be shortly discussed, afford a perfectly analogous case, and yet localised cerebral ganglia are certainly present there. If, as I shall endeavour to show, nuclei actually form central nervous organs in multicellular animals, it is evidently highly probable that they do so also in the unicellular. In the cases referred to of multicellular animals in which a circumscribed brain is not yet present, the exertion of will is only to be conceived as proceeding from circumscribed central points (nuclei) which are in connection with one another, and to which the impressions received by the rest of the cell are conveyed. Since all the action of will depends on the employment of impressions which must be conveyed to the organ of will along separate protoplasmic conducting paths, therefore the organ of will can scarcely be the whole protoplasm of the body, but rather the very presence of a will implies in itself

tion of life in snails. Dugès and Moquin Tandon obtained this result (cf. Milne-Edwards, *Leçons sur la Physiologie et l'Anatomie Comparée*, Tome viii. p. 304), and recently J. Carrière has confirmed it (*Studien über die Regenerationserscheinungen der Wirbellosen*, I. *Regeneration bei Pulmonaten*, Würzburg, 1880.

division of labour in the province of nervous life, and it is impossible to see why the nucleus should not be considered as central nervous organ in unicellular forms, if it can be shown that it plays this part in the multicellular.

However, the question can only be decided in the future. Possibly special fibrils for the conduction of nerve stimuli will yet be demonstrated by more exact researches directed to this point,[1] and will at the same time lead us to the central organ. In any case, nerves like those of the multicellular forms are not present, and yet the animals evidently receive stimuli by means of the cilia which are connected with the protoplasm, which stimuli must be conducted along definite paths to the central organ. These paths can be nothing but strands of protoplasm ; probably they are formed,

[1] Cf. Engelmann, *Zur Anatomie und Physiologie der Flimmerzellen*, *Pflüger's Archiv*, Bd. xxii. 1880, p. 505. This investigator believes he has seen such fibrils in Stylonychia passing into cilia which vibrate at different rates. I may also remark that I have repeatedly found the ciliated cells of the surface of the mantle and gills in Anodon to be most distinctly connected with nerves which obviously fulfil the same purpose. After treatment with bichromate of potash the cells can easily be isolated still in connection with the nerves. I have hitherto failed to see a connection between the nerves and the fine protoplasmic threads which I was able in these cells to trace downwards below the nucleus, and which are continued above into the cilia, but such a connection is very probable (cf. Eimer: *Weitere Nachrichten über den Bau des Zellkerns, nebst Bemerkungen über Wimperepithelien, Archiv für mikroskop. Anatomie*, Bd. xiv. 1877). I consider these threads as nervous paths, as nerve-plasma, and the knowledge of these was the basis of my view that in the Infusoria also the protoplasmic threads form nervous paths. From the analogy of the relations of the fibrils from the cilia in the ectoderm-cells of Medusæ which I have described (cf. *Die Medusen*, Taf. iv.), it may be assumed that these threads in Infusoria also pass into the nucleus. On the other hand, that such threads pass into nerve-fibrils, I have shown in the ectoderm-cells of Carmarina hastata (Cf. *Medusen*, Taf. xii. fig. 8, 19 ; Taf. xi. etc.) These cells appear distinctly striated longitudinally ; their protoplasm is clearly transformed into nerve-plasma, and the fibrils representing the latter, diverging from one another below, are continued directly into the nerve-fibrils of the ring-nerve (" brush cells "). The fact that each of these ectoderm-cells is thus in connection with numerous nerve-fibrils, would seem to show that they still act as central nerve-cells, although ganglionic nodules have already been developed beneath the epithelium (cf. below : the first development of the central nervous system, especially Hydra).

as already remarked, by fibrils of protoplasm having the ordinary appearance.

We shall see that even in the lower multicellular animals the conduction of nerve stimuli takes place without the presence of actual nerves. I have only touched upon the much less definitely understood relations in the ciliated Infusoria in order to indicate that there is a parallelism between these and the important facts to be shortly discussed, and to lead up to the latter.

In the preceding I have principally referred to muscles and nerves as evidence in support of my propositions, and shall follow the same course hereafter, because these organs of so-called animal life are specially suited to bring into view the effect of the relations of the organism to the outer world.

The essential distinction between the formation of organs in the unicellular and multicellular forms is that in the former the organs appear as parts of the cell, which is itself the animal, while in the latter each organ is formed of numerous cells. For instance, in the unicellular Vorticella a part of the cell forms a muscle-fibre, in the multicellular many cells form a muscle, which is physiologically equivalent to that muscle-fibre. Morphologically, therefore, the two can be no more compared than any other organs in the two divisions of animals. But that they, like other organs in the two groups, are physiologically equivalent, that in each case, from a quite different starting-point, an apparatus acting in a similar way in relation to the outer world has been produced, this alone warrants the conclusion that it was this very relation to the outer world which modified the protoplasm in both cases into an equivalent though not homologous structure, that the inheritance of acquired characters is here exemplified.

The Evolution of Organisation in Multicellular Animals

The Evolution of the Germinal Layers as Primitive Organs

Multicellular animals must have arisen from colonies, formed in consequence of the advantage of association, of unicellular forms still at a low grade of development, of protozoa which had yet developed no kind of permanent organs except, perhaps, the nucleus. The way in which these colonies were formed was, probably, that the new protozoa produced by division did not separate, but remained connected until the colony had a sufficient number of members. But a solid sphere (Morula) could not have been the most suitable arrangement of those members, although they may at first have assumed that arrangement, as some do still; the best arrangement must have been that which enabled all the members to exercise their powers for the benefit of the whole and the parts, and as these powers were the same in each, this condition could only be fulfilled when all the cells were equally exposed at the surface, so that they formed a single layer constituting the wall of a spherical vesicle containing a central cavity. Thus all the individuals were in the best position to obtain equal information as to the conditions of the outer world, and for common defence, as in our square of infantry bristling with bayonets. There are also living forms still showing this arrangement of the "blastula," *e.g.* the colonial Volvocidæ, consisting of unicellular flagellate Algæ, each of which directs its flagellum outwards, as the soldier his bayonet. Before long, to effect locomotion in a definite direction, certain more powerful cells must have gained a superiority over the rest and assumed the leadership, so that now the several cells co-ordinated the movements of their processes, and these became synchronous on all sides, and thereby the locomotion of the whole forwards or backwards took place.

Those cells which were situated at the pole of the blastula must have specially developed their power of food-inception, because they had most opportunities of exercising it: they came first and most frequently into contact with food. As several neighbouring cells thus devoted themselves more and more exclusively to food-inception, the first step was taken, functionally, in the evolution of the first common organ, of the digestive endoderm, and also in the formation of an invagination of one pole of the blastula, which gave rise to a second common organ, the primitive gut. Every increase in the degree of this invagination was an advantage to the nutrition of the whole, for more food could be taken up in it. In this respect, therefore, selection may have been actively at work. Thus arose the gastrula, an organism of the form of an egg-cup, with a wall of two layers of cells lying one over the other, of which the inner is derived from the invagination of an original single layer.

With this invagination we get at once two chief or primitive organs of the multicellular animals, the ectoderm and endoderm. While the latter serves only for the reception of food, the former has functions of three kinds :—

(1) The protection of the body on the outside, as its covering; (2) to act as the organ of relation, instead of a nervous system, to place the body in communication with the outer world; (3) locomotion.

With regard to the latter, it is to be observed that it was probably a definitely-directed activity, namely, locomotion in a definite direction, which led to the development of permanent flagella on the cells in place of the irregular, variable pseudopodia. And this modification must have taken place before the completion of the endoderm, for the latter itself still bears flagellate cells, while at the same time it can also send forth pseudopodia to take up food; but its flagella are retractile, as all flagella, being derived from pseudopodia, must have been originally, and as in many other cases they are still, *e.g.* in the Flagellata.

Again, it must have been in consequence of function (use) that the endoderm cells gradually came to differ from those of the ectoderm in size and internal structure, and when the gastrula developed into a fixed sessile animal, disuse caused the disappearance of the flagella.

As there are now living forms which retain the morula and blastula form, so there are others which exhibit the gastrula form, although the gastrula has undergone various modifications. Hydra is such an animal.

In Hydra it is also evident how a third primitive organ of animals arose, namely, the mesoderm.

In Hydra and many allied forms the ectoderm cells consist of two parts, an external, which receives stimuli, and an internal, which is muscular (neuro-muscular cells). This differentiation also cannot possibly have arisen except as a result of function, of activity. In other animals the separate layer of supporting and muscular cells forming the mesoderm arose by growth and multiplication towards the interior, by the separation of special cells from the ectoderm and endoderm.

It must be described as a special peculiarity belonging to the organisation of a portion of the mesoderm cells, that there "crystallises" in them calcareous or siliceous or horny matter, the first especially in perfectly definite, and often extremely complex, shapes.

The formation of the supporting skeleton of these lower animals, therefore, has probably not been determined by use or function, but rather by definite laws of development, and selection; but this applies, not so much to the origin of definitely shaped calcareous or siliceous spicula, as to the solidity of the skeleton in general.

Origin of Muscles

A striking proof of the importance of activity, function, in the evolution of the organism is afforded by the mesoderm in the muscles of that layer.

In unicellular forms and the larvæ of the multicellular free and rapid movement in the water is effected by flagella. Muscles evidently first arise in consequence of ever-repeated changes of form, of the bending, extension, and contraction of the body, by which it is diminished in order to avoid danger, or enlarged in order to seize food, or moved in various ways in fighting and in locomotion.

It is well known to histologists how difficult it sometimes is to distinguish smooth muscle-cells from certain connective-tissue-cells. Both arise in the mesoderm, from the same fundamental layer. I showed in a particular case years ago that, in fact, connective-tissue and muscle pass directly one into the other, the latter developing first in those directions of the body in which contraction takes place, while in the rest only connective tissue arises. This is the case among the Ctenophora in Beroë ovatus, where the gelatinous tissue of the body is traversed by connective-tissue-fibres and very simple muscle-fibres, which pass into one another so gradually that a boundary between the two cannot, in spite of all endeavour, be determined. The muscle-fibres consist of long tubes of contractile substance, which in the fresh condition seem homogeneous, but can be separated into longitudinal fibrils by reagents, and which contain unorganised protoplasm containing nuclei at intervals. These fibres are surrounded by a sarcolemma. "The typical connective-tissue-fibres are cylindrical, strongly refracting, delicate threads running a straight or contorted course, and are usually enlarged to a spindle-shape at places by nuclei, which, however, lie at very long intervals from each other. In the coarser fibres a clear streak can be seen in the centre, which in many cases appears as a canal, so that then the whole fibre forms a tube." Between the typical connective-tissue- and muscle-fibres there occur "forms perfectly intermediate, and there is no means of defining a distinction between the fibres which contain the smallest proportion of contractile substance and those which contain none at all, for even staining with carmine supplies no criterion when only a small quantity of this substance is present. The first unmistakable trace of contractile substance appears in the fibres as a doubly refracting layer on the inner wall of the central cavity, surrounding the axis of the fibre as a delicate envelope.

"The connective-tissue-fibres traverse the gelatinous substance of the animal from within outwards, as well as from above downwards; very often, however, in a direction at right angles to that of the muscle-fibres.

" . . . In the lower Medusæ, and also in the Acraspedota, the gelatinous supporting tissue of the body is not yet traversed by muscle-fibres, but only by connective-tissue-fibres. It seems here not to be itself contractile, or only by the aid of the musculature lying upon it. In the Ctenophora the gelatinous tissue is abundantly traversed by muscle-fibres passing from wall to wall, and it can therefore be contracted in itself.

"Here, from the cell layer which produced both connective-tissue- and muscle-fibres, the latter have developed in those directions in which the body made the greatest efforts to contract. In those directions in which such effort was made not at all, or only slightly, the evolution went no farther than the formation of connective-tissue-fibres, or at most there arose muscle-fibres which contain but a small proportion of contractile substance.

"In a direction at right angles to that in which, at a given part of the body, contraction exclusively occurs, the contractile substance will be least developed, and hence we shall find the most typical conuective-tissue-fibres at right angles to the muscle-fibres."[1]

When I speak of an effort to contract, I do not of course intend to support Lamarck's view, especially not in the sense that anything can be produced by the "will."

Contractility is one of the fundamental properties of protoplasm, and by excessive exercise of this property, according to my explanation, the characteristic muscular substance is produced from the protoplasm. That effort is exerted in this process is self-evident, but this produces an effect not directly, but only indirectly.

Only thus can be explained the fact that musculature is everywhere most developed in those parts of animals in which its presence is most necessary to their requirements, *e.g.* in Medusæ, which move by the contraction of the bell, on the under side of the bell; in worms, whose locomotion is due to contortions and creeping and twisting, in the "dermo-muscular tube"; in snails, in the foot; in bivalves, in the adductor muscles, etc.

How could selection, or even sexual mixture produce muscles at particular parts of the body where previously none were present? Selection and mixture can only deal with what is in existence.

But the muscles of multicellular animals afford in yet another direction evidence in support of my view of the modification of organisation by functional activity.

Origin of Striation in Muscles

The text-books of histology still continue to divide muscles into the smooth and the transversely striated. On

---

[1] *Vid. Beroë ovatus*, 1873, *op. cit.* p. 30, *seq.* Similar results have been obtained by my friend Flemming with regard to the connective-tissue- and muscle-cells in the urinary bladder of Salamandra maculata. He says: "There are per-

the whole the former means the muscle-cells, the latter the muscle-fibres. But from the comparative standpoint this division is quite unjustifiable, for there are transversely striated muscle-cells as well as striated muscle-fibres, and there are also unstriated, smooth muscle-fibres (Ctenophora). Moreover, every one who has carefully pursued comparative studies of muscles will agree with me that the transverse striation arises gradually, so that transitions occur between striated and unstriated musculature. Thus the distinction between muscle-cells and muscle-fibres can only be admitted as in accordance with fact, if it is understood that the former are uninucleate, the latter multinucleate, since there are simple uninucleate muscle-cells which are so elongated that they might equally well be called fibres in the ordinary sense of the word.

On the other hand, recently histologists speak more than formerly of the "muscle-prisms" (*muskelkästchen*) as the fundamental element of the striated muscle-fibre. But the fibril is rather to be considered as the fundamental element, and comparative observation leads necessarily to the conclusion that the subdivision of this fibril into single successive rods, which seems to exist in the most highly developed condition of striation (the fact is still not absolutely decided), can only be something acquired at a late stage of evolution, and that at earlier phylogenetic stages it does not occur at all.

Thus I have already in my work on the *Medusæ* advanced the proposition, and endeavoured to prove it, that the striation of muscle is produced through its own activity, is therefore an acquired and inherited character.

The evidence for this lies in the following: (1) Striated

manent intermediate stages of form between uninucleate muscle-cells and connective-tissue-cells, and therefore no sharp distinction between the two."—*Zeitschr. f. wiss. Zoologie*, Bd. xxx. Supplement, 1878, p. 473.

muscle only occurs, but also always occurs, where very active muscular contractions take place. Thus the visceral musculature of Vertebrata is everywhere composed of smooth muscles, with the exception of the heart, which is composed of striated muscle-cells—the former, therefore, occur at every part of the viscera where movement is slow, the latter alone where it is extremely active. A no less significant fact is the following : leaving aside the smooth muscle-cells which occur in the mesoderm of sponges, the neuromuscular cells form the commencement of the development of the musculature in the zoophytes. But in spite of their primitive character the muscular elements of these cells, in cases in which they produce vigorous and rapid motion, are transversely striated, while in other cases they are smooth. They are smooth in Hydra, but striated on the lower surface of the bell in the Medusæ, which, in order to effect locomotion, respiration, and to obtain food, is in constant motion.

The inactive Mollusca have generally smooth muscle-cells—striated muscle occurs only in one position, namely, in the adductors of those Lamellibranchs which can swim swiftly in the water by the rapid flapping of their valves, as butterflies fly by the flapping of their wings, *e.g.* Pecten. But in relation to this question the adductors of Anodon are still more worthy of attention : in these I find the first traces of striation, but it is not yet a permanent morphological character ; it appears only temporarily and only at separate parts of the fibre, merely as the outward sign of activity.

It is usually stated that the Arthropods have striated muscle-fibres only. It is true that in these actively-moving animals the striation is especially well marked, and it is in them that the subdivision of the fibril into "prisms" has been described, this supposed constitution having been afterwards attributed to all muscle-fibres as though the "prisms" were the elementary parts of the fibrils everywhere.

Even the intestinal musculature of the Arthropods is transversely striated; and the intestine obviously in Crustacea as well as in Insecta has in a pre-eminent degree a respiratory function, and on account of this, as well as in many cases in consequence of the ceaseless action of the digestive machinery, it is in constant movement.

But if we examine the muscles of the thorax of flies in the beginning of spring when they have just been revived by the warmth, and have not yet flown, we find that the fibrils are almost all smooth without any striation, an observation which has been for years regularly repeated in my histological demonstrations. This spring in these classes we made the experiment of repeatedly shaking up the flies so as to make them move and fly, and thought, even after this slight motion of the wings, we could notice that a much greater number of the fibrils were striated than before. When we examined the thoracic muscles of flies this summer we found all the fibrils most perfectly striated.

Evolution of the Various Sense-Cells from a Common Cell Layer

The history of the nervous system proves if possible still more conclusively than the origin and development of muscle how completely the origin and elaboration of organs are governed by external stimuli and by exercise.

It is actually required by my theory that the nervous system should, as it in fact does, proceed from the epiblast. For the nervous system consists of the organs of relation, that is, the organs by which the body is placed in relation to the environment, by the reception of stimuli and the excitement of a response to them. If, as we must assume, not only according to the biogenetic law, on account of all the facts of embryogeny, but also on account of the structure of the lower multicellular animals now living, all multicellular animals have been derived from forms whose only organs were two or three germ layers, then the external germ layer, the ectoderm, must naturally have originally been the medium of those relations to the external world. And in these multicellular forms which still consist only of these "primitive organs" the ectoderm must be still the medium of relation. This most external of the three layers of cells of the larva or of the body came, and comes, first into contact with the outer world. All external stimuli acted first upon it. By the repeated incidence of the stimuli this layer must have become more and more fitted for their reception, and subsequently for dealing with them, must have had these capacities exercised and strengthened. Only thus can it be explained that the nervous system, that brain and ganglia wherever they can be distinguished, arise from the epiblast.

Again, the most important parts of the nervous system, the brain and higher sense organs, the organs of sight and hearing, and also of taste and smell, are always situated in those parts of the body which come most into contact with the outer world, which are so placed as to be oftenest affected by stimuli—in worms, molluscs, arthropods, and vertebrates at the anterior end, in Medusæ on the edge of the bell and so on.

The sense organs are not only derived from the epiblast, but usually remain situated in the ectoderm. The sensory cells moreover are developed from ordinary ectoderm or epiblast cells.

At first only simple epiblast or epidermis-cells were present everywhere; from these touch-cells were developed. All the higher sense-cells have been developed from touch-cells— this is irrefutably proved by comparative anatomy.

The origin of the various sense-organs from the epidermis-cells is, as I have shown, seen in a wonderfully beautiful way in the sense-organs of the toponeurous Medusæ, *e.g.* of Aurelia aurita.[1] Here at particular spots in the margin of the bell, on the so-called marginal bodies and in their neighbourhood, lie in close proximity, visual, auditory, and probably olfactory organs, as well as tactile organs, formed by groups of epidermic-cells specifically modified. A pavement epithelium covers the greater part of the surface of the bell. The cells of this epithelium at the edge of the marginal bodies change first into flagellate cylindrical cells, which also line certain depressions of the surface near the marginal bodies, and here probably are subservient to smell or taste, but also, like the ordinary flat epithelium, are doubtless tactile. Such flagellate cells are modified here and there into a kind of visual rod, and, the cells surrounding these acquiring pigment, the eye of the jelly-fish is formed. Close to these eyes on the clavate extremity of the marginal body the epithelium again becomes flat, and a vesicle containing crystals being developed beneath it, forms the auditory epithelium.

The history of the sensory-cells, their origin from one and the same primitive form, even if nothing else did, would indicate that all sensory stimuli may be but different qualities of one and the same stimulus, different forms of motion of external media—even the stimuli of taste and smell.

The coarsest of these stimuli is that of touch. To this the epidermis-cells were in any case susceptible from the beginning, and the protoplasm of even the lowest protozoan is also affected by it. Whether or how far this is also true of the other stimuli is not yet established. The remarkable

[1] Cf. *Die Medusen*, and my address to the Congress of Naturalists at Munich in 1877: "On Artificial Divisibility and the Nervous System of the Medusæ," published in the Proceedings of the Congress, also printed with additions in the *Arch. f. mik. Anat.* Bd. xv. 1877.

facts brought to light by Pfeffer open a far-stretching vista of possibilities with regard to the effect of stimuli on the lower organisms. Pfeffer has shown that chemical stimuli affect certain vegetable spermatozoa; that, for instance, they are attracted in water by malic acid, and that they collect in fine tubes filled with that substance, as though impelled by a secret force, so that they can be captured in such tubes.[1] I have already referred to a similar remarkable sensibility to stimuli on the part of protoplasm, at p. 312.

It seems therefore certain that the sensitiveness of protoplasm to mechanical stimuli, like that of contact, and to chemical stimuli, is one of its fundamental properties. Since the other stimuli, termed sense stimuli, are merely more delicate kinds of one or other of these, the question arises whether protoplasm is from the first sensitive to these, although only in a very slight degree, or whether it has only become sensitive to them by practice under the coarser stimuli—has been educated to perceive them. With regard

---

[1] W. Pfeffer, *Lokomotorische Richtungsbewegungen durch chemische Reize*, *Berichte der deutschen botanischen Gesellschaft*, Bd. i. 1883, p. 524: "Malic acid is the specific stimulus which attracts the antherozooids of ferns into the open archegonia. In like manner, when this substance is unequally distributed in water the antherozooids of Selaginella move towards the spots where it is most concentrated. In a similar sense cane-sugar is the specific stimulus of the antherozooids of the leaf-moss. . . . Not one particular, but every nutritious substance is, when unequally distributed in solution, a stimulus to motile bacteria, which move towards the richer or better nourishment. I have also found that an extract of meat is able to attract the swarm-spores of Saprolegnia and Trepomonas agilis, one of the Flagellata. All these are cases of chemical stimulation, by which the motile organism is excited to travel by its ordinary means of locomotion towards the more concentrated medium, that is, in the opposite direction to diffusion."

Pfeffer attracted the antherozooids of ferns with malic acid into capillary tubes, and with exceedingly small quantities of the substance: the effect was marked even with a solution of ·0001 per cent.

We have here, then, a chemical stimulus which ensures fertilisation—the malic acid conducts the antherozooids of ferns. Other antherozooids are influenced by other such stimuli, *e.g.* those of mosses by cane-sugar. For further details see W. Pfeffer, *Lokomotorische Richtungsbewegungen durch chemische Reize* in *Untersuch. aus dem. Bot. Inst. zu Tübingen*, Bd. i. 1881-85, p. 363.

to which question we may refer to the fact that the stimulus of light affects very primitive organisms, and that bacteria travel towards the water that contains most oxygen. We can also assert definitely that in many of the lower multicellular animals simple epidermic, or at any rate tactile-cells, must be capable of perceiving various kinds of stimuli, since touch and taste-cells and touch and olfactory-cells pass into one another by gradual transition, and since touch-cells become in some cases visual-cells, in others auditory; while, on the other hand, the so-called end-organs, which usually serve as organs of taste, have in some cases become eyes, as Grenacher[1] has shown in the Arthropoda. The evolution of organs of touch, or of touch and taste into eyes is also beautifully shown in leeches. It is obviously in the lower animals, whose senses are not acute, more difficult to determine whether or how far they perceive gustatory or olfactory stimuli, and still more, whether they distinguish either from touch. But after numerous researches and observations of my own, so much seems to me certain, that originally one and the same cell perceives different sensory stimuli, as, for instance, many animals which have no eyes, such as insect larvæ and worms, are evidently affected by light through their epidermic cells, and indeed it seems to act upon them as a painful tactile stimulus. Probably all so-called sensory stimuli act upon protoplasm originally in the form of tactile stimuli, hence the fact that all sensory-cells first appear morphologically as tactile-cells. Sensitiveness to diverse kinds of stimuli must therefore have been gradually evolved in different ectoderm-cells as a division of labour.

Animals thus probably originally perceived light and sound, and, to judge from anatomical relations, smell and taste also, not as such, but as tactile sensation, and many still so perceive them, by means of one and the same kind of

---

[1] Grenacher, *Unters. über das Sehorgan der Arthropoden*, Göttingen, 1879.

cell. At a later stage the functions of the cells diverge, while the cells themselves develop specific forms. Evidently touch, smell, and taste continue longest to be perceived by the same cells, and for this reason we look in vain in many of the lower multicellular animals for separate organs for these sensations.

The sensory stimuli, as various stages of one and the same stimulus of motion, have, according to my view, been the ultimate cause of the origin of various specific sense organs, a view I previously expressed with special reference to eyes.

This evolution of the characters of sense-cells could not possibly have been produced by sexual mixture and selection, —by variation of the germ-cells—although I do not deny to the first two processes a share in the accomplishment of the result. Not variations of the germ-cells which occurred by chance in a definite direction, but a definite capacity of modification in the protoplasm of the ectoderm—the property of the latter of becoming altered in a definite way under particular stimuli—has determined the modification.

The best evidence of the truth of this, and at the same time of the effect of use, lies in the fact that the higher sense organs have always developed on the parts of the body best adapted for the reception of the respective stimuli, and in different animals on different parts, while the larvæ had originally an ectoderm all of one kind, as the lower multicellular animals have in their adult condition.

A proof, moreover, that the organism under the action of external influences can only undergo particular definite modifications, that it can only yield to external demands by modification in a definite and limited degree, lies, as I have already urged, in the "Medusæ,"[1] in the fact that the higher sense-organs

[1] P. 220, *seq.* I refer here to my description of the auditory organ of the Medusa Carmarina as compared with the ordinary structure of the auditory organ of the cycloneurous Medusæ and many worms. Cf. my previously cited address to the Naturalists' Congress at Munich, 1877; also *Die Medusen*, p. 222.

(auditory and visual) often have a quite similar structure in animals by no means closely related, and this in cases where they must have arisen independently, because the common ancestors of both, and even the ancestral forms of both derived from the latter, were entirely without such organs.

Certain Medusæ and worms, for example, have perfectly similar auditory organs, although these cannot have been derived one from the other—indeed the auditory organs of many Medusæ and worms are more closely similar to one another than to those of other Medusæ or other worms. Consider the eyes of vertebrates and cuttle-fishes. In the latter is repeated the whole plan of structure of the retina of the former, only in the reverse order: in the vertebrate eye the expansion of the optic nerve lies on the inside of the retina; in the invertebrate on the outside. In the slug Onchidium, however, as Semper has shown, we find the same relations as in the vertebrate. All these three kinds of eyes must then have arisen independently, and yet they are constructed of the same parts.

"This fact," I insisted further, "is of quite peculiar importance. It shows in the clearest way . . . that in consequence of the relations of organisms to particular influences of the external world perfectly similar forms may arise, even without any immediate blood-relationship between them, not, on my view, because the material provided in the animal organism had little capacity for evolution, but because with this material organs can only be constructed within narrow limits of variation to fulfil in the best possible degree a perfectly definite and constant requirement from without."

I said further that the anatomy of the sense organs offers specially numerous examples of this in two directions; that notwithstanding the absence of all direct relationship of descent (1) resemblance, or exact similarity of form, has arisen; (2) that similar forms, but in different combination, have been produced. Further, " the more definite, the more powerful and continuous, and the less modified by other influences the actions of the external world on given organised materials, the more similar will be the structural forms which they produce, even when the blood-relationship of these materials is by no means close.

" Such an action is present in those definite unchangeable physical influences which determine the origin of the sense-organs in a given formative material, in consequence of the necessary requirements of adaptation. The power of adaptation, in comparison with that of heredity, here comes into extraordinary prominence. The influences of the former are very powerful. Even where the manifold slighter relations of life between individuals come into play reciprocally, these relations after repeated action have a great effect in changing forms, and when long continued are important agents in modification. But the influences which primarily govern every organism in consequence of the necessities of its existence, the physical influences of the media in which it lives, constitute forces which, though simple, are counteracted by no opposing factors, which are ever powerfully acting in the same way, and to which, from the given material, only a limited variety of structural modifications can be adapted. And therefore, in spite of the endless variety in detail, in particulars, a certain uniformity of organisation in the gross must exist: thus the same plans of structure may appear repeatedly where there is no close blood-relationship to account for them. Since the continuous influence of elementary physical forces necessarily exhibits itself most in the organisation of the sense organs, a comparative anatomy of these organs, founded essentially upon heredity, can only be established, usually within the smaller divisions of the animal kingdom, and among all others their similarities of form are the least trustworthy guides to phylogenetic relations."

These arguments are directed against the prevailing fashion in comparative anatomy of referring all similarities of form dogmatically to blood-relationship, and of leaving the direct and indirect influences of the external world out of consideration. The view here urged that the nervous system and sense organs have been formed in the multicellular animals from the original body covering, from the epiblast, because the epidermis has naturally from the first been the medium of relation to the outer world, because it received impressions and required to be rendered capable of giving the impulse to reaction against those impressions, to defence and attack, this view will be opposed by no embryologist or physiologist. But perhaps the most important support for this view will be afforded by tracing the first appearance of a morphologically distinguishable nervous system in the animal series, as I have traced it in the Ctenophora and Medusæ.

The Origin of the Central Nervous System

Embryology has long since taught us that not only in the Vertebrata, but also in the Arthropoda, Mollusca, and Vermes, the central nervous system is developed from epiblast-cells, which separate from the superficial layer whence they are derived and pass to a deeper position, and has also shown that such animals develop from similar embryos which consist of germinal layers, and which are evidently genetically related together. It has also long been known that lower forms of the Metazoa exist, which remain throughout life essentially in the morphological condition of these embryos, and which although no separate nervous system can be recognised in them, clearly manifest the power of sensation and even of volition. Thus the question necessarily arose whether these powers in such forms do not reside in the ectoderm.

Accordingly, N. Kleinenberg has explained the cylindrical

epidermis-cells which give off internally a transversely-placed fibrillar process, so that a fibrous layer is formed between the ectoderm and endoderm, as neuromuscular-cells,[1] that is, as epidermis-cells whose outer part is capable of receiving nerve stimuli, while the internal process is of muscular nature, and transforms directly the stimuli received by the former into motion.

The Hydra is simply a gastrula-sac consisting of two layers, these neuromuscular-cells (the ectoderm) and digestive cells (the endoderm), and provided with prehensile arms. When a direct reaction ensues upon stimuli falling on the ectoderm-cells, the movements of the animal are reflex actions, and it unconsciously performs movements on stimulation, just as the Mimosa folds its leaves when they are touched.

The question how the Hydra can exercise volition as it certainly does, without a separate central nervous system is not discussed by Kleinenberg. Yet, when we observe the motions and general behaviour of many free-swimming ciliated larvæ, the conviction is forced upon us that although they consist only of epiblast and hypoblast, they pursue definite purposes, that their actions are in some degree directed by a will. It must therefore be assumed, whether nervous cells are somewhere present in Hydra or not, that ectoderm cells in the lower Metazoa are the seat of volition. But a morphologically recognisable nervous system consisting of separate nerve-cells and nerve-fibres must, according to these considerations, in its most primitive form lie immediately beneath the epidermis, connected on the one hand with the latter, on the other with muscles. Moreover, it probably extended at first as a layer all over the body.

These predictions were completely fulfilled by my researches upon Beroë ovatus.[2] The further conclusion was

---

[1] Kleinenberg, *Hydra*, 1872. [2] Cf. *Beroë ovatus, loc. cit.* 1873.

also verified, that the formation of a definitely circumscribed brain will take place by the gradual concentration of nerve-cells at one or several points of the body which are most exposed to contact with the external world, and on which therefore the sense organs develop. These conclusions were verified both by the anatomical and the physiological investigation.[1]

The central nervous system of Beroë ovatus consists of nerve-cells which lie beneath the delicate epidermis of thin flat epithelial-cells, which are distributed over the whole surface of the body, but are accumulated in larger numbers at the closed end of the sac-shaped body in the neighbourhood of the sense organ there situated. At this spot the nerve-cells form externally a sense-ganglion, and beneath that the rudiment of a definite brain formed of aggregated nerve-cells. But the nerve-cells scattered over the whole surface are connected by delicate nerve-fibrils on one side with the epidermis-cells, on the other with the muscle-fibres which traverse the gelatinous tissue (neuromuscular fibres).

The nervous system of the great Scyphomedusæ, the toponeurous Medusæ, is, according to my anatomical researches, first published in 1877, but suggested in 1873 by my section experiments,[2] similarly constructed; but the accumulations of nerve-cells which represent brains are here found chiefly on the edge of the bell, again in the neighbourhood of the sense-organs. Brains connected by a ring-nerve are formed in the same position in the cycloneurous Medusæ, and their cells and fibres are demonstrably formed of ectoderm

---

[1] Cf. Th. Eimer, *Versuche über künstliche Theilbarkeit von Beroë ovatus, angestellt zum Zweck der Controle seiner Morphologischen Befunde über das Nervensystem dieses Thieres*, Arch. f. mik. Anat. Bd. xvii. 1879.

[2] *Sitzungsberichte der physikalisch-medicinischen Gesellschaft zu Würzburg für das Gesellschaftsjahr* 1873-74, Part ii., erste Sitzung am 13 Dez. 1873; and *Ueber künstliche Theilbarkeit von Aurelia aurita und Cyanea capillata in physiologische Individuen*, etc., Verh. der phys.-med. Ges. zu Würzb. N.F. Bd. v. 1874.

cells, which are connected with the ectoderm, but which have come to lie beneath it.

Thus we have in the Medusæ a laminar central nervous system extending over the body, whose cells are commencing to concentrate at spots particularly suitable for communication with the external world, and there to form definite brains. In the higher Metazoa, in Vermes, Mollusca, Arthropoda, and Vertebrata, these brains, or ganglia, are completely formed, but their embryonic condition still indicates the epiblast as their original place of origin.

But in the former case the first appearance of a morphologically recognisable nerve-system in the animal series, its formation from the ectoderm, can only be explained as the effect of the constant action of external influences upon the organism, and by the inheritance of this effect—by the inheritance of acquired characters with the aid of selection.

Accidental variability of the germ-plasm as a determining cause seems here also completely excluded.

A more detailed account of the researches above mentioned on the nervous system of the Medusæ, and especially of my experiments upon the subject, is contained in the address, "On the Idea of the Individual in the Animal Kingdom," which forms the Appendix to this work ; for a full discussion I must refer to my original papers.

Only one point further I must draw attention to: the nerve-cells in the Ctenophora and Scyphomedusæ are so far from being morphologically differentiated and recognisable, that I have been accused of mistaking connective-tissue-cells for nerve-cells. It is, however, obviously, on my view of the matter, necessary that nerve-cells should at the commencement of their evolution be similar to other cells. It is a known fact that even in higher animals the nerve-cells in the embryonic condition cannot be distinguished from other embryonic cells. Only function could impress upon nervous, as on other cells, a definite morphological character. Thus I sought at first in vain for nerve-cells or brains in the Scyphomedusæ and could only discover the spots at which the latter exist by the section-experiments above mentioned. Such experiments afterwards completely confirmed my description of the nervous system of Beroë. In both cases, in Beroë as in the Scyphomedusæ, the presence of a number of nerve-cells, or of brains, could be recognised by the fact that the parts in question, when separated from the rest of the animal, alone, or at least in a pre-eminent degree, exhibited life (movement).

In Beroë I had found anatomically a gradual decrease in the abundance of the nerve-cells from the closed end towards the aperture. In power of movement the parts of the animal when divided transversely exactly corresponded to this result. The separated polar portion moved immediately after its separation, in all respects like an entire uninjured animal. The nearer to the aperture they were taken, the more the parts were affected by separation, the longer time elapsed before they showed traces of movement, the less were they capable of life.

VICARIOUS NERVE-CENTRES.

In the course of my section-experiments I observed a remarkable fact which at present stands alone, and which has an important bearing on my whole argument.

The Medusa Aurelia aurita, when I had removed from it all the primary nerve-centres, became completely motionless. But after the animal thus mutilated had lain several days in clean sea-water, it gradually commenced to exhibit movements again: one day, movements at first trembling and irregular appeared in the umbrella. These movements evidently started from a definite spot which might be at any part of the umbrella, and extended thence over the whole, just as in the uninjured animal they proceed from the nerve-centres on the margin. Gradually the movements of the animal became regular, they succeeded one another in a definite rhythm, as in the entire animal, and the animal thenceforth behaved completely like an uninjured specimen. Thus a new nerve-centre had been formed in place of the old from the nerve-cells scattered over the body-surface, and had taken upon itself the movement and direction of the animal. I have repeated this experiment many times with the same result, and have observed mutilated animals living in this way more than eight days.

The movements of Medusæ are, as I have shown, involuntary, but they can be retarded or hastened, diminished or intensified voluntarily. As involuntary movements they are respiratory, as voluntary they are the means of locomotion.

The only possible explanation of the reappearance of the movements is that the need of respiration in the still living but motionless animal first produces convulsive contractions, which gradually become rhythmical. But how it happens that they become rhythmical, and that this rhythm may be controlled by some portion of the nerve-cells scattered over the body, and how these cells are able to influence the movements by volition remains a mystery, unless we assume that all the nerve-cells of the Medusa still possess, as an inherited and originally acquired character, the faculty of directing motion, even after the function of direction has been assigned to the nerve-centres of the margin of the umbrella. In order to explain the facts, it must be further assumed that some of the nerve-cells of the surface have retained this faculty in a higher degree than others, or that they happen to be more vigorous at a given moment, and are therefore able to exert their inherited faculty more powerfully than the rest. It would thus only require external stimulation to bring this faculty into action: Medusæ which were left after the operation in unchanged water deficient in oxygen, did not recover their power of motion.

I must here refer to a particular experiment which bears upon this subject.[1] It must be explained that in Aurelia aurita, on which the experiment was made, there is an anus between every two marginal bodies.

On 29th August, at 12 noon, of two specimens of Aurelia 18 cm. in diameter, I cut away from one the marginal bodies, from the other the parts around the anal openings. The first remained after the operation entirely motionless, the second contracted as usual. In the former, twitchings occurred only when I pricked it with a needle—slight imperfect contractions.

Second day. 30th August, 11 A.M. The animal (O) deprived of its marginal bodies lies flat, extended, and motionless, the remaining parts of the margin somewhat turned upwards. Even when pricked with a needle it reacts extremely feebly. The other specimen (M) still contracts actively. The separated pieces of O contract vigorously, as they did yesterday immediately after their removal; the separated portions of M have, on the contrary, lost their fresh appearance, they are no longer elastic, but flabby and thin—they seem to be dead, for not even the tentacles on them react to stimulation.

Third day. 31st August. M contracts most vigorously (in the same water) at least ten times per minute, but generally much oftener. On O still not a single contraction is to be perceived.

Fourth day. 1st September, 12 noon. M contracts (in the same water) vigorously. O makes at intervals of $2\frac{1}{2}$, $1\frac{1}{2}$, $1\frac{1}{2}$ minutes slight contractions, but only with two particular anal lappets out of the eight, and these two contract simultaneously. When the animal is pricked in the centre with a needle, the same two lappets which exhibit spontaneous movements contract strongly, and some of the others

---

[1] It is described as Experiment D at p. 81, *et seq.*, of the *Medusæ*. Cf. fig. 5 below. The dotted lines in the figure terminate on the margin at the anal apertures.

contract feebly, while yesterday no spontaneous movement, and scarcely any reaction to stimulation were observed in this specimen. The Medusa, therefore, has recovered itself in certain parts only; in these parts vicarious centres of contraction must have been established, the action of which does not extend beyond their boundaries. The specimen still lies flat and extended, while M when contracting assumes completely the bell-like shape. Three-quarters of an hour after the addition of clean water, O contracts at intervals of 2, $1\frac{1}{2}$, $1\frac{1}{2}$ minutes the same two anal lappets as before. M contracts very actively. Both behave almost in the same way as in the old water.

Fifth day. 2d September. M contracts actively in the old water; O lies still and begins to decompose, only the two anal lappets which yesterday contracted spontaneously are still entire. The other anal lappets have dissolved away, except two, pieces of which are still left.

Sixth day. 3d September, 8.30 A.M. Only some of the separated pieces of O which contain marginal bodies are left alive, all the rest is in process of dissolution. Putrefaction in the vessel, which has also caused the death of M.

Seventh day. 4th September. The fragments containing marginal bodies alive yesterday are still alive, but are reduced to narrow ($\cdot$75 cm. broad) strips, and each is merely a marginal body with a morsel of the umbrella margin connected with it.

Eighth day. 5th September. One fragment of the margin with its marginal body still alive and contracting, although the water is completely putrid from the decomposing Medusa left in it all night.

In this case, therefore, two new nerve-centres came into action in two portions of a Medusa while the rest was dying, and these centres, when the living portion of the body was stimulated, produced contraction in the anal lappets controlled by them, just as do the " brains" of the intermediate parts in the uninjured animal.

It is also remarkable that when the Scyphomedusæ die gradually the brains with the parts surrounding them remain alive longest—Aurelia thus gradually perishes, until at length only eight small fragments containing the nerve-centres are left,[1] and in like manner Beroë dies towards the aboral pole which contains the greatest number of nerve-cells. But when I cut out one of the eight antimeres of an Aurelia, containing a marginal body in the middle of its lower edge, its death proceeded in the manner explained by the accompanying figure.[2] It began in the middle and proceeded upwards and downwards, until only the lowest portion containing the marginal body remained. And this piece went on diminishing towards the marginal body.

The results of section in Beroë still more strongly support my hypothesis of the development of vicarious nerve-centres.

The locomotion of the Ctenophora is effected by the strokes of the small paddles formed by the concrescence of cilia, again a motion which can take place both involuntarily and voluntarily.

Separated portions of this animal all behave after some time exactly like the whole: no difference in the movements can be recognised—voluntary motion of the paddles is evident even in the fragments.

FIG. 2.

Thus certain nerve-cells here also must assume the function of direction. But this occurs the more quickly, as follows from the facts already described, the greater the number of such cells present in the fragment—only the polar fragment

---

[1] *Die Medusen*, p. 80, *et seq.*
[2] From *Die Medusen*, fig. 10, where p. 61, *et seq.*, contain further details of the process of dying. The stalked knob represents the marginal body.

moves immediately after its separation in the same way as the uninjured animal.

The preceding results lead to the inference that a definitely circumscribed nervous system is not absolutely necessary for the exercise of volition. Co-ordinated movement might also occur if the nerve-cells were uniformly scattered over the whole surface of the body, without a permanent centre. This was the original condition, and in a still earlier stage the seat of volition was in the ectoderm cells. The latter is even now the case in some two-layered, free-swimming larvæ which exhibit voluntary action, and the presence of special nerve-cells is not indispensable to the activity of *e.g.* the Hydra. Quite recently attempts have been made to demonstrate the existence of such cells in Hydra, but it seems to me the evidence is at present insufficient. The fact that each fragment of a Hydra, however it is cut up, behaves like an entire animal, and that it grows into an entire animal, that this is true even of a fragment of one of its tentacles, indicates that morphologically expressed centralisation can scarcely exist in this animal.

That a single will can exist even with a number of brains, is also beautifully shown by the Medusæ. I have shown how the eight brains of Aurelia produce single movements: the impulse proceeds always from one of the brains, and passes immediately to the others—sometimes one, sometimes another originating the impulse—but this does not exclude the possibility of all eight acting at the same time, indeed, this is probably necessary for the production of certain movements.

Now, whether in a given animal there are eight such central points of nervous activity or thousands—as many as there are nerve-cells or ectoderm-cells on the body surface—makes no difference. If we assume that Hydra has no special nerve-cells, then the impulse which produces action may start now from one, now from another ectoderm-cell, or group of such cells, and spread from thence—by simple contact since no nerves exist—to the neighbouring cells, and so communicate itself to all the muscular processes.

Brains, or a brain, as the case may be, could only arise in consequence of the fact that certain ectoderm-cells, or groups of such, came more frequently into contact with the outer world, and accumulated experience, or that were from their favourable position adapted to form the middle point for the activities of a larger number of neighbouring cells—cerebral ganglia could only be developed in consequence of the inheritance of acquired characters.

The Cell-Nucleus as a Central Nervous Organ.

But if the individual cells of the ectoderm are capable, as the view I am advocating supposes, of giving rise to voluntary action, we must seek in them some apparatus which serves as a central nervous organ ; and this apparatus can only be the nucleus.

Thus I am brought back to a question previously suggested, Whether the nucleus of the unicellular organism is not also to be regarded as its central nervous organ?

For the larval Blastula is certainly, like the Volvox colony, only an aggregation of unicellular beings. And an affirmative answer to the above question is supported by numerous facts which I have already brought forward in my papers on Beroë and on the Medusæ. In these I have described the nucleus in general as the central organ of the cell, as its organ of life, in the sense that it originates and governs the processes of life in the cell, while in the animal cell I have considered the nucleus as the central nervous organ.

I was first led to this view by considering the great importance of nuclei in every nervous system. The nucleus and not the cell-plasma is the essential part of a nerve-cell: the latter serves only for conduction. Hence the extraordinary magnitude of the nuclei in the ganglion-cells of all animals which possess a somewhat highly developed nervous system.[1]

Secondly, this supposition is strongly supported by the peculiarities of the nuclei in the nerve-fibres of the lower multicellular animals. In these fibres, as I have shown for instance in Beroë, the nuclei, surrounded with very little protoplasm, are placed at intervals in such a way that a nervous fibril passing through the nucleolus connects one nucleus to another. Thus these primitive nerves look like chains of nuclei connected by conducting fibrils, and resemble a series of telegraph stations connected together for the purpose of renewing or strengthening the current at the beginning of each stage. Obviously the ganglionic swellings along the nerves in higher animals, *e.g.* the spinal ganglia, form a similar apparatus, but with the added function of crossing the conducting fibres.[2]

Thirdly, I find that also in the sensory-cells of Medusæ nerve-fibrils pass through the nucleolus, and are continued, as already mentioned, into the cilia, or terminate in the nuclei of the epidermic-cells.[3] The same thing occurs in the sensory-cells of higher animals.

Fourthly, nerves frequently terminate in the nuclei of cells which are not sensory, and this mode of termination will, I believe, be found in future to be very common. It occurs, for instance, in the epidermis, in muscle-cells, etc., and in all probability, as before mentioned, in ciliated cells.

---

[1] Even in the nerve-cells of Medusæ (Carmarina hastata) the cell-plasm obviously consists of conducting fibrils (*Medusen*, Taf. viii. figs. 8, 10); and the same thing is easy to observe in the nerve-cells of the cerebral ganglia of Helix pomatia.

[2] Cf. *Beroë*, Taf. viii. fig. 72; *Medusen*, Taf. xi. fig. 9.

[3] *Medusen*, Taf. iv. figs. 1, 7, 21; Taf. xiii. fig. 9, etc.

In the latter the cilia are connected at their bases with fine fibrils, which probably pass through the nucleus and form the continuation of nerve-fibrils.

Lastly, it is evident that the nerve-fibrils start from the nucleoli of the nerve-cells, for they pass in a radiate manner from the nucleolus and form a fibrillar network in the nucleus. On the other hand, the same histological relations are seen in nuclei in which nerve-fibres terminate, most clearly in Medusæ, but also in the higher animals. I consider the so-called "Eimer's ring of granules," which is also most clearly to be seen in many nerve-cells, to indicate the points where the fibrils radiating from the nucleolus bend round to join the complicated network of the nucleus.[1]

This fibrillar network in the nucleus—and this is the particular bearing of the preceding upon our subject—must, like the fibrils in the cell which are connected with it, be of nervous nature, and this is undoubtedly the condition of things in the ganglion-cells of the higher animals: it can be seen, for instance, in the cerebral ganglia of the Helix pomatia. Even in Medusæ (in the marginal bodies) definite fibres of extraordinary delicacy serving exclusively as nerves have arisen [2]—the "dotted substance" in the brain of higher animals is the indication of a fibrillar network connecting the cells, and the structure of this is so delicate that it will probably never be possible to follow out the course of the fibrils. But these nerve-fibrils must have been derived from simple strands of ordinary protoplasm, and such they are still in cells which are not specifically nervous.

From such protoplasmic strands true nervous substance with infinitely delicate and regular connections must have been formed by the constant passage of nervous impulses and

[1] Cf. *Die Medusen*, Taf. viii. fig. 6, 7; Taf. ix. figs. 10n; Taf. xi. fig. 10; *Beroë*, Taf. viii. bes. fig. 82; and *Weitere Nachrichten über den Bau des Zellkernes*, etc., loc. cit. Taf. vii.

[2] Cf. *Die Medusen*, Taf. iv. figs. 2, 6, 13, 14, etc.

by the inheritance of the properties so acquired. Similarly, only by modification of the original ordinary protoplasm, in consequence of the constant exercise of nervous action, can conducting nerve-fibres and nerve-cells have arisen. Spontaneous variations of the germ-plasm cannot possibly have produced a system of organs so wonderfully delicate, both morphologically and physiologically, exclusively as a means of relation between the organism and the external world. This must have arisen through external influences and exercise.

The fibrillar network above described occurs also in the germinal vesicle of the egg-cell, even with the radial arrangement round the germinal spot. That which may afterwards become the path of nervous impulse is originally the path of nourishment, and the radiation of the strands to and from a central point is the most effective for the latter purpose. In the streaming of the protoplasm of vegetable-cells, *e.g.* in the hairs of Tradescantia, we see the commencement of this typical arrangement of protoplasmic threads. The protoplasm flows as though driven by a secret force—as also in the Foraminifera—towards the central point and then again from it: the central point being the nucleus which originates and governs the vital processes in the cell. Such paths, at first fluid, became fixed, and finally, where they had to convey nervous stimuli, became nervous fibrils.

The same fibrillar network in the nuclei of the germinal cells, in the germinal vesicle, is considered by Weismann, on the other hand, as the idioplasm, *i.e.* the firm substance which conveys the characters of the species from generation to generation.

I have already in the *Medusæ* laid weight on the fact that only in the ova of animals are the nuclei so extraordinarily large as in the ganglion-cells, and have attempted to explain "the prominent part played by the nucleus in ova, and also in spermatozoa," the "great importance possessed by these elements as the instruments of an enormous development," by the suggestion that the nuclei originate this development, "a supposition which is considerably supported by the most recent observations" (1878).

Originally even the multicellular animals had no special paths of nervous conduction. Possibly in the ectoderm-cells of Hydra a network of ordinary protoplasmic threads still constitutes the communication between cell and cell, nucleus and nucleus.

Origin of Nerve-Fibres and their Vicarious Action.

But a more distinct morphological effect of the exercise of nervous action is seen in ectoderm-cells, in which the course of nervous impulses is through the cells lengthwise to the nerves or muscles connected with them. For instance, in the ectoderm-cells over the nerve-ring of Medusæ, which are continued into nerve-fibrils, and also in the sensory portion of neuro-muscular cells, the cell-plasm is, as I have already mentioned, longitudinally striated in consequence of its transformation into extremely fine fibrillæ. In the former instance these fibrillæ pass directly into nerve-fibrils; in the latter, they pass into the muscular portion of the cell, after having in both cases come into connection with the nucleus. These relations are quite similar to those described in ciliated cells—that is, the fibrillæ are nerve-strands which have been produced by functional activity from the protoplasm of ordinary epithelial-cells.[1]

It follows from the preceding that the nerves of Zoophytes originally consisted of chains of cells derived from the ectoderm. This is also the most probable origin of the nerves of the higher animals. In the gelatinous tissue of Scyphomedusæ, in Aurelia aurita for example, we find that the amœboid cells

[1] Cf. *Die Medusen*, Taf. xii. figs. 8, 12; Taf. xi. fig. 6, and others.

2 A

of that tissue, connected by definite but extremely primitive nerve-fibres, form chains passing through the tissue from one surface to the other. I believe that we have here a survival of the most primitive nervous communications in multicellular animals. The course of nervous impulses, originally variable, has probably become fixed in consequence of their continual passage in the same direction; and from the same cause, to judge from the relations above described of neuro-ectodermic-cells, has been developed the axis-cylinder, composed of fine fibrils, of the nerves of higher animals.

The nerves of the cycloneurous Medusæ, which are also chains of nerve-cells, are, on the other hand, developed in such a way that they remain dividing ectoderm-cells connected from the first by nerve-fibrils, which are processes from themselves. Thus arise the ganglion-cells interpolated in these animals in the course of nerve-fibres.

When making experiments on the section of Medusæ, I made an observation on the vicarious action of nerve-fibres, which finds its proper place here. I was trying at the time to ascertain whether in the Scyphomedusæ (toponeurous Medusæ) a nerve (ring-nerve) is present at or near the margin of the umbrella. With this object I made a cut in an Aurelia aurita, with scissors, through the margin for a distance of 1 cm. between every two marginal bodies, in order to see whether the interdependence of the movements of the eight portions of the umbrella on one another, in other words, whether the co-ordination of the motion excited by the activity of the several ganglia, would continue or not. As the co-ordination was completely maintained, I next, in another Aurelia, cut out the whole central portion, leaving a continuous marginal ring about $1\frac{1}{2}$ cm. broad, and containing all the eight marginal bodies.[1] This ring began, after a pause such as always occurred after similar operations

---

[1] The following figure is also taken from the *Medusæ*, p. 31.

in consequence of the shock, to contract rhythmically like an entire specimen, while the central portion sank to the bottom as if dead. Then I made cuts through the outer edge of the ring inwards for 1 cm. between every two nerve-centres, so that the portions which each contained one of the latter were only connected by eight bridges, ½ cm. wide, at the inner edge of the ring. The co-ordination of the movement of the several parts continued still as before. It continued also when I prolonged half of the cuts till only a thin connecting strand was left. I also prolonged the other four cuts, and then made alongside each of these another cut in the opposite direction from the internal edge of the ring nearly to the natural margin. Thus three of the portions were connected together and with their neighbours only by a flat piece of tissue a few mm. broad, and only directly connected in the direction of one of the diagonals of these flat pieces. Now, the co-ordination of the contractions of the several portions became uncertain the more uncertain the smaller the connecting bridges of tissue. Thus it is proved that the several nerve-centres are connected, not by a ring-nerve, but by the delicate nerve-fibrils which pass in various directions through the gelatinous tissue, that is, by nerves which in this case again have by no means the morphological character of ordinary nerves, but which have the appearance of connective-tissue-fibres, and which are not yet united into bundles, into proper nerve-cords.

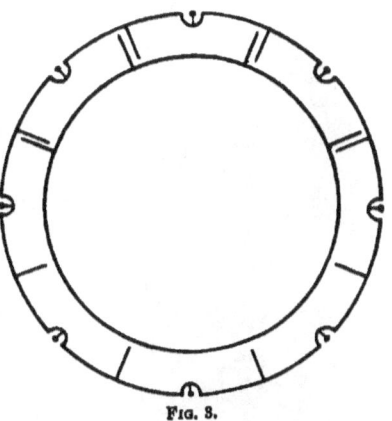

FIG. 3.

I repeatedly observed that the co-ordination of the movements of the several portions, after having been destroyed by the cuts, after a time was re-established. Communication must therefore have been effected by nerves which previously did not perform this function; some nerve-fibrils must have taken up the function of others, have acted vicariously in place of these.[1]

Experiments on Beroë gave me similar results.[2] I have already briefly stated that in this animal also when parts are separated a new centre of action appears in them. I now return to this subject, and give a short account of my experiments.

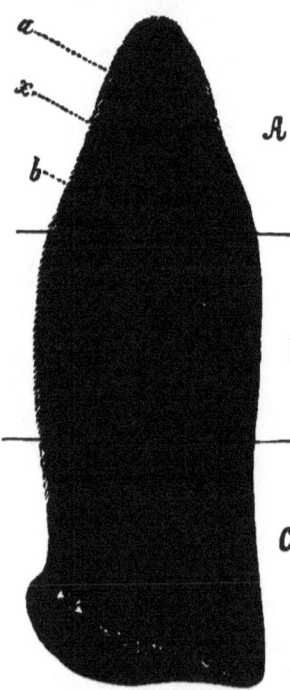

Fig. 4.

Experiment A.—I cut five Beroës into three parts transversely so as to form from each three parts of equal height, of which the upper, *A*, contained the aboral pole, with the largest aggregation of ganglion-cells; the lower, *C*, contained the mouth; while the third, *B*, was the middle portion of the body.

After the division, the movement of the swimming-plates ceased completely in all the pieces, but recommenced after a short time in all those pieces which contained an aboral sense-organ. Shortly afterwards, movement began again in those portions distinguished as *B* and *C*, but in these ceased again after a time, while in the *A* pieces it continued.

When I examined the pieces again after four hours, I

---

[1] Cf. *Die Medusen*, p. 31. Romanes's results also proved the same vicarious action of nerve-fibres in Medusæ, *Phil. Trans.* vols. clxvi., clxvii.
[2] *Archiv. f. mikr. Anat.* Bd. xvii.

found active movement going on in almost all the pieces; in
A, however, the plates were most active, and some of the *B*
and *C* pieces showed no motion.

Then I cut off small fragments from these various portions
of the animals. The swimming-plates of these fragments
were immediately after the operation in all cases motionless,
but after two hours I found active movement going on in
them also.

On the second day the plates were still in active motion in
all the pieces, the small fragments included, excepting a few
of the latter. On the third day all showed active movement.

On the piece forming the middle portion of one Beroë was
exhibited the peculiar phenomenon, that the plates of one row
vibrated in the opposite direction to that of the others, and
the movement in this row was no less active than elsewhere,
and, as in the others, proceeded in waves in rapid succession.
Everywhere else the movement in the separate pieces proceeded in the same direction in which the plates vibrated in
the entire animal: the waves passed from the aboral pole
towards the mouth.

This reversion of direction in the swimming-plates of a
single row of one piece of an animal while the rest vibrated
normally forms a very remarkable instance of independent
nervous action in a separate piece of an animal, an instance
of reversal of function. Moreover, I observed repeatedly on
the separated tube-shaped middle portion of a Beroë that the
movement in one and the same series of plates could proceed
now in one direction, now in the other.

Many repetitions of such experiments gave the same
results: the movement of the swimming-plates always ceased
after the division for a moment or for some minutes, sometimes even for some hours, but in all cases recommenced,
reappearing always earliest in the pieces which contained the
sense-organ, and later in the others. Usually also in the

former the motion after its reappearance attained sooner than in the others to the rapidity which it had in the entire animal—in fact, in many of the pieces without sense-organ, or in fragments of these, the motion remained permanently slower, as well as more irregular, than it was in the entire animal. Most of the pieces, however, recovered completely, and halves of Beroë containing no sense-organ usually soon swam about exactly like an entire animal, and also in the same direction as the latter, the mouth forwards, they reacted to stimuli exactly as entire animals, and seemed to be not at all inferior to these in psychical capabilities. Such a piece had also the power, like an entire animal, of stopping at pleasure the motion of any one of the rows of plates. As in the entire animal, the movement in the pieces as a rule commenced at the edge nearest the sensory-pole, and proceeded towards the mouth. But in rare cases I saw the movement commence somewhere in the course of the row of plates.

The difference between the results of experiments on Beroë and those on Medusæ shows that in the latter the localisation of nervous action has proceeded further than in the former, a conclusion with which my histological observations are in perfect harmony. In Beroë the total body-surface is in a still higher degree to be regarded as the brain, the accumulation of ganglion-cells in the sensory-pole in Beroë is more gradual and less distinctly defined than in the marginal bodies of Medusæ. In the latter I was able to stop all movement by pricking with a needle any one of the eight nerve-centres, in Beroë this is not possible. In Beroë separated portions soon begin to move again, in Aurelia aurita it was first necessary that a vicarious nerve-centre should be formed from cells which had either lost or almost lost the function of central nerve-cells. Hence motion reappeared in Aurelia rarely, and always in a manner which exhibited most clearly the clumsiness of the action of the apparatus at first. In Beroë, on the contrary, the independent activity of the pieces followed very quickly and from the first with greater certainty. Beroë shows therefore, with respect to the nervous system, a much more embryonic condition than the Medusa investigated.

Experiment B.—I cut through one of the rows of plates of a Beroë and the subjacent tissue, with a pair of scissors, at a point about 2 cm. below the aboral pole, at the level of x in Fig. 4. The movement of the plates ceased for a moment in the whole animal. Then the movement recommenced first in the uninjured rows of plates, next in the upper portion of the divided row, a, and lastly in the lower portion of the same, b. After it was re-established everywhere, it appeared that the motion in the two portions of the row operated upon, above and below the cut, went on independently. Both in a and b it was rapid, in b it could even attain greater rapidity than in a. In both portions the direction of the motion was towards the mouth, and the movement consisted in the same rapid succession of waves as in the uninjured animal. The fact that the waves in the two sections were independent was established not only by mere observation, but more certainly by the following experiment: when I lightly touched the upper portion with a needle the movement ceased in it for a moment, while in b it went on as before—and similarly the movement could be stopped in the lower portion while it continued in the upper. Immediately after the operation, the waves in a proceeded from the aboral pole to the incision, while in b waves quite independent of these in a began at the incision and passed towards the mouth; but eight hours afterwards it appeared to the eye as if the connection between a and b was re-established, as if continuous waves beginning at the aboral pole were propagated through a across the incision to b and then to the mouth. It was found, however, when b was touched with a needle that the movement in this portion was interrupted, but not in a. Thus a certain degree of independence in the activity of the two portions still existed. On the other hand, if I touched a in the same way, the movement was no longer interrupted in this portion only, but in both a and b. It follows that the connection was almost completely re-established. Three hours before it had still been possible to interrupt the movement in a alone by a touch.

Towards evening I divided the other seven rows of swimming-plates as well, by making a shallow circular cut round the whole body, from the cut already made. The movement of the plates thereupon ceased all over the animal, but recommenced very soon on the row previously operated on, and after some time on the oral segments of all the rows, last of all on the oral portions of the rows—but in these it was irregular and not very vigorous.

On the following day at 10 A.M. I found all the plates of the animal operated on in active motion. But a complete connection between the oral and aboral portions of the rows was not everywhere established. In four of the rows the waves seemed to be continued from one portion into the other, they were propagated apparently continuously across the cut; but in the other four they evidently originated in each portion independently, only occasionally they had in two of these the appearance of continuity—however, in the latter it was still possible to interrupt the movement in the lower portions while it continued in the upper, but the converse was not possible.

Experiment B I have made with different variations on numerous other Beroës, and with results essentially the same: gradual recovery of the movement and of its continuity in the several rows always took place, until the movement went on in the animal after the operation in the same way as before.

But the following phenomenon also was repeatedly observed: when a circular cut was made in a Beroë in the manner above described some centimetres below the aboral pole, and after the movements had again become regular and vigorous in both the oral and aboral portions of the rows, but before their connection in the two portions was completely restored, the animal was completely divided into two parts by deepening the cut already made, then the movement in the two parts immediately after the complete separation went on exactly as it had before the separation. In such a case the larger tube-shaped oral portion of the animal swam away exactly as if it were an entire uninjured animal, just as a branch of an oleander tree, which has been made to send forth roots while still attached to the stem by being surrounded at a particular spot with earth contained in a pot, goes on growing independently when it is separated from the mother-plant by a cut below the roots.

Thus, while the movements of the swimming-plates is stopped for a time by dividing a Beroë into two parts or into several, and usually also by merely making a superficial cut in the animal, yet when the separation is preceded by the preliminary operation above described, it may be completed without producing any effect on the divided portions. This proves that the two portions acted quite independently before the complete separation, and that the reappearance of the continuity of the movement of the plates must be the consequence of a subsequent concrescence, or must be caused by the establishment of new nervous paths which take the place of the old.

Facts are also known which prove the vicarious action of nerve-fibres in Vertebrata, for example, in the results which follow from alternate section of the two halves of the spinal cord.

The facts above brought forward not only show how rapidly during individual life parts of the organism can become adapted to special functions; they enable us also to understand how in course of time such functions have become localised, and have modified the structure of the parts devoted to them more and more, as each generation inherited the alteration produced in its predecessor.

I will now add some further arguments concerning this question, derived from a subject to which I have already referred.

THE ACQUISITION AND INHERITANCE OF PECULIARITIES OF VOICE AND SPEECH, AND THE SPEECH OF ANIMALS

I have already opposed Weismann's argument against the inheritance of acquired characters, which depends on the assertion that neither the faculty of speaking nor of reading is inherited.

In one of his latest writings[1] he says: "When we adequately realise how energetically and how uninterruptedly we practise speaking throughout our whole lives, whether we are speaking aloud, or thinking silently to ourselves, and when we consider that in spite of this continual practice which has been performed by all human brains and vocal organs for centuries, the faculty of speech has not in the least degree become established as a hereditary character, we shall be inclined ever afterwards strongly to doubt that any acquired characters in the true sense of the word can ever be inherited."

I have already remarked that we ought not to expect the faculty of speech, or of reading, to be inherited, because these are very complex accomplishments, not simple faculties of the organism, which alone we bring with us into the world at birth.

Apart from the fact that the development of human speech dates from a period geologically recent, we cannot argue about its heredity from the length of time which has passed since

[1] *Ueber den Rückschritt in der Natur* (Retrogression in Nature).

its origin, because it has not been constantly the same from its beginning. Thus our living languages are only about two thousand years old, and even in this time have been modified. It is surely enough, as I before insisted, that we have acquired and inherited the faculty of perceiving tones, and producing sounds of various pitch. The rest depends on conventions and practice of comparatively short duration.

It is certainly impossible to contest that the faculty of producing particular vocal sounds has evidently become hereditary in the members of different nations. While I was writing the preceding during a sojourn on the island of Spiekeroog, an Italian organ-grinder came before the window of my room and ground out his shrill tunes. I asked him where he was going to-day. To " Orik," said he. I knew no such town. At last I found from his residence-permit that he meant the town of Aurich on the mainland. I endeavoured in vain to make him pronounce the name correctly. " Orik " he continued to repeat ; it was perfectly impossible for him to pronounce the " au " and " ch." It may be objected, that in this case it was simply want of practice ; that if the Italian practised for some years he would certainly at last be able to pronounce the word " Aurich " quite correctly, and the children of Italian parents would certainly have no difficulty if they were brought up in Germany. Granted, but not unconditionally. A trace of the tendency to speak in the Italian manner would probably remain even in the latter case.[1] How is it to be explained that the Jews, wherever they are found in Germany or elsewhere, pronounce

[1] In consequence of the efforts of the German Scholastic Association young people of German race, Cimbrians and Goths, but speaking only the Italian language, are frequently brought from South Tyrol to Germany in order that they may learn German. It has been noticed in such people that " the structure of their vocal organs being of the Germanic type, they learn High German with peculiar facility." Cf. *Fünfter Jahresbericht des Frankfurter Vereins zur Unterstützung deutscher Schulen im Auslande,* 1887. An accurate investigation of such cases would be of great value.

certain sounds of the German or other native language in a peculiar manner, even when they are scattered in small numbers among a population of German or other nationality, unless we attribute it to peculiarities of the larynx, of its muscles and nerves, that is, of the powers of movement in the organ? The native pronunciation must as undoubtedly depend on the congenital structure of the larynx as, conversely, an artificially acquired pronunciation must in time have an influence on that structure. We can most easily form a judgment in this matter when we consider the dialects of our native language, because here we are able to appreciate the slightest variations. What variety we find in the pronunciation of the German "ch," from the harsh throat-sound of the Alemannic people in Switzerland and Upper Baden to its complete conversion into "sch" which frequently occurs among people of Frankish race! I have special personal grounds for the conclusion that the faculty of producing the Alemannic throat-sound is inherited, that it depends on peculiarities of the larynx. But as we know almost every village has its dialectic characteristics, and these evidently depend in many cases on peculiarities of voice. The strangest and ugliest peculiarities of voice and speech that can be found among Germans are those of the inhabitants of the rural part of the town of Tübingen: in their speech the shrillest falsetto alternates with the harshest bass in all possible transitions, which it is impossible for an ordinary German to imitate. It seems as though the vocal muscles of these people had been by the most laborious practice so distorted as to pronounce the simplest sounds of speech in a perverted and barbarous way. Any one who doubts that this is due to peculiar anatomical and physiological relations of the vocal organs—although the variations may be too minute to be demonstrated—must believe it possible to produce with a fiddle the tones of a brass instrument out of tune. And there can also be no doubt that this pronunciation, which occurs only among a limited population, and moreover only in one part of a single town, has been inherited. It seems to me that such a peculiar pronunciation could only have been developed in one of two ways : either some man possessed it accidentally in consequence of laryngeal peculiarities, and transmitted it to his descendants ; or some influential personages introduced it by a deliberate perversion of pronunciation, and others imitated them, and thereby the larynx gradually acquired peculiar characters, which in course of time were more strongly developed and inherited. These are probably the two ways in which dialects in general and ultimately languages have arisen.[1]

My conclusions concerning this subject are strongly supported by the facts presented by the voice and speech of

[1] It is certain that the accurate investigation of the order in which sounds appear in children, of the development of the powers of pronunciation and of forming words, such as Professor Preyer carried out on his own children (cf. *Die Seele der Kinder*), might throw a great deal of light on this subject if it were applied to the most various nations. We might by its means discover the primitive sounds which formed the basis of subsequent evolution. It has struck me that many children use the sound "eng" at a very early age to express any feeling or the desire for anything. In the latter case it is often very vigorously uttered, the head, arms, and hands being at the same time stretched out in sudden jerks towards the desired object. Idiots sometimes have recourse to the same sound all their lives. I have a remembrance dating from my early youth of such a man in a village, who was for this reason nick-named "Eng-eng." Possibly there is a profound connection between these phenomena and the fact that the anthropomorphous apes of Africa have the same or similar names in their native habitat. The gorilla is called "engena" or "ingjina," the chimpanzee "engesego" or "ingjisego." On the speech of children compare also K. Vierordt, *Deutsche Revue*, Bd. iii. p. 29. Herodotus (ii. 2) relates that the Egyptian king Psammetich (670-616) endeavoured to discover which was the most ancient of all nations and languages in the following way : He gave two new-born children to one of his shepherds to be brought up without hearing any sound of human speech. They were to lie in an isolated hut, and the shepherd was to take she-goats to them at proper intervals that they might suck milk. After two years the children came to the shepherd with outstretched hands and cried "Bekos" (evidently an imitation of the bleating of the goats). As that was the Phrygian word for bread the Egyptians concluded that the Phrygians were older than they. A similar experiment was made by the Emperor Frederick II, but the children under trial died too soon (*Vide* Raumer, *Geschichte der Hohenstaufen*).

birds. That the peculiarly-constructed larynx of a singing-bird must owe its peculiarities and capabilities entirely to acquirement and inheritance, scarcely any one can doubt. In these animals the faculty of singing is actually hereditary: it is a fact that song-birds kept in captivity, even when they have had no opportunity of imitating old birds, begin to sing, and learn, though but imperfectly, to sing like their ancestors. We may therefore expect that speech in animals, when it is very simple, when it consists of only a limited number of sounds and sound-combinations, will be inherited more completely than in ourselves. Hitherto so little attention has been paid to these matters that even at the present day some zoologists and zoological text-books still regard not only reason but articulate speech also as exclusive possessions of the human race. What knowledge have we of the speech of animals to justify us in maintaining such a general distinction? It is certain that the language of many birds possesses a copious vocabulary, and among these are some of comparatively little intelligence, for instance, our domestic fowls. These creatures converse with one another with their hoarse voices often for hours together, as do also, among others, the swallows by their twitterings. Although these conversations are to us unintelligible, it is not extremely difficult gradually to learn something of the language of fowls and other birds, to understand their expressions of surprise and alarm, of pleasure, invitation, and warning. After much careful study it seems to me that such modes of expression are universal among fowls, and that they are hereditary, for one finds them used in the same way by individuals from the most various districts, and even when chickens are taken from the fowl-house when quite young they develop the same language. In this case it is the same as in that of instinct: the simpler the language, the fewer the definite requirements which the powers of vocal expression have to satisfy, so much the more completely are they inherited—so much the less variable is the mechanism of the vocal organs which is developed.

Nevertheless, the languages of animals have their dialects, possibly even that of the same species differs in different regions. Any one who has passed sleepless nights in Naples in the oppressive heat of August, will have discovered that the crowing of the cocks there is different from that of ours at home: they shriek in the most unmelodious fashion with a force which goes to one's marrow. Possibly the creatures have developed their great power of voice in order to make themselves heard above the noise of the streets, and the peculiarity has been gradually confirmed by heredity in course of time.

In our gardens the blackbirds give warning to all other birds when a cat makes its appearance. As soon as one of these beasts of prey shows itself they follow it from branch to branch, uttering the cry "dag, dag, dag," in rapid succession, the greater their alarm the quicker the *tempo*. Then the smaller birds come and join them with a confusion of anxious cries. At evening, when twilight comes on, the blackbirds utter the same cry for a time, but then it is not a warning cry, and the other birds do not so interpret it. At this time the blackbirds fly about as though quarrelling, scolding, and pursuing one another before they go to sleep. The meaning of this behaviour I have not been able to discover. When the blackbirds first see anything which makes them anxious—as when they first catch sight of a cat—they utter once only, flying from one place to another, a short succession of sounds which, as far as I remember, is something like "diridiridiridirollo." Now, I have noticed that the blackbirds of different districts, for instance those of the mountains at certain places where I listened to them, in comparison with those of Tübingen, present in these cries very considerable, even surprising differences. Local variation of the note is also known to occur in the chaffinch (Fringilla cœlebs) and in the nightingale (Luscinia Philomela).

I have made the latter remarks chiefly with the object of recommending the study of a subject, which although it ought to supply science with some remarkable facts, has hitherto received very little attention from zoologists.

G. Jäger [1] says with reference to this subject, "By means of looks, gestures, and sounds animals speak a very plain language, and it requires only a somewhat persevering attention to learn this language. . . . This sound and gesture-language reveals to us completely the feelings of animals, and their desires sufficiently disclose to us their intellectual powers. The sound-language which most mammals and birds, and some reptiles, fishes, and insects possess, consists of cries expressing feelings, like the utterances of a child in its earliest years; these cries are more or less prolonged tones, *i.e.* vowels, or noises, *i.e.* consonants, which are uttered once or several times in succession, while human words are combinations of tones and noises arranged according to definite laws, articulate. The interjections of our word-language are the most closely related to the cries of sensation of animals, for the former are in fact nothing but cries of sensation scattered among our vocabulary of words. The cries of animals, however, have not merely the value of interjections, they are something more. Thus the animal can express several sensations by modification of its voice, by modulation of its tones. Thereby animals are able to communicate their sensations and condition even during the night when they cannot see each other's gestures."

On the language of monkeys I find some remarks by J. von Fischer,[2] who quotes the above passage from Jäger. He says

---

[1] *Zoologischer Garten*, Bd. iii. p. 268, 1861.
[2] *Zoologischer Garten*, Bd. xxiv. p. 294, *et seq.* 1883.

concerning Macacus erythræus, s. Rhesus, as follows: "In fact, one learns to understand the language of animals in a very short time. I understood the sound-language of each of my monkeys, and knew exactly thereby the state of its feelings at any particular time. As the requirements of the life of animals are much simpler and less variable than ours, so is their sound-language much more limited. But the homely peasant speaks a language much poorer in words than that of the cultivated member of town society. The vocal expressions of the Rhesus were very simple, and consisted of vowels, so that they most resembled the interjections of human speech, although they had not exclusively the significance of the latter. The utterance at different times varied much in pitch, in force, and in quality, according to the feelings of the animal, so that the monkeys were well provided with means of expressing the temporary condition of their emotions."

The Rhesus expressed his feelings partly by voice and partly by facial expression. When he desired a thing, he cried "oh," or "o-oh," in the latter case the second syllable being higher than the first. At the same time he laid his ears close to his head, drew back his brows, and pointed his lips. Joy and pleasure he expressed by a grunting or gurgling throat-sound, which sounded like a hoarse "äh." At the same time he laid back his ears for a short time, drew back all the skin of his head for a moment with a jerk, his eyebrows being thus also drawn back, and stretched out his mouth with the lips narrowed. In extreme pleasurable excitement, when he laughed, he disclosed his teeth as far back as the middle of the molar series, and uttered a slight tittering sound like "kikiki," and so on.

Valuable results would certainly be obtained if students of language would take up this subject, and would endeavour to investigate what I think ought to be recognised as the speech of animals. One of them, Schleicher,[1] remarks: "Speech, that is, the expression of thoughts by words, is the only character exclusively peculiar to man"; in support of which conclusion he appeals to Huxley's well-known essay "Man's Place in Nature," in which that investigator comes to the conclusion that speech alone separates man from the anthropoid apes most nearly related to him.

But even though these anthropoid apes have really no power of speech, it does not follow that the faculty is not possessed by other animals.

Let us learn of those who study language what speech consists in.

Schleicher says on this point: "Sound-gestures, in some cases highly-developed sound-gestures, for the direct expression of its feelings and desires, the animal possesses, and by means of these, as by means of other gestures, animals are able to communicate their feelings to one another. Accordingly it is usual to talk of the speech of animals. But the faculty of directly expressing thought by means of sound is possessed by no animal, and this alone is the meaning of speech. How fully this is in fact recognised in our ordinary consciousness is shown by the consideration that an ape endowed with speech, or even an animal utterly different from man externally, if it possessed the power of speech, would be regarded by us as a man."

The zoologist and the anatomist, be it remarked by the way, would leave the attitude of mind described in the last sentence to the philologist, and likewise the subsequently-expressed conclusion that microcephalous idiots are not to be regarded as really human, because in these speech, and even the capacity for it, is wanting.

But apart from this, if the definition which Schleicher

[1] A. Schleicher, *Ueber die Bedeutung der Sprache für die Naturgeschichte des Menschen*, Weimar, 1865.

gives of speech is correct, then animals speak; for, as is shown by the instances I have given, and as the simple observation of nature teaches, many of them indisputably possess "the faculty of directly exchanging their thoughts by means of sound."

I go even farther, and maintain that even what is called "articulate" speech does not constitute a distinction between man and the lower animals. And in this the remarks of Schleicher himself afford me assistance. For in the same short paper he says: "Scientific investigation proves clearly that speech is something which has been quite gradually developed, something which once did not exist. The comparative anatomy of languages discloses that the more highly organised languages have been evolved quite gradually from simpler languages, probably in the course of very long periods of time: at least the science of language finds nothing to contradict the conclusion that the simplest modes of expressing thought by sound, the languages of simplest structure, have gradually been derived from sound-gestures and imitations of sounds such as animals employ. . . . But if speech alone makes the man, then our primitive ancestors were not at the beginning what we now call men, for they only became men with the development of speech. But the development of speech is for us equivalent to the evolution of the brain and the vocal organs. Thus the results of the science of language lead necessarily to the conclusion that man has been gradually evolved from lower forms, a conclusion at which the natural science of the present day is known to have arrived from other premises. For this reason alone the investigation of speech is of great scientific importance, especially in relation to the evolution of man. . . . Those languages which have hitherto been analysed into their simplest elements, and those which have remained at the simplest stage of evolution, show that the oldest forms of all languages were essentially the same. The oldest sounds of which languages consist are those which express perceptions and ideas. Expressions of relation (distinction of parts of speech, declension, conjugation), are at this stage wanting; all this is a later development, to which many languages have not attained, and which the others exhibit in different degrees of perfection. Thus, to indicate one example in Chinese at the present day, there is no distinction in sound between the parts of speech; true verbs, as distinguished from nouns, among all the languages I have studied, I have found only in the Indogermanic. Morphologically, but only morphologically, according to our results, all languages are originally essentially similar; but on the other hand, even these primitive languages must have been different in sounds, as well as in the ideas and perceptions which were reflected in the sounds, and must have further differed in their capacity for evolution. For it is positively impossible to derive all languages from one and the same primitive form."

In his earlier and more widely-known work, in which he explains the origin of language on Darwinian principles,[1] Schleicher gives an example in support of these views. He says there: "The oldest form of the words which now in German appear as *That, gethan, thun, Thäter*, was at the time when the original Indogermanic language arose, *dha*, for this syllable *dha* . . . is found to be the common root of all these words. At a somewhat later stage of development, in order to express definite relations, the roots which then performed the part of words were pronounced twice and another word, another root was added on to them; but each of these elements was still independent. For instance, in order to indicate the first person of the present tense, the speaker said *dha dha ma*, from which, at a subsequent period in the

---

[1] A. Schleicher, *Die Darwin'sche Theorie und die Sprachwissenschaft*, Weimar, 1863.

history of the language, by fusion of the elements together, and by virtue of the tendency to alteration in the roots, was produced *dhadhâmi* (in Sanskrit *dádhâmi*, Old Bactrian *dadhâmi*, Greek τίθημι, Old High German *tôm, tuom* for *titômi*, modern High German, *thue*)."

I think the above completely justifies me in speaking of the sounds uttered by blackbirds as their language. "Dag, dag, dag," is as good a sound as "dha, dha, dha," not distinguished from the latter as an "imitative sound" or a "sound-gesture." We may conclude that it means in general, "Take care! Danger!" but with a special tone and in a certain succession it probably means distinctly "Cats!"—uttered in another way, perhaps "Owl!" or "Crow!" and in any case the variation of the same "word" which the blackbirds utter every evening has a meaning quite different, for then the other birds take no notice of the cry.

Thus, just as in the lower stages of human speech, the development, the complexity of language in birds evidently just proceeds from the repetition of sounds. By different modes of repetition and by differences of tone in addition a variety of things can be expressed. But in the language of blackbirds, as in the higher stage of evolution of human language, the further step has been taken of adding another sound at the end of a repetition, in the cry, "diridiridirirollo." Thus all the elements necessary to the development of the most perfect language are present. The drum-language of the negroes of the Cameroons shows how much can be said merely by variations in the repetition and in the loudness of a single tone, which is less than an uttered sound. Thus flexibility of speech is not always so complete a criterion of mental capacity as we are accustomed to assume without further reflection. The want of flexible speech is wanting in the anthropomorphous apes, not because their brains are too lowly organised, but because their laryngeal mechanisms are not sufficiently delicate. It is true of course that the development of speech causes modifications in the brain and nervous system, but primarily only such modifications as relate directly to speech, and only at a later time those which are connected with the higher mental development made possible by speech.

If not only is our highly-evolved language derived from a simple sound-language, but there are at the present day tribes who have never advanced beyond the latter stage, then no fundamental distinction exists between the language formed by the voices of animals and human speech. It depends, as I have said, merely on the structure of the larynx that the animals most nearly related to man have no well-developed vocal language. Such animals contrive to make themselves understood in other ways; the apes convey a great deal of meaning by facial expression, while insects, *e.g.* the Hymenoptera, especially the ants, have evidently a highly-evolved means of communication in the tactile language of their antennæ.

But if the development of speech depends on the structure of the larynx, we ought to be able to point out evident anatomical differences in this organ even among different races of men. This objection I have often heard expressed by eminent philologists, with the addition, "But nothing of the kind has ever been found;" whence it was then concluded that the structure of the larynx has nothing to do with the origin of dialects or with the evolution of language. It is quite true that nothing of the kind has been found; but to anatomists and physiologists such a discovery is not necessary to enable them to ascribe the modifications of language essentially to anatomical causes, in other words, to the acquisition and inheritance of anatomical and physiological peculiarities depending on habitual use. Schleicher takes exactly the same view of this question as myself, and in quoting his words I am brought back to a subject already discussed, the learning of foreign languages, my view on which is further developed by Schleicher. On the latter subject he says :[1] "If language really depends on particular adjustments in the brain and vocal organs, how is a man able to acquire any other or even several other languages besides his own ? To which I might . . . answer that a man can learn to walk on all fours, or even on the hands alone, and yet no one will doubt that our natural mode of progression is determined by our bodily structure, and is the expression of that structure. But let us examine into the objection more minutely. The first question to be asked is, Whether a foreign language is ever perfectly acquired and appropriated ? I doubt this, and at most will only admit it to be possible when a man exchanges his native language for another in early childhood." In that case, however, the person in question would be a different person, for his brain and vocal organs would develop in another direction. It is further to be considered, in talking of the acquisition of European languages, that all the Indogermanic languages belong to the same family, and, regarded broadly, are species of the same language. " But show me the man who thinks and speaks with perfectly equal facility in German and Chinese, or in the New Zealand and Cherokese tongue, or in Arabian and Hottentot, or in any other two languages differing in their fundamental constitution. I do not believe such a man exists, any more than I believe that any individual will ever be able to walk with equal agility and comfort on all fours and on his two feet. It is often even impossible to us to pronounce sounds peculiar to foreign languages, or even to distinguish these with our ears correctly and exactly." He concludes, therefore, that a given vocal organ, like every other kind of organ, has a definite function, which is, and remains, always natural to it.

[1] *Ueber die Bedeutung der Sprache*, etc.

The postulate from which Schleicher derives these conclusions is the belief: "That the activity, the function of an organ is, so to speak, only one of the qualities of the organ itself, although it is not always possible to the scalpel and the microscope of the investigator to exhibit the material causes of that quality." As with gait, so it is with speech. Speech is the symptom, perceived by the ear, of a complex of material relations in the structure of the brain and vocal organs, with their nerves, bones, muscles, etc. Lorenz Drefenbach had already expressed the same ideas: "The material basis of language and its varieties has not yet been anatomically demonstrated, but to my knowledge a comparative investigation of the vocal organs of people speaking different languages has never yet been undertaken. It is possible, even probable, that such an investigation would lead to no satisfactory results; nevertheless, this would by no means be enough to destroy the conviction of the existence of material conditions of language in the structure of the body. For who would deny the existence of such conditions, although they are still concealed from direct perception, and possibly can never be made the object of direct observation?" The results of infinitesimal quantities and relations are, he continues, in some cases wonderfully striking; as, for example, in the phenomena of the spectrum, of colour and smell in plants, and so on. As light is to the sun, so is sound to speech; as in the former the character of the light is evidence of a material condition, so in the latter the character of the sound.

With these arguments of the philologist I fully agree, and here repeat the proposition which forms the subject of the whole of the present section, and embraces that of the preceding: That functional activity—exercise—universally precedes and determines the higher evolution of organs. The larynx of our ancestors was at first quite incapable of producing language

—as it was exercised in the production of various kinds of sounds, it produced them more and more perfectly, and its capabilities were gradually improved. These capabilities therefore are acquired, they were transmitted from one generation to another; but as they were developed ever greater changes must have taken place in the powers of contraction of the muscles and the efficiency of the nerves of the organ, and above all in the vocal chords. These changes, on account of the great delicacy of the mechanism, were able to produce great effects, although they are not anatomically demonstrable —any more than the special structure of the fingers which helps to determine a peculiar handwriting can be exhibited by dissection. Accidental variations of the vocal chords under the action of selection may have promoted the progress of the evolution.

Concluding Remarks.

In the second section I gave reasons for the conclusion that the external stimuli which act directly upon the protoplasm are the primary causes of the manifold variety among organic forms. On the other hand, as I there explained, I regard the influences of use, of functional activity, which have been discussed in the last and preceding sections, as the indirect or secondary causes of growth and of the modification of structure. As the direct influence of stimulation, which I have previously named "impression," promotes growth, so also does exercise, which modifies the protoplasm and, by producing an increased inflow of the nutritive fluids, adds to its bulk.

But in many cases the effects of direct and indirect stimulation cannot be separated. When a stimulus acts constantly in a certain way upon the protoplasm, the latter is exercised in a corresponding manner, and is thereby modified. In fact, when we speak of use as the cause of growth or of the modification of structure, we mean spontaneous use, the active motion of the organism; and the influence of exercise in causing growth is therefore principally observed in animals. But when subsequently we consider for a short time the causes of modification in the vegetable kingdom, we shall see clearly how impossible it often is to separate the direct effects of external stimuli from those of exercise, how the two frequently coincide. And from a physical point of view this requires no explanation.

It is, however, a necessary consequence of my conception of growth that I should regard the effects of indirect stimulation of the living substance also as growth.

SECTION VIII

THE IDEA OF ORGANIC GROWTH—THE LAW OF ORGANIC FORM—RECRESCENCE

The Idea of Organic Growth

We have to distinguish from one another (*a*) individual (personal) growth, (*b*) the growth of the race (the species) or phyletic growth. The latter is, however, merely the sum of the modifications due to growth which the individuals of a line of descent have undergone in course of time, by which modifications, together with the separation between it and its allies, the species is constituted. Usually personal (individual) organic growth means the regular changes which take place in a given organism under external influences (food, warmth, light, gravity, etc.), and which are connected with an increase in size; for we assume that this growth depends on the multiplication of the particles of the body, and accordingly the assimilation of nutritive material enters largely into our idea of the process.

On the other hand, by organic growth I mean every physiological change of structure which is naturally produced in a given organism by external influences or by constitutional causes, which is not morbid and not accidental, and which is permanent, or only temporary because it precedes a further stage of modification.

Growth, therefore, does not necessarily produce visible alterations—the changes which precede an increase in size are growth, and so are alterations in the position of particles (molecules) which produce no increase in size; such a change of position may only cause an alteration in the form of the body, but it may even produce a diminution in size.

Growth is by no means necessarily the result of the assimilation of food: the action of any external stimuli is capable of causing changes in the position of the particles of the body, and thereby of causing growth in my sense of the term. The idea includes every inheritable alteration either of material or form in the organism, or (as underlying these) every alteration of the interaction of forces in the organism; and therefore the result produced by stimuli influencing the protoplasm indirectly is growth. Therefore, also, even diminution in size and degeneration is growth. Thus plants and animals grow to a smaller size in the north and on mountains than in the south and in valleys, and the exclusive influence of one condition, abundance of food, leads to the degenerate processes of growth which parasitic forms exhibit in organs related to other conditions.

Thus two things are necessary to produce growth (1) the given composition of the organism; (2) the action of stimuli (food being considered as a stimulus). It was in reference to the former that I described in previous works some of the causes of the modification of forms as "internal" or "constitutional causes."

Since living beings differ from one another in the composition of their bodies, so stimuli act differently upon them, produce in some changes different from those they produce in others: they grow in different ways.

The constitution of the body, which has so important an influence in determining the course of growth in an organism, is to a very great extent the result of the inheritance of characters from ancestors; to a small extent it is due to acquirement, *i.e.* modification, during individual life, or is the consequence of the mingling of the characters of the parents. This latter small element in the constitution of the body is the cause of individual variation; but it is of the greatest importance, because on it depends essentially the continuous modification of forms.

If the individual growth of the organism is, as according to the preceding we may briefly express it, in the last result nothing but the effect of external stimuli (including food) upon the tissues, and if we assume that there was a primitive organism from which all succeeding living beings have been derived, then the variety of growth which took place in the latter must of necessity have been originally due to the variety of the external influences (stimuli) under various conditions. Sexual mixture was not at the beginning in operation. But the peculiarities thus acquired were inherited by the organism in its whole and in its parts, and selection increased them. The greater the number of modifications thus by continued inheritance acquired by a succession of organisms, the greater will be the peculiarity of constitution produced.

Thus the individual growth of every plant and every animal is a brief and rapid repetition, under the continued influence of similar stimulation, of the series of effects produced by external stimuli in the course of vast periods of time on the tissues of its ancestors.

The character of the individual growth of every living being therefore depends essentially on phyletic growth, the individual growth includes phyletic growth in itself. For even the peculiarities of constitution of different individuals which are due to inheritance are really a consequence of phyletic growth.

Since the individual growth of every living being is thus a

stage of phyletic growth, since the latter, wherever we contemplate it, represents a sum of individual growths, both are traced back to one and the same process—fundamentally they cannot be separated.

Phyletic growth, or the evolution of the organic world ever into higher and more complex forms, or at least into forms of different structure, is, as I have said, merely the sum of the processes of growth of the ancestors—together with the result of external influences on the forms during their development and their existence. This additional modification which the individuals as such undergo is—together with the influence of crossing—the very cause of the constantly progressing evolution. All that the members of a series of individuals directly connected by descent acquire constitutes together the material for the formation of new species.

Individual growth was described above as a process which according to constant laws forms permanent conditions, or conditions which when they are transient constitute stages in the further development in the same direction. The same was previously asserted of phyletic growth; it forms permanent stages which serve as the basis for further development, and which thus render progression (or retrogression) possible.

The variety of the external conditions, *i.e.* of the external stimuli (including food) and of the constitution, together with crossing, necessarily led to variety of growth, to variety of structure, necessarily led to the formation of a variety of species, provided that separation of the continuous chain of forms into separate links took place.

Thus we can appropriately speak of the organic growth of species.

The reader should compare the previous propositions with the arguments of Section II.[1] I have there made special use of the biogenetic law as evidence that the world of

---

[1] Pp. 21, 25, 51, etc.

organisms has *grown* in course of time out of cells. I described the individual development as an abbreviation of phylogenetic growth.

Growth, in the sense of continual modification under the influence of stimuli is a fundamental property of protoplasm.

Since the manifold variety of the forms of the organic world is due to growth, it is seen to be a necessary consequence of that fundamental property.

As I have already insisted in a previous passage,[1] reproduction also is a fundamental property of protoplasm, because it is indeed nothing else than continued growth ; and, I have to add, since it essentially consists in the transmission of the properties of one protoplasmic unit (person) to another, it is on this fundamental property of protoplasm that the immortality of life depends.

Crossing and Selection as Indirect Causes of Growth

Besides growth as the result of the action of stimuli, we have to consider in relation to the production of the heterogeneity of the organic world, and as indirect causes thereof, crossing (sexual mixture) and selection. Of these two crossing alone is capable of creating anything new, of contributing by its own action to the growth of the organic world. This possibility is due to the fact that by the mingling of two forms a third new form may be produced. But this does not occur to so great a degree nor so commonly as is frequently assumed. Nevertheless the influence of crossing in the modification of forms is an important one. Natural selection can, as I have repeatedly remarked, create nothing new. It only so far contributes to the growth of the organic world that it selects the forms which are most fitted for life, and preserves them for the future action of new stimuli and of

[1] P. 24.

crossing. Its influence is highly important, but it produces its results principally by the aid of crossing: by sexual mixture various directions of growth already determined may be united and strengthened or modified in the most various ways. Whatever in the result of mixture is useful, has a greater claim to exist and is capable of giving rise to new, altered directions of growth. Thus the power of selection lies chiefly in the promotion and diversification of organic growth. It is like crossing, only an indirect cause of the evolution of living beings. As was previously pointed out, sexual differentiation, the necessary antecedent of crossing, has itself only been developed in course of time—is itself the result of peculiarities of growth in two different directions, and these peculiarities of growth are again primarily conditioned by peculiar external influences (nourishment), and preserved by selection.

The Law of Organic Form: Its Application to the Form and Structure of Plants

In accordance with the preceding, we may formulate a law of the form of living beings in the following terms: The external form of every individual, every variety, species, genus, family, etc., is the resultant of a number of processes of growth which have taken place in its ancestors, together with the effect of external conditions which have acted upon it during its individual development and life, and of spontaneous internal modifications.

In other words: The external form of every organism is the result of the action of external influences on all its ancestors, together with the effect of such influences and of spontaneous internal modifications during its individual life.

The influence of selection is included in this definition.

Moreover, by "form" I mean here essentially the form which grows from the germ by natural necessity as the result of regular growth. But with the words "during its individual life" reference is made also to the alterations of form which organisms undergo during life and which may be inherited. These inheritable modifications produce at first little or scarcely any visible external alteration; they are internal, dynamic. Nevertheless, as I have stated, external features due to age, and even mutilations, as well as diseases which affect form, may be inherited.

By the expression "spontaneous internal modifications" I mean the alterations produced by constitutional causes, among them some senile changes, which help to determine form.

Vegetable physiology supplies the most palpable evidence that the forms of organisms are determined by the influence of external forces on protoplasm, by acquired and inherited properties.

Is it not the influence of light which determines the direction of growth in plants? Do not light and air produce the direction, expansion, and position of leaves on which nutrition depends? Is it not the force of gravity which determines the form of roots and causes them to grow towards the centre of the earth? Has not nutrition the most profound influence on the form of all parts of plants? Is not the influence of temperature equally important?

To which the reply will be made: that if all external stimuli were removed all these effects would disappear, that they themselves are not inherited, but only the predispositions towards them. I have already remarked previously that every "predisposition" presupposes a molecular modification. With the predispositions are inherited the modifications which alone render possible definite directions of growth under the influence of external stimuli. It is certainly true that if all

2 c

stimuli are removed all organic form disappears. If only the necessary warmth is wanting, there is an end of it. Life itself, and with it all the forms in which it resides, are only the effects of stimuli, and therefore the above objection is entirely delusive. Indeed, the definition of species, genera, etc., is rendered possible only by the characters produced by external influences. Take away the latter, and with them the characters conditioned by them, and we have no species, genera, etc., no "kinships" left. It follows from this consideration that, as already maintained, the question of the formation of species is by no means so profoundly affected as systematists, and, curiously enough, physiological botanists suppose, by the fact that plants—as, for instance, species of cereals—revert to an original form when the external conditions which determine their peculiarities cease to act.

We employ colours as distinguishing characters. Chlorophyll is a most important factor in the life of plants, and yet in the absence of light the green colour disappears—it is not inherited, but the materials which are its basis are inherited, and these materials have been acquired. If we removed from plants during their development the stimuli which act upon them, without causing their death, so that we obtained abnormal forms, these would still retain by inheritance certain established properties which would cause them, when the action of the stimuli was renewed, to develop only in the definite directions which were natural to them.

In all organisms definite tendencies of growth are inherited, which alone constitute the specific qualities of the various species.

The stimulus of gravity having acted upon plants during endless ages, so that roots were developed in a downward direction, the parts which became roots by this particular kind of growth have necessarily developed a tissue of a special character, a tissue whose cells tend to grow towards the centre of the earth, and in that direction only. The roots also developed in consequence of their peculiar growth special morphological characters, a cell-structure which is more or less different from that of other parts of plants. All such speciality of structure has certainly arisen simply as a result of the inheritance of acquired characters, and it constitutes the type, the "kind" of the plant. Similarly the tendrils of climbing plants—structures which owe their origin to the action of external stimuli, to the inheritance of acquired characters—are essential to the definition of their species; and equally so are the position and colours of the leaves, etc., which are due to the action of light and air. In all such cases special morphological and physiological characters have arisen in consequence of the continuous action of stimuli, and the functional activity in definite directions thereby set up in the cells of plants—direct stimulation and functional action cannot here for the most part be separated. In many such cases the direct effects of stimuli and function, and the inheritance of these effects, are so obvious, that it seems to me we must recognise them unless we refuse to recognise cause and effect altogether.

Must not the water which a desert-plant stores up within itself, in order to keep itself alive in spite of drought, have had an indirect and direct influence on its whole internal constitution? And when a seaside-plant like Salicornia herbacea takes so much salt from the sea-water that it has a salt taste, must not this salt have had an influence on its whole cell-structure—an influence which enters into its whole specific character and which is inherited? Who would substitute for such obvious action of the simplest causes the accidental variation of the germ-plasm and selection?

I will not here present the anatomical evidence of my view, since every text-book of botany will supply it. I will only adduce some especially remarkable and cogent facts.

The parenchyma of the leaves of plants which grow in the light differs completely in cell-structure from that of those which grow in the shade : the cells in the two cases are of quite different shape,—in the latter long and prismatic, in the former short, and so on.[1]

The so-called compass-plants place their leaves in sunshine so that their edges are directed to the north and south, whereby the leaf is least exposed to the sun. This peculiarity was first observed in the North American Silphium laciniatum, but is found equally developed, according to E. Stahl, in a native species of lettuce, Lactuca scariola.[2] The vertical leaves turn their largest surface to the rising sun. In proportion as the sun rises higher, the angle at which its rays strike the leaves becomes smaller, until finally at mid-day all the leaves when regarded in the direction of the sun's rays present only their edges to the eye. In the afternoon the incident angle of the sun's rays upon the leaves again gradually increases, so that towards evening the light again strikes them at right angles. Silphium laciniatum belongs like Lactuca to the Compositæ. But the leaves of many Papilionaceæ, *e.g.* of the beans, in strong sunshine place themselves edgewise : by the twisting of the articular protuberance the leaves are brought into the position which presents the smallest surface to the sun. By this means excessive heating and illumination are avoided.

Plants show this property especially in dry regions, and it is probably much more widely distributed than it has yet been observed to be. However, some other plants are known in which it occurs in a less degree.

This, therefore, is a case in which the action of stimulation is useful, as it is for instance in the carnivorous plants, in

[1] Cf. E. Stahl, *Ueber den Einfluss des sonnigen oder schattigen Standorts auf die Ausbildung der Laubblätter, Jenaische Zeitschr.* Bd. xvi. 1883.

[2] E. Stahl, *Ueber sogenannte Compasspflanzen, Jenaische Zeitschr.* Bd. xv. 1881.

which extraordinarily minute quantities of material often cause the reaction. Sensitiveness of this kind in relation to certain stimuli is inherited. There can, however, be no doubt that light and air have had an influence in the manner supposed by me also on the permanent inherited position of leaves.

Indeed, I believe that leaves in general owe their origin partly to the action of light and air. That adaptation has had some influence in their production, both generally and in particular in the evolution of their more minute structure, is to me self-evident, but that, *e.g.* the variety of form of the leaves of our native trees is essentially due to adaptation, is a belief for which I can find no basis. I consider the differences rather to be for the most part the simple results of growth as affected by direct external stimuli, principally by nourishment, a conclusion supported by the changes which occur when the nutrition is altered. Of course the forms are partly determined by the veins of the leaves, which are the channels for the passage to and fro of the nutritive material, and which have the additional function of keeping the leaves expanded. But it is not possible to say that just this or that distribution of the veins is a necessary requirement of adaptation—this also is an obvious consequence of definite directions of growth.

THE RECRESCENCE OF LOST PARTS AS AN EXAMPLE OF ORGANIC GROWTH

The discussion of the influence of definite directions of growth on the form of organisms leads me to the subject of recrescence, which it seems to me can only be explained by the aid of my theory of organic growth.

Recrescence is, in fact, nothing but an effect of the same causes which condition growth in definite directions. Among cases of recrescence two kinds can be distinguished (1) those in which obvious external stimuli directly excite the new process of growth; (2) those in which no such excitation occurs. Among the first group is the growth into new plants of parts of a plant when cut off and brought under favourable external conditions—such as cuttings which send forth roots when they are set in the earth, and grafts. Grafting can with certain limitations be performed on the bodies of animals, including man. I allude to the transference of pieces of another person's skin to the human head, or to other parts of the body-surface (transplantation).[1]

A similar phenomenon, although it does not fall under the definition of the term grafting, is the readhesion of noses, ears, and even finger-joints which have been cut off, when they are immediately brought into contact with the exposed surface—a process of growth which like grafting is so far connected with recrescence that to effect the readhesion new parts must be formed—therefore the last remnant of recrescence in man.[2]

Vöchting has shown that the action of gravity alone is sufficient to produce roots on cuttings from plants, and that it is not even necessary to set them in the earth: if twigs of willow bearing buds of similar age along their whole length are laid in a chamber which is kept dark and saturated with moisture (or in moist earth) roots appear on the under side only. Still more striking is the effect of an external stimulus in Lepismium radicans, a plant belonging to the Cactus tribe, on the production of roots, for these are produced on any part of the sprouts of this plant from which the light is excluded.[3]

---

[1] Recently Thiersch has even transplanted pieces of the skin of a negro on to a white man, and conversely; in the first case the pieces soon became white, in the latter black.

[2] For other examples see O. Weber, *Die Gewebserkrankungen im Allgemeinen und ihre Rückwirkung auf den Gesammtorganismus.* I. Bd. i. *Abschnitt* of V. Pitha and Billroth, *Handbuch der allg. und spec. Chirurgie,* 1865.

[3] H. Vöchting, *Sitzungsb. d. niederrhein. Gesellsch. f. Natur-u. Heilkunde in Bonn, Sitz. v.* 3 Jan. 1876.—On the divisibility of plants and the influence

Cases belonging to the second of the above-defined groups are much the more important for my argument, for in them the recrescence is evidently entirely the result of the definitely directed powers acquired by the ancestors and transmitted by them to their descendants.

The processes which underlie recrescence are no others than those which condition asexual multiplication. Indeed, in lower animals, where recrescence leads to the production of entire animals, the two phenomena coincide in every respect. Whether we cut a small annelid, a water-worm (Nais proboscidea), into two parts, with the result that each part grows into a new perfect individual, or whether the worm divides itself spontaneously and grows into two new animals, in both cases we have exactly the same process.

In the recrescence of the water-worm we have an example of the second kind of recrescence; there is no external stimulus which directly prompts it. Of the relation of nutrition to the process I say nothing—the possible objections to be made on this ground are obvious of themselves, as also the considerations by which they can be extenuated or overcome. The parts of the worm, without any special external stimulation during the process, grow in a definite direction in a perfectly constant manner again into an entire animal—that is for me the principal fact. New eyes and a new tentacle are produced on the anterior segment, and a new nerve-ring and cerebral ganglia within it. From this anterior segment, by its repeated division, the longitudinal growth of the new animal takes place, that is, from before backwards, just as in natural division. This recrescence evidently repeats the process which originally led to the formation of the worm with its eyes and proboscis, when it was evolved from a uniseg-

of internal and external forces on the formation of their organs, see Pflüger's *Archiv f. Physiol.* Bd. xv. (Extract from *Ueber Organbildung im Pflanzenreich*, I. Bonn, 1878.)

mental organism, as we have grounds for assuming it once was.[1]

The same process repeats itself in the individual development of the worm—in all these cases we have definitely directed processes of growth, which can only be the result of inheritance.

The remarkable fact that in an animal thus divided organs, and sometimes very complex organs, develop in the proper positions, without any aid from the direct action of external stimuli, requires to be illustrated by further examples before I enter more minutely into its explanation.

Everyone knows that when the tail of a lizard is cut off another grows in its place—even Pliny dilates upon the fact. Aristotle mentions its occurrence in salamanders and snakes. But the recrescence of such a complex highly-evolved organ as the eye of vertebrates is much more remarkable. Blumenbach recorded the occurrence of this recrescence as early as the last century, first in newts,[2] and afterwards in Batrachia[3] (according to Merkel it had previously been recorded by Brühl). But this recrescence only takes place when some remnants of the various layers of the eye are left in connection with the uninjured optic nerve.[4] In Batrachia, according to Blumenbach, the lower jaw is also reproduced. The legs of newts, when renewed, grow again in the same way as when they develop in the larvæ.[5] The spinal cord is also reproduced in Amphibia and reptiles, as, for instance, in tadpoles, in newts, in Pleurodeles Waltlii—in the latter animal very completely in five months. The spinal ganglia likewise are renewed, apparently growing out from the spinal cord.[6] That

---

[1] Cf. p. 62.

[2] Blumenbach, *Specimen physiologiae comparatae*, 1787, p. 31.

[3] Idem, *Kleinere Schriften zur vergleichenden Physiologie*, 1800, p. 31.

[4] J. Ch. Eggers, *Von der Wiedererzeugung*, Würzburg, 1821.

[5] Götte, *Ueber Entwicklung und Regeneration des Gliedmassenskelets der Molche*, Leipzig, 1879.

[6] P. Fraisse, *Die Regeneration von Geweben u. Organen bei den Wirbelthieren*,

the legs and claws of Crustacea, the tentacles of snails, and even the heads of the latter, if the brain is not destroyed, are reproduced is well known, and has already been partly discussed. Equally well known is the recrescence of the arms of star-fishes. The legs of insects also grow again. I have only brought together here a few out of the large number of instances of recrescence.

How then does it come about that such organs, many of them exceedingly complex, grow again from the body at the places where they were originally present, without the direct action of stimuli?

The old explanation was that it was due to the "formative tendency," and this was deemed sufficient. We may still use the same term, provided that we get rid of the meaning which the earlier school attached to it, namely that of a spontaneous impulse more or less independent of matter. If we guard against this, if there be any need to do so, and employ the term in the sense of the action of definitely directed forces contained in the material of the organism, then there is nothing to be said against it. We apply the word "shoots" in an entirely passive sense to parts of plants. And in the same sense recrescence consists in the production of shoots, of normally-constructed parts which are, as it were, shot out from the organism by natural necessity. Thus we may use the word tendency or impulse in an active sense, though with certain limitations, in speaking of recrescence. But we conceive the formative tendency not as an independent force by the action of which everything is explained, but as a force of which we have to discover the causes.

My view is that the recrescence of lost parts of the organism must be ascribed to heredity, that the latter is the

---

besonders Amphibien und Reptilien, Cassel and Berlin, 1885. In reptiles (lizards and geckos) the recrescence of the spinal cord takes place only imperfectly.

mechanical cause of the restoration of the mutilated organism as a whole to its previous form.

Ever-repeated inheritance not only causes every organism to repeat in its individual development the directions of growth by which its form is determined, but, even after that development is completed, causes those directions of growth to manifest themselves in the restoration of lost parts. For recrescence not only depends on exactly the same causes as ontogeny, it is to be regarded as a continuation of the latter. By the ever-repeated inheritance of the characters of an organism, and therewith of its organic unity, it has come about that heredity not only reconstitutes this unity by development, but also endeavours to reconstitute it after injury.

In order to understand this we must previously grasp the fact that all the several particles of which the organism is composed, that all the cells, all their molecules, stand to one another in perfect correlation, so that each is influenced by the fate of the others, and that also in each such particle there is something of the tendency to form a whole—to unite together with the other particles to form a whole, just as inorganic particles crystallise into a whole from the mother liquid. The formation of crystals is in fact scarcely less mysterious than the development of organic forms—the latter also, to return to a previous comparison, is a kind of crystallisation. Thus I also come back to the conception of correlation, which is nothing else than the expression, the consequence of the relations of all particles of the organism with one another.

I have explained correlation also as organic crystallisation, and recrescence is in fact an instance of correlation.

My conception of the nature of the process of recrescence becomes more intelligible when the attention is fixed, not on the morphological parts of the body, but on the forces which govern them, when the organism is conceived as the expression of a sum of forces which mutually supplement one another, which govern the particles and restore their equilibrium when it is destroyed—restore equilibrium in the sense of reconstituting the previously existing whole.

That a tendency to develop into the whole by growth is impressed upon the several particles of the organism by heredity is best shown by the fact that many multicellular plants and animals can be subdivided almost into their constituent cells with the result that each portion grows again into an entire organism. Since Trembley's time it has been known that the smallest parts into which our common Hydra is divided become new entire individuals. The question, however, arises whether such fragments must not consist of at least two connected cells, an ectoderm-cell and an endoderm-cell, if recrescence is to ensue [1]—because the ectoderm and the endoderm even in these lower animals respectively have special characters established by heredity. In any case, recrescence from a single cell must be possible in those lower organisms which consist of simple colonies. It takes place in this way even in higher plants. A very remarkable instance of this is recorded by Vöchting:[2] "A piece from the middle of the surface of a vigorous thallus of Lunularia vulgaris was cut up with a sharp knife on a smooth plate of cork until the fragments were so small as to form a coarse-grained pulp. The largest fragments were about half a cubic millimetre in size, while the smallest were considerably smaller. This pulp was spread out on moist sand and protected as much as possible from external disturbing agents. After some time young sprouts showed

---

[1] According to Th. W. Engelmann this is in fact the case. Cf. *Zool. Anz.* I. p. 77, 1878.

[2] H. Vöchting, *Ueber die Regeneration der Marchantien*, Pringsheim's *Jahrbücher für wissensch. Botanik*, Bd. xvi. heft 3, pp. 15, 16, Berlin, 1885.

themselves, at first one or two, but afterwards an ever-increasing number, and at last quite a forest of young fronds were growing from the pulpy mass. Examination showed that by far the greater number of the particles had remained fresh and been able to produce adventitious sprouts.

"This experiment clearly shows how remarkably the thallus of this plant is able to endure external violence, what energy of life resides in even the smallest aggregate of its cells. This is almost sufficient proof that the unity of the organism is contained potentially in each single vegetative cell; indeed, it ought not to be impossible under suitable conditions to prove the truth of this proposition by direct experiment."

By experiments on the subdivision of willows, Vöchting obtained further results, which led him to the following conclusions: "In whatever direction we divide the organism (the willow), and to whatever degree we continue this subdivision, in every fragment the whole organism lies as it were concealed, provided that the fragment contains cambium-cells. If the subdivision could be actually carried so far as to isolate an uninjured cambium-cell, this cell would doubtless be able to reproduce the whole organism."

Such facts can only be explained in this way: That to each cell of an organism possessing such a power of recrescence, as to a germ-cell, the properties which the whole organism inherited have been transmitted also by heredity; and that, further, every such cell, like a germ cell, by inheriting the tendencies of growth and the general formative powers of the ancestors, has acquired the capacity of growing again into a whole organism.

The higher such an organism stands in the scale, so much the more formative capacity must it have inherited—the willow more than the Lunularia,—and in a series of species directly descended from the same ancestors in a straight line, the more highly-organised species must always contain in itself the sum of the most essential of the tendencies of growth of the lower, and must be capable of exhibiting those tendencies.

Thus, as it seems to me, as by individual development, so also by recrescence, complete proof of the truth of my theory of the organic growth of the living world is supplied.

I must make my argument complete by entering further into details.

The reason why the power of recrescence is greater in simple organisms than in more complex, why in general it is greater in proportion as the grade of organisation is lower is obvious: for, as was stated above, the power of recrescence depends upon the retention by the cells of the organism of the properties of the germ-cell, or mother-cell. The lower we descend in the scale of multicellular animals and plants, the more these approximate to colonies of unicellular beings—the more is each of the component cells similar or equivalent to every other, similar or equivalent to the germ or mother-cell. Therefore recrescence in a simple cell-colony such as constitutes any of the lowest multicellular plants or animals is not surprising—it almost explains itself that every cell of such a colony can grow again into the whole. Every cell in such colonies possesses every capacity required for independent life. The higher the grade of the organism, that is, the greater the degree to which division of labour, the differentiation of the formative tendency into various directions, has been carried, so much the less is this the case, so much the more do its cells depend on other cells, its parts on other parts, finally on central organs, without whose aid they cannot live and multiply, and therefore cannot produce recrescence. Again, the power of recrescence is greater in youth than in age. It is greatest indeed in the embryonic or fœtal condition, as is shown, for instance, by the development of incomplete rudiments of fingers after fœtal self-amputation in man.[1]

The reason of this is that the cells have still in the young state—the younger, the more—the same general character as in lower organisms; further, that the protoplasm has not been completely used up in forming definite structures, and is in every respect less exhausted, more capable of life and growth than in the adult condition. The more minute explanation of the differences will be given in what follows.

In organisms also which have no fixed external form, as, for instance, many sponges, the recrescence of the whole out of any part is not very surprising, and yet it is only a step from this to the recrescence of the definite regular forms of such multicellular animals as the Hydra. However, the differentiation of the cell-layers even in sponges makes the explanation of the process less self-evident. The greater the differentiation of function and the constancy and complexity of structure the more difficult it becomes to understand recrescence. For from the formation of the germ-layers onwards the functions of the cells of multicellular animals increasingly diverge: the cells of the various layers soon begin, if I may so express it, to meet with different experiences, they must acquire different capabilities. They must therefore gradually be restricted to the power of forming only organs or parts of organs which proceed from the germ-layer to which they belong.

It is obvious, however, that in considering the question our inquiries cannot stop at the germ-layers, the simple or primitive organs which retain their original character in zoophytes throughout life: they must be extended to every portion of a germ-layer which is destined with others to form some part of the body, and therefore to every complex organ, and in plants

[1] Cf. Spiegelberg, *Lehrbuch der Geburtshülfe*, Lahr, 1878, p. 356.

to every group of cells which is to form a part of the organism. The question is in every case, To what degree the formative possibilities of the cells in question have become unchangeable and definitely restricted, to what degree they are still of a general kind. That at the lower stages of development these possibilities are still general we have already seen. We know that in plants even of a high grade of organisation roots may grow from the most various parts if these are brought under suitable conditions. Indeed, according to Vöchting's experiments, any fragment of a willow may, as we have seen, grow into a perfect tree,[1] provided only that the fragment contains a certain formative tissue, namely, cambium. These facts might suggest the question whether a special cell-tissue is not present also in the adult animal from which all parts can be produced, a kind of animal cambium, a general formative tissue which retains general formative powers throughout life. The blood will first occur to us as most likely to fulfil these conditions, and especially the white blood-cells, which take so prominent a part in constructive and destructive processes,—but we have no evidence that these can produce any but mesodermic tissues, no evidence that endodermic or ectodermic tissues can be produced in higher animals from other than their own primitive layers. In fact, it follows from the nature of heredity that they cannot. To

[1] I have already pointed out how impossible it is to reconcile such facts as this with the theory of the continuity of the germ-plasm. From the fragment of willow is formed a willow tree with its reproductive organs, that is with germ-plasm, and thus the latter is contained in the fragment—in Lunularia is contained in any particle whatever. Van Bambeke (*Pourquoi nous ressemblons à nos parents ?* Bruxelles, 1885) also draws attention to the same conclusion, pointing out that from a piece of a Begonia leaf a whole plant with its flowers may proceed. But his assertion (p. 46), that Weismann assumes the presence of germ-plasm in somatic-cells is evidently due to a misunderstanding. Weismann only supposes that small quantities of germ-plasm are present in the somatic-cells of hydroid polyps, "afterwards passing through an endless succession of cells till they reach those remotest individuals of the colony in which the sexual products are formed," etc. (*Continuität*, p. 61). But here the germ-plasm only exceptionally occurs in somatic-cells, and migrates from them to others.

explain the recrescence of complex tissues and organs in the higher animals, it is necessary to assume, not merely that the parts have a general relationship to one another, a general inherited formative tendency, but that they have at the same time a special formative tendency directed towards the structure of the part to which they belong. This special relationship shows itself in the fact that these parts are specially correlated with one another which are derived from the same germ-layer, epidermis, hair and horns, for instance, in Mammalia.

The higher the grade of organisation in the animal kingdom, the greater the division of labour, the more complex the organs, so much the less are all cells in a condition to reproduce different parts or the whole body, so much the more are their powers restricted to the redevelopment of particular parts.

Thus it is evident that the cells of certain regions of the body of the higher organisms are endowed by heredity with particular tendencies of growth : they have a tendency by union with one another to constitute, to complete, particular parts. This is an instance of the establishment by functional action and heredity of definitely directed forces.

This is shown to me in the simplest way by another experiment of Vöchting's : he hung up a twig of willow in a position the reverse of the natural, in a vessel where it was exposed to moisture and the other conditions favourable to growth. This twig produced roots not only at the lower end, where their development was favoured by gravity, but also and in superior numbers at the upper end—evidently there in consequence of the established hereditary tendencies of growth.

Subsequently Herr Vöchting drew my attention to some remarks of his which well express the significance of the examples I have previously given of definite growth-tendencies due to acquirement and inheritance, although for his part he puts forward the view I maintain as only hypothetical. He speaks of the explanation of the gradual increase in height of plants on the earth, and says: "If we start from the conception that the geotropic and heliotropic organs were first produced under the influence of gravity and light, it follows according to the views now current upon the evolution of organisms, that wherever other conditions were also favourable, more and more lofty forms must have been produced, till at last the existing giants of the vegetable kingdom were developed." And in a previous passage: "After I had demonstrated the influence of gravity and light on the development of roots and sprouts, the idea naturally offered itself that the fundamental contrast between the two was to be regarded as a gradually accumulated effect of these two forces, especially of gravity. In fact, it would appear very strange if these forces, notwithstanding their constant influence, had produced no hereditary properties. When we reflect that the pace to which a mare is trained is inherited by the filly, we are logically compelled to look for the hereditary result of those forces in the organisms on which they have acted in the same way for inconceivable periods of time. If such a result were not present it would be in the highest degree surprising."[1]

Vöchting thus meets the arguments by which Sachs opposes his view, that the results of his experiments on the recrescence of plants are to be regarded as phenomena of heredity. Sachs looks upon these phenomena as the result of the influence of gravity on the young growing organs. The opposing action of light and gravity does not, he says, depend on heredity, but the contrast represents only a "predisposition," which is implanted in the organs of every individual by light and gravity during its development.[2]

[1] Cf. H. Vöchting, *Ueber Spitze und Basis an den Pflanzenorganen*, Botan. Zeitung, pp. 596-97.
[2] J. Sachs, *Stoff und Form der Pflanzenorgane*, Arbeiten des bot. Inst. zu Würzburg, Bd. ii. p. 469, *et seq.*

But this is to dispute heredity altogether, it is maintained that the variety of plant-forms is due to external stimuli only, which act upon plants during their development. Whither does such a view lead? To the denial of the very nature of embryonic development, which incontestably depends principally on heredity. It leads to the assumption that all germs are equivalent, and that the fact that from a cherry-stone a cherry-tree, from a pea a pea-plant arises, is due only to the difference of external stimuli.

In a work [1] in which he defends himself against Sachs's attacks, Vöchting uses language which even more distinctly than the previous quotations shows the agreement between his views and mine, an agreement which I value the more because it was only after the preceding pages had been written that I became accurately acquainted with the contents of the work referred to. Amongst other things he says that every organ, of whatever kind it may be, contains internal determining conditions, and these in the cases recorded by him primarily determined at what points roots should arise in a fragment of willow. Such internal conditions "primarily determine the position of new formations. The effects of the action of the external stimuli are thus modified by internal factors."

These "internal conditions" are the same as those which I have called constitutional causes.[2]

Like Sachs, Pflüger also denies heredity [3]—and inasmuch

---

[1] *Ueber Organbildung im Pflanzenreich*, ii. p. 192, Bonn, 1884.

[2] Sachs has since (J. Sachs, *Stoff und Form der Pflanzenorgane*, ii., Arb. d. botan. Inst. zu Würzburg, Bd. ii. p. 689), not only assumed that there are such internal causes, but even declared that he never denied their existence. It may be remarked that Sachs (*Naturwissenschaftliche Rundschau* 1886, No. 5) also maintained that he had put forward the theory of the continuity of the germ-plasm before Weismann (Sachs, *Pflanzenphysiologie* 1882, ch. xliii.) But as Weismann replied (*Zur Annahme einer Continuität des Keimplasmas. Ber. d. naturf. Ges. zu Freiburg i. B.* 1886), Sachs only referred to the fact that from the germ-substance of one generation that of the following is derived.

[3] Pflüger, *Ueber den Einfluss der Schwerkraft auf die Theilung der Zellen*

as he repudiated the inheritance of acquired characters he must be allowed the credit of preparing the way for the theories of Weismann and Nägeli. His view depends on his discovery that gravity has an influence on the development of the embryo even in animals, since the rudiment of the young frog appears, not as usual in the dark, but in the light pole of the egg when the latter is kept uppermost. Pflüger proceeds therefore from the argument, that if heredity existed the dark hemisphere of the frog's egg would have retained by inheritance alone the function of forming the principal part of the embryo. But this interpretation of the facts cannot well be sustained, because as Pflüger himself admits, the whole segmentation certainly proceeds from the nucleus, and the artificial inversion of the egg possibly causes a simple change of position of the nucleus, and thereby of the seat of its influence on the surrounding protoplasm. It is possible that the protoplasm follows the nucleus in its change of position. The movement of the nucleus may be due to its having a lower specific gravity than the rest of the egg. Experiments on the action of gravity would therefore be much better evidence for Pflüger's contention if gravity produced the same effect after segmentation was completed.[1]

In connection with his results concerning the influence of gravity on development, Pflüger discusses recrescence, seeking

und auf die Entwicklung des Embryo, Arch. f. die gesammte Physiologie, Bd. xxxii. 1883.

[1] According to Rauber and Roux (Rauber, *Schwerkraftversuche an Forelleneiern, Sitz.-Ber. Nat.-Ges.* Leipzig, Feb. 1884 ; Roux, *Ueber die Entwicklung der Froscheier bei Aufhebung der richtenden Wirkung der Schwere, Breslauer Aerztl. Zeitung,* 22 März 1884), under the influence of a centrifugal force the embryo is formed in the direction of that force, a result which harmonises with the explanation I have given above. Cf. also Born (*Ueber den Einfluss der Schwere auf das Froschei, Arch. f. mikrosk. Anat.* Bd. xxiv. 1885), who formerly maintained the explanation given above, but afterwards, it seems to me on insufficient grounds, abandoned it. O. Hertwig also (*Welchen Einfluss übt die Schwerkraft auf die Theilung der Zellen, Jen. Zeitsch. f. Naturw.* Bd. xviii.), after experiments on Echinoderm ova denies any profound importance to Pflüger's results.

to use this phenomenon also as an argument against heredity. Since a part of the body taken from the whole grows again from the remainder—is formed anew—it is evident that it is formed again notwithstanding the absence of all hereditary connection between the lost part and its substitute. At the same time he endeavours to give an explanation of recrescence in the following fashion:[1] "When just what was lost is replaced, it is clear that the newly-produced member does not arise from a pre-existing germ of the member. The sectional surface of the stump of the arm draws nutritive material towards itself, and organises the molecules of that material into an arm. The organising force is, however, a molecular force, which cannot extend its influence from the living substance of the stump to a distance, but acts only by attracting the nutritive molecules which come within the sphere of activity of the molecules of the stump, driving them into definite positions, and so gradually produces a new living layer over the exposed surface. The organisation of this new layer evidently depends on the law of the organisation, *i.e.* of the molecular arrangement and the chemical condition of the surface over which the new layer is formed. The condition of this layer is, in a word, the mathematically necessary consequence of the condition of the older generating layer. But since in the embryonic development this latter layer was also in existence before that now reproduced arose, so this must then have arisen in exactly the same way as it now arises for the second time. Thus layer is formed upon layer, the younger always the child of the older, until the organ is again entire. The process of recrescence therefore consists in this, that the exposed surface of the stump of the arm acts, as it is always acting, on the molecules of the layer next to it, directing, arranging, organising them, so that every nutritive particle which comes within the region of that surface at once

[1] *Op. cit.* pp. 65, 66.

obeys the laws prescribed by it. . . . Since therefore the most superficial layer of living molecules in the mutilated stump, an almost imponderable minute quantity of substance, produces the whole limb with mathematical necessity, very much as a snowflake forms an avalanche, and since this is true of all members, therefore it is not difficult to conceive that the whole head and trunk can be produced from a very much smaller surface, a kind of ellipsoid, if adequate nutritive material is supplied to this surface." Farther on it is particularly insisted that the regenerating action takes place always in a definite direction. If a nerve-trunk is cut through, the surface of the portion still connected with the nerve-centres exhibits an enormous regenerating activity, while the other surface, together with the peripheral nerve, perishes, although otherwise it is under the same conditions as the former surface. The organising surfaces of a body show therefore a polarisation, in that the one of their sides does not exhibit the same properties as the other. The direction of the polarisation of the regenerating surfaces is the cause of the direction of growth, and for this reason a part of an animal cannot in general by growth regenerate the whole animal. The peculiarity mentioned by Pflüger of the recrescence of nerves seems to me to demand a different and obvious explanation. It is easy to understand why only the central portion of the divided nerve remains alive and exhibits the power of recrescence, while the peripheral dies, for only the former remains under the necessary influence of all the vital conditions of the whole body, is still active as a nerve ; the other has given up its activity as a nerve, and must therefore as such perish.

Pflüger thus supposes that just as in the embryonic development of an organ its growth at a given moment takes place always from the already formed "polarised" terminal surface, so also in recrescence. That, he says further, in a dog when the gall-duct or a long piece of nerve is cut out it is regenerated is intelligible, because the piece removed is the result of the direct and indirect organising action of the surface exposed by the section, and is continually (in the restorative processes of metabolism) being reformed by that action.

In reply to this it may be remarked that a part serving a particular purpose may be regenerated sometimes in the interior of the body in consequence of simple mechanical conditions and simple stimulation—possibly, for instance, a duct conveying a liquid, the contents of which took the same course after it was severed as before. But the idea that after the removal of any part the polarisation of the proximal exposed surface is the cause of its recrescence seems to me sufficiently disproved by the fact that parts of animals can grow again into entire animals. This fact can only be explained on the assumption that all parts of the body, in consequence of an acquired definite growth-tendency, have a share of a definite formative power in the way I have maintained, although the higher the grade of the animals the more heredity has produced a certain degree of peculiarity in the formative powers of the several parts. All the facts of recrescence, it seems to me, are incontestably in favour of, not against, the heredity of acquired characters. Moreover, Pflüger himself, at the conclusion of his arguments, has a sentence which at least in a certain sense acknowledges the influence of the entire organism in recrescence. "If," he says, "in certain regions of the body, by poisons or injuries, the groups of molecules are disturbed out of their normal arrangement and thrown into disorder, then almost the whole body, all of it that remains normal, exerts a constant organising influence until from layer to layer the normal arrangement of the particles is re-established; each layer imposes a definite law on the one succeeding it and works without intermission until it gradually again impresses its own stamp upon its neighbour."

To my thinking, on the contrary, the recrescence of lost parts falls under the same laws of acquirement and inheritance as ordinary growth : it is nothing else but growth proceeding with increased vigour under peculiar conditions.

How natural a process it is, and how justified we are in asserting its subjection to the ordinary laws of growth, and how dependent it is upon the contemporary condition of the whole body, is excellently shown by the recrescence of the horns of the Cervidæ, and by the periodic renewals of hair and feathers. These are instances, not of pathological, but purely physiological recrescence, and examples of recrescence in general than which none better can be conceived. In the Cervidæ, with the exception of the reindeer, the horns are developed only in the males, and after castration they cease to be developed in them. They develop according to perfectly definite laws every time they are renewed, becoming more complicated every year : they grow. But this growth, their normal recrescence, no longer goes on when the testes are removed ; the horns degenerate and diminish with the cessation of sexual activity.

This example at the same time clearly exhibits the relation of recrescence to correlation, on which I have already insisted.

Recrescence is only the increased action under special conditions of a process which constantly goes on in our bodies as long as we live. Like recrescence, the constant renewal of the parts of the body, even of the highest animals, throughout life depends upon acquired and inherited growth-tendency, on acquired and inherited formative power. This renewal is essentially nothing but a gradual process of recrescence. To explain it, it is necessary to assume what I have assumed for the recrescence of lost parts, and what all growth resulting in a definite form presupposes, namely, that by the inheritance of acquired characters there is established in every developing and every adult organism a relation of the particles to one another, which finds its expression in their striving to form themselves into the whole, and to maintain or re-establish the co-ordinated whole; and further, that the higher the degree to which division of labour is developed, so much the more are special growth-tendencies developed in the several parts.

Another consideration which I must emphasise on account of its importance to the above view and to my whole theory of the organic growth of the living world is this: that the renewal of the cells of the body which goes on constantly throughout life, that is, the division of the cells, depends on the same processes which accompany the division of the original egg-cell, and which take place during the whole of development. This continued multiplication or recrescence of cells in the adult body thus appears as in some sense an after-effect of fertilisation, as a continuation of the process of growth which constitutes the development of the organism. Thus the whole metamorphosis of organisms, their whole life, depends upon the acquisition and inheritance of properties, and on the growth thereby determined. If the capacity for this ceases the organism dies.

CONCLUSION

I CONCLUDE the first part of this book with the Address following, not only because it contains some of the most important of the views here expounded, but especially because it expresses the fundamental conception from which all my views arise—the doctrine of the unity of organic nature.

My view of the causes of the origin of species is nothing but the logical application of that conception to the explanation of the manifold variety of the living world. It might be replied that this conception equally influences the minds of all who endeavour to explain the origin of species by the gradual modification of organisms. It is true that no evolutionist at the present day contests the fact that varieties, species, genera, families, etc., more or less plainly pass by transitions into one another, and that these are in the last result artificial conceptions. Indeed, it is the most confident disciples of the doctrine of evolution, of Darwinism, who most loudly complain, not merely of the manufacture of species, but particularly that systematists regard varieties, not as a kind of transition from species to species, but as a comparatively unimportant deviation from the established type of the species. Nevertheless, in the more recent attempts to explain evolution, this principle of the unity of organic nature has been more or less disregarded. These attempts show that transmutationists make no less distinction between variety and species than the systematists. For the former recognise that certain influences of the external world are the causes of the origin of varieties, but deny that the same causes have led to the evolution of species. They also abandon that principle when they maintain that the modification of unicellular forms is governed by other laws than those which regulate that of the multicellular.

And it is certainly astonishing that this really self-evident truth has never yet been followed to its conclusions, namely, that if the same fundamental laws hold throughout the organic world, and if species are only a collection of individuals, and genera a collection of species, families only a collection of genera, and so on, groups which have differentiated themselves from a number of originally similar individuals, that then the same laws must hold for the development of these groups as for that of the individual, that the causes which modify an individual within the boundaries of a species must be the same as those which modify it beyond these boundaries. Finally, that therefore also the special laws of growth which influence and modify an individual during its life, must be essentially those which underlie the manifold variety of the whole world of organic forms.

Any one who thus completely renders allegiance to the supremacy of the principle of the unity of the organic world, who rejects everything which contradicts this principle, cannot help admitting that in truth, as I assert, the ultimate origin of the various kinships in the animal and vegetable kingdom is to be traced to individual differences, and that the differences between the former, like the latter, must be essentially determined by external conditions, by the modifications of organic growth.

Only in this way also is it possible to explain in a natural way, as I have attempted to do, the origin and unity of all forms of mental action.

APPENDIX

APPENDIX

ON THE IDEA

OF THE

INDIVIDUAL IN THE ANIMAL KINGDOM.

An Address delivered at the 56th Congress of German Naturalists and Physicians at Freiburg, i.B., on the 21st September 1883,

BY

Prof. Dr. G. H. THEODOR EIMER of Tübingen.

From the stenographic report published in the daily proceedings of the Congress.

Man loves to isolate himself from Nature.

He is reluctant to confess his kinship with beings which he thinks beneath him.

He alone will be lord, alone wise; he the crown of creation.

But who gives him the right to assume this pre-eminence?

How small is his power!

Man the mighty—is powerless against the infinitesimal organisms which seek to enter his blood and destroy him.

A wave kills him, while the ocean is teeming with life that mocks his sovereignty.

Does not the consciousness of our weakness, of our imperfection, steal upon us when we see the bird of prey circling in the air? He is scarcely visible to our dim eyes; yet sees the smallest creature on the earth which he has chosen for his prey.

Does he think more humbly of himself, this king in his own realm, than the lord of creation beneath him?

Yet no less has every animal perfect right to feel himself lord in his own kingdom; for each is perfectly adapted in its own way—if it were not, were it not lord in its own fashion, it would not exist.

When we look at the smallest Infusorian under the microscope we see that it feels itself supreme in the drop of water which is its universe—actively and confidently it feels about, moves like an arrow hither and thither, unconscious in whose hand it is, and that in a few minutes the drop of moisture on which its life depends will have disappeared.

And if any one replies that man, though not in every detail, yet on the whole is lord of the earth, let him consider that this lord, so full of pride one moment, in the next becomes conscious of his weakness, and humbles himself before a higher power.

For this self-subjection, which resigns our fate into higher hands and implores their aid in time of need, is first prompted by our helplessness under the blows of life, under the resistless powers of the outer world, under the might of death.

Thus man, the lord of creation, the contemptuous despiser of the Nature surrounding him and its creatures, bows himself in humility before this Nature, degrading himself to idol-worship, placing in the hand of a feeble fellow-creature the sceptre of infallible wisdom and power.

How can we find a way out of this state of contradiction, how discover the limits of the claims which man can with dignity make upon the external world, and thereby lay the foundation for a more harmonious development of his mental life ?

I believe by the universal recognition of the actual relations towards one another of organic beings in nature.

In the choice of my method of treating this question I have been guided by the desire of showing by special investigation how mistaken are those who are never tired of maintaining that the doctrines of Natural Science, especially the newest, are in antagonism to morality, and to idealism in general.

I will point out how rather the contrary is the case, how these very doctrines, rightly understood, coincide with the laws which make the noblest claims on life.

And how could it be otherwise since these laws have sprung from the original demands of Nature herself.

To carry out my purpose I intend to attack the apparent independence of the animal organism which is implied in the word individual, *i.e.* indivisible, to dispute the claim of the organism to separate isolated existence, and will endeavour to prove that the conception of such an indivisible entity is unable to withstand a more exact investigation.

I shall endeavour to prove this (1) by a consideration of the question of the isolated unity of individuals themselves; and (2) by the consideration of the direct connection of the individuals, even of different species, with one another, as shown by new evidence.

I

Among the differences which have been stated to distinguish animals from plants, it has been insisted that one of the most prominent is that the animal is a complete distinctly-defined unity from which no part can be separated without injury to the whole, and without the probable destruction of the separated part, while plants on the contrary are tolerant of the most varied injuries, and can be artificially divided, multiplied by division. This conception is connected with and is indispensable to the other conception, that the animal as distinguished from the plant possesses an indivisible soul.

In consequence of this conception, the greatest astonishment was excited when in the middle of the last century Trembley showed that a small insignificant but genuine animal, the Hydra or fresh-water polyp, which lives in our waters, could be cut up into quite small pieces in the most various directions, with the result that each piece grew into a new perfect individual. These experiments attracted the greatest notice, not only from naturalists, but also from philosophers and theologians, on account of their bearing on the question of the indivisibility of the soul.

The philosopher Bonnet and others repeated the experiments of Trembley, and extended them to other animals, especially to a small worm which likewise lives in our fresh waters, the Nais proboscidea, an animal which resembles the earth-worm in form and is allied to it. It was found that when this worm is cut transversely into several portions each part goes on living after the division. But it was also found that this worm multiplies by dividing itself in exactly the same way as it can be divided artificially.

But it must be pointed out that the conditions are different in the two cases. The whole Hydra can be cut up in all directions; but in the worm there are a number of segments in linear series which are to a certain degree independent—the rings of the body,—and subdivision cannot go beyond the separation of these without endangering the life of the parts.

But the physiological treatment of zoology has in recent times been extraordinarily neglected, in my opinion, to the great injury of the science; and such physiological experiments were not for a long time continued. It struck me as very desirable to try experiments of a similar kind, for it seemed to me that by operating on animals intermediate in organisation between the Hydra and the worm it might be possible to discover traces of the commencement of the nervous system in the animal kingdom.

For in the worm we have—and this is the reason why we can divide it with the result mentioned into its component parts (segments)—a nervous system which is repeated in every segment—a nerve-cord which runs along the ventral side of the animal, and in every segment enlarges into a ganglion, and each of these ganglia can in some sort be regarded as a brain of the animal. It is the same with the other organs—each is repeated in every segment of the worm, so that the animal is to be regarded as a composite of several parts, each of which is a whole complete in itself. But this worm was developed from a simple unsegmented larva.

For my first experiments I took various kinds of jelly-fishes. I will here only mention those which I made on the large kinds. Many of you know these creatures, at least those who live at the sea-side or visit watering-places. Some kinds of them are at times a great annoyance to bathers, for when they come in contact with the skin they cause great irritation by means of small glands they possess which send out minute stings charged with an acid. Such a jelly-fish may be compared to an umbrella if we suppose that the latter is made of a watery gelatinous substance, and that instead of the wires which form the ribs of the umbrella there are canals running in the jelly which meet in a common chamber at the apex of the umbrella, the stomach of the animal, and open to the exterior through the stick of the umbrella by an aperture at its lower end, the mouth. Round the margin of this umbrella, at regular distances, there exist several small bodies, which have long been conjecturally regarded as sense-organs, the so-called marginal bodies, each one at the end of a canal.

When I divided one of these animals into two halves by an incision passing through the summit, each half went on swimming independently. It must be borne in mind that the locomotion of the large jelly-fishes is produced by dilation and contraction of the umbrella. Then I cut the animal into four, then into eight parts; I could cut it into as many parts as there were marginal bodies—and each separate piece endeavoured to go on moving independently.

When I took one of these pieces—a radius of the whole creature—and cut off pieces from this radius, each piece when

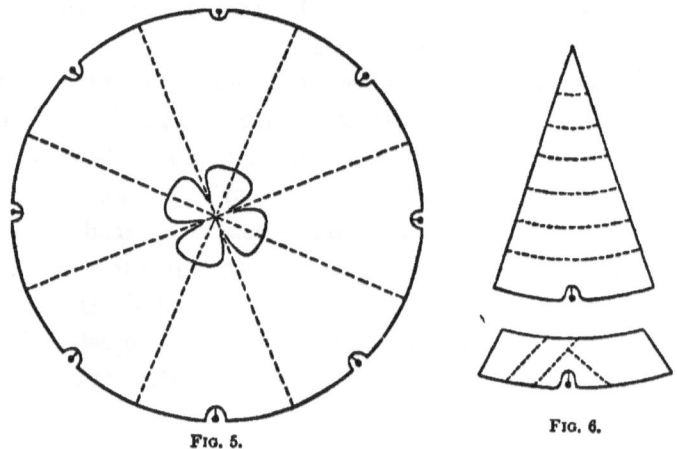

FIG. 5.   FIG. 6.

separated was dead and motionless if it did not contain the small body on the margin, the so-called marginal sense-organ. But the smallest piece containing this organ continued to move, pulsated rhythmically in a perfectly constant manner, somewhat as the heart of a frog when dissected out from the body. A prick with a needle in a particular part of the marginal organ, and the motion in this small portion also ceased. In this organ then was the starting-point of the motion, here was the nerve-centre.

When I cut out the central portion of the umbrella so that only the margin was left, the latter went on moving quite cheerfully in the water like an independent and uninjured animal, but the central piece seemed dead. When from an entire specimen I cut out all the marginal organs except one, the animal went on moving as if it were perfectly uninjured; but if I cut away all the marginal bodies, the animal seemed dead, no longer moved.

If one carefully observes how the uninjured animal moves in the water, he will notice that the motion starts always from one of the marginal bodies and spreads in various directions as if it were propagated through telegraph wires. But at one moment the motion begins at this organ, the next at another, the next again at another.

Where then are we to look for the individual in such an animal? It is not an indivisible whole, but a divisible, and divisible into so many parts. But neither can we describe the smallest part of the whole which is capable of life as an individual, for such a part cannot continue its independent life indefinitely, it perishes after a time.

If we compare these phenomena with those which I have described in the case of the worm we see a difference. In the latter case we also obtain by section a part capable of independent life, but it only remains an individual for a moment; immediately afterwards it begins to develop again into an entire worm, which does not happen in the jelly-fish.

But to proceed further. When I made a jelly-fish motionless and then kept it in conditions favourable to life in pure sea-water, then after a time, at some spot or other on the summit, twitchings appeared; these movements extended to wider and wider areas, became more rapid and vigorous, and after a time a regular pulsation was re-established in the mutilated animal—it continued to live. Thus a new centre of life has arisen at some point in the animal, and this new centre now accomplishes the movements of the whole—a most marvellous phenomenon, which, however, is explained by the peculiar relations of the nervous system. These on further inquiry were revealed to me by the minute microscopical investigation of these animals.

This nervous system, in fact, represents the beginning of all nervous systems. It consists, besides nerve-fibres, of nerve-cells, which are scattered over the surface of the animal, and at various points are aggregated so as to form the beginnings of central organs, a number of separate brains. But so slight is the differentiation of the peripheral from the central nervous system, that, under certain necessary conditions, a rudimentary brain of this kind can to some extent develop itself from the scattered nerve-cells in any other position, that is, some other group of nerve-cells can begin to act as a brain.

Subsequent microscopical researches made by others have confirmed my observations and extended them to other zoophytes.

Time will not allow me to describe the experiments and researches which I have made on the ribbed Medusæ (Ctenophora)—experiments which would of course afford a deeper insight into phenomena of this kind; I will therefore pass on to another consideration which may throw some light on the question of the definition of the individual in the animal world.

There are animals which are usually not very appropriately described as colonies. The corals are of this class; they all arise originally from single beings, each from a simple larva, which begins to divide itself, and ultimately comes to form a colony of numerous individuals, as it is usually expressed. But a coral colony is by no means a co-ordinated whole; it can be divided up in every direction into these single creatures (polyps), and each will grow into a new colony, a new animal-tree.

It is much the same with star-fishes and sponges. These

also are composed of single parts, each of which can, when separated from the whole, under certain conditions, grow into a new animal—in sponges this takes place in any fragment separated from the whole and thrown again into the water. We have thus in all these cases, in the adult animal, not something simple, and this is what I want to emphasize, but an aggregate of similar members.

Such an aggregation occurs also in animals where it is not usually suspected—for instance, in insects and their allies, in the group of Arthropoda generally. A fly or a bee is not an originally simple whole, but it is clearly evident from its developmental history that it was evolved from a worm-like creature, *i.e.* one which consisted of a linear series of similar parts. We recognise this condition still to some extent in the caterpillar, and also in the maggot or similar larva of other forms. Thus the bee has been evolved from a series of segments originally equivalent. But a unity has been developed in this animal in consequence of the division of labour which has taken place among the segments. The various organs are no longer distributed equally among the several segments of the body, but collected together, united in different regions. This applies also to the nervous system, although closer observation shows that here there is not yet the unity which might be expected. For the several nerve-centres which are present in the body of an insect have more or less different functions: stimulation of one group of nerve-centres affects the sexual organs, of another the locomotive organs, and so on. I only say therefore that in the body of an insect we have a composite structure which has arisen from a number of equivalent segments; in consequence of division of labour these segments have been united into a single body, differentiated into several groups composed of different organs and lying one behind another.

The same is the case in our own bodies. Even the

mammals, including man, are not simple individuals; our bodies are transversely segmented. In the course of evolution, it is true, amalgamation or concentration has proceeded to a very high degree, at least in the highest animals; but in the lower forms of the vertebrata this is scarcely carried further than is the case in the Arthropoda, especially in insects. The central points towards which the segments have been united appear plainly when we attempt to divide these animals also in different directions. Insects can be divided with the result that the separate parts, *e.g.* the posterior part of a stag-beetle, retain for a long time an independent life — a very long time if they are protected from desiccation. The same holds also for vertebrates; for we can cut off the head of a frog and the remaining part of the nervous system retains its functional activity. It seems to me certain, notwithstanding the contrary view which still prevails, that voluntary action in this animal proceeds also from parts of the spinal cord; for after the head has been cut off the creature evidently acts quite consciously. A tortoise, it is also known, can survive decapitation for a long time, and apparently likewise responds voluntarily to external stimuli. The higher we ascend in the animal scale, the less do we find separated parts capable of surviving, and the more does that possibility of recrescence disappear which still exists in the lower vertebrates, in which, for instance, legs or tails are reproduced when lost. In ourselves, traces of this power of recrescence still remain in the formation of scars, and in the reattachment of certain parts which are restored to their place immediately after they have been separated from the body.

Our subject can also be illustrated by facts of the reverse kind. As in the cases mentioned, a differentiation and union of equivalent parts produced a compact and co-ordinated whole —individuals, as it were, being specialised into organs—so in other animals organs become externally separated from the body, so that locomotive organs, digestive organs, sexual organs, remain connected with the body only by a peduncle.

This occurs in Zoophytes, for instance in the Siphonophora, in lower forms of animals generally. It also happens that organs of this kind separate from the whole and swim about independently; that sexual organs, for example, leave the common body, and, provided with visual and auditory organs, move freely in the water; and that from the mingling of the sexual elements of these isolated sexual organs new animals arise. It even happens that in the course of evolution such sexual organs come to be the principal form of the species, while the original stock gradually dwindles away—just as (and the comparison might be pursued far) if the separate blossoms of a tree, male and female, separated and moved about freely, and gradually in the course of evolution came to exist as independent species, while the original tree gradually, as such, disappeared.

The Medusæ afford instances of both these kinds of phenomena.

Again I will illustrate the subject by a brief reference to another class of facts, namely, those of the social life of some animals. In a bee-hive there are, as is familiarly known, different kinds of individuals: there are the queen, the drones, and the workers. In a community of animals allied to these, the ants, and also in Termites, we have still another class of specialised members, called, because they serve for the protection of the social body, soldiers. These different classes all work together for the good of the whole community in the wonderful manner known to all of us.

How can we account to ourselves for the origin of these different forms? Only by this explanation: that they have all been derived from a single primitive form, the organs of which have been atrophied or specialised, as the case may be, in one direction or another; that in this way the classes have been differentiated, and at the same time alterations of the organs of nutrition and of mental life have been produced. We can, in fact, describe the different members of the bee-community as organs of the state, and a wide outlook is opened for us when we devote even a brief consideration to the mental life of this animal society in relation to the question of the existence of real individuals in organic nature.

We know that mere "intelligence" alone is ascribed to animals as distinguished from man; it is not admitted that they are also possessed of "reason," and it is remarkable that this distinction is upheld to the present day even by zoologists. But it is my belief that no perfectly definite distinction between intelligence and reason is possible. We distinguish, to put it briefly, intelligence from reason essentially only by this, that the latter takes account not only of the immediate requirements of the single creature, but considers also indirectly, deductively, everything which affects that creature; that reason does not merely live from hand to mouth, looking only to immediate needs, but considers future needs depending on the time and also the society in which the individual lives. Reason will therefore always take account of the future and of the community in which the individual lives, because it must reflect that the good or ill of the community is the good or ill of the individual.

Now, when we regard the mental life of a bee community from this point of view, we necessarily find that it shows reason in a high degree. The animals act throughout according to the requirements of a well-organised state in various directions. They act in complete accordance with the common interest of all the single members of the state, and with the requirements of the future. It is true that it seems to the observer as if this action was in many cases mechanical, as if the animals' conduct was in particular instances not the direct result of mental reasoning, but as if they performed reasonable actions mechanically. But in other instances we are compelled to admit that, after fully considering the particular circumstances, they do what is best for the future and for the common weal.

If we suppose, for example, that the collection of honey has become mechanical, that the bees no longer reason consciously in performing this labour, yet we must assume that originally they began to collect honey from reflection and reasoning; for otherwise they would never have come to do it mechanically. We can also say, on the other hand, that by inheritance actions originally due to reason have gradually become mechanical, automatic, instinctive, so that the animals continue to practise them mechanically as the inheritance of ages. It might be thought that to regard the social life of the bees in this way as merely mechanical degrades it in our estimation. But such a judgment would be mistaken. Since these actions are performed automatically, the animals are left free to exert themselves in other ways : they have time, opportunity, and energy to apply their mental activity in other directions ; and we can even see in the development of such a state of things the ideal condition of the state, since the individual performs his part for the good of the whole more and more mechanically, and as a matter of course, and is no longer compelled to consider on each occasion whether he must do it, and how.

Regarding the phenomena in this way, we see a marvellous unity resulting from the combination and co-operation of the mental activities of separate units, since these now work for the good of the community in a perfectly constant manner, like the parts of a machine.

Let us consider yet another side of the question. I said

previously that even the development of male and female forms among the members of the bee community was a case of division of labour. This proposition may be extended to the rest of the organic world, and it may be said that wherever a male and female sex exists there is no real perfect individual. The two parts absolutely belong to one another, and only form a whole together, not each for itself.

But still further. The whole multitude of higher animals are all composed of cells. Only the lowest animals are unicellular—in this sense also all higher forms are compound aggregates. We might accordingly, in seeking a basis for the idea of the individual, describe the lowest forms which are not sexually separate as individuals, as indivisible wholes. But even this view on more minute investigation proves false, for the multiplication of every such individual and its origin from another are found to be the results of the same process as the multiplication even of the highest forms: of the separation of parts from the whole, of a division or budding, as the case may be.

Yet another point of view. In the process of metabolism there is a constant assimilation of nourishment and excretion of effete material taking place in the body, and therefore a constant change of the constituents of the body. No organism continues the same, and even if we were able to speak of an isolated organic being, we could only do so by speaking of the shortest given moment in the existence of an animal, for at every moment its substance is changing.

We can, of course, artificially establish individuals in the most various ways; we can say that by an individual we understand this or that, if we define our conception in each case in a particular way. But, as I have shown, in the last result we come to the conclusion that the contemplation of single organisms makes it impossible for us to give a perfectly logical definition of "individual," for the simple reason that such an indivisible organic unit does not exist: it does not exist because all single forms are directly or indirectly connected with one another.

Herewith I come to the consideration of the second part of my subject.

II

The doctrine of evolution, we all know, assumes that the whole of nature forms a great whole—in particular, that the animal kingdom is a connected whole. It is very remarkable that the knowledge of the grounds on which this doctrine rests has not become so widely extended among us in Germany as might have been expected. I believe the reason of this lies in the circumstance that German research has almost entirely confined itself to the investigation of anatomy and embryology; while Darwin, who infused a new life into the doctrine of the evolution of species, took account chiefly of the external forms of animals and plants, and particularly of their biological relations. It is for the greater number of men absolutely impossible to follow the researches into details on which the doctrine of evolution among us, in the German way of treating it, is supported. The study of systematic zoology and botany, which principally depends on the investigation of external form, has even been expressly and publicly tabooed by the principal representatives of Darwinism, and it has been declared scientifically unallowable to occupy one's self with such things at all. I have long been of a different opinion, and believed it would be profitable to fix the attention keenly for once on a living creature commonly accessible, to consider the external relations of some single animal or some single plant, and in this way to seek material for the discussion of the connection of forms with one another. And, in fact, I obtained in this way some very remarkable results.

Hitherto the significance of the markings, the spots and

stripes, which occur on the skins of animals has been undervalued in a quite extraordinary way. It has been the custom to consider all this as of no importance, and no one has thought of looking in any way for the laws which govern these phenomena and the connection between them. But by attentive examination I found that these characters followed most wonderfully definite laws, that there is no point in the marking of any animal which has not a quite special and definite typical importance, and that among all the markings which occur on the surface of the animal-body a high degree of connection is to be recognised. Thus there are different types of marking which pass by gradual modifications one into another, so that, when the extreme forms of such markings are connected by the intermediate forms, those which are apparently most different can be brought into a connected relationship.

For example, if we place together all the Lepidoptera which belong to one group, the butterflies, we shall find that there are surprising connections between one species and another, between this and a third, and so on; connections which point to the perfectly gradual modification of the species according to the plan of arborescent ramification, and to the fact that the cause of this depends on the variation of the species, in other words of the individual, in perfectly definite directions. The species varies here and there, but only in a few perfectly definite details, and usually in one direction only. There appears on the wing, for instance, first a new minute streak as a darkening in a perfectly definite position; on a second specimen of the same species it is larger and shows whither the evolution will lead. In the most nearly allied species it forms a constant conspicuous diagnostic character, in the next it is still more altered in the direction indicated, perhaps accompanied by another new variation appearing for the first time. So we pass from species to species; the last compared with the first shows scarcely any resemblance, but the intermediate forms connect them completely. Thus we pass from Papilio Podalirius to P. machaon, thence to the species of Vanessa, of Hipparchia, of Apatura, and so on, and finally we recognise the indirect connection of all the species.

But the matter may be made much clearer by another example. I am able to demonstrate that dogs and cats are indirectly related. The evolution of the carnivora proceeds from the civet cat. The markings of this form lead to those of the cats, of the hyenas, and of the dogs, and if you look at any dog which approximates to the original form of the Canidæ, that is, which has the jackal or wolf form, you can discern in very faint dark marks on its skin the stripes of the hyena, and certain corresponding markings of the domestic cat. Even the most conspicuous variations of the markings of the dog which have gradually been produced, the spots of the most varied kinds which are apparently irregularly scattered over the body, can be traced back to these original lines. Again we have always original darker areas which extend over a larger area, then vary in different ways, divide into separate portions, coalesce, and so on, and thus produce markings between which it is apparently impossible to trace any connection. I notice, for instance, that the markings of hunting and sporting dogs can be traced to a simple plan: you find in most cases a spot on the hindquarters, two on the back, one spot on the neck, and again another on the head.

These examples must suffice for the present occasion. I have next to point out that the males always are the first to assume new characters. To this law I have given the name of "male preponderance." The males then are always at a more advanced stage of evolution than the females. But there is at the same time a difference depending on age, for the older males assume new characters first and transmit them to the whole species. Of this also very instructive examples might be given; but I will not enter upon these, but will only mention that the females usually exhibit the markings of the young males, that they therefore—if the ladies will permit me so to express it—stand at a lower stage than the males. On the other hand, the young always exhibit the original markings—therefore in any race the young in respect of marking stand at the level of a long past generation. It is known, for instance, that in many deer certain markings on the side of the body occur only in the young, in others in the adult females also, and only in rare cases in the adult males. The whole modification of the marking proceeds from a longitudinal striping into a spotted condition, and finally into transverse striping, till the marking disappears altogether. If we take the roe-deer or the red-deer as an example, we have in the young a longitudinal striping, which is, however, already resolved into spots, or very soon is so resolved. In the old roe-deer this kind of marking is no longer seen, in the female red-deer it lasts longer. Cervus axis, on the other hand, still stands at the lowest stage of evolution, for even the adult male retains the longitudinal striping. Dama vulgaris, the fallow-deer, is in an intermediate condition, for it shows the fully-developed longitudinal striping only in the young and the females; but even here it is but feebly defined. Similarly we find that in our wild cat the marking in the young animal is at first still more or less of a longitudinal striping, afterwards resolves itself into spots, then becomes a transverse striping, until in the old animal, and especially in the old males, the marking has almost disappeared. We have here a good illustration of the so-called biogenetic law, the law that the development of any animal repeats the stages which were successively reached by its ancestors.

But there is another point to notice. We find that the evolution of marking proceeds also in a definite direction in this respect; that as a rule modification extends from behind forwards, so that when the male assumes a new stage of marking it appears first at the posterior end of the body. Thus we may see in birds of prey that the tail first shows the transverse striping, while in the middle of the body we have spots, and on the anterior part, the head and neck, there are still longitudinal stripes. When we place a whole series of allied species alongside one another we can see which is the most advanced in evolution, and we may possibly detect in this species the traces of a new marking on the most posterior part of the body of the male. In the course of generations the new marking advances forwards, then still a new one begins behind and passes forwards, and at last a trace of the first is only to be found remaining on the most anterior part of the body of any descendant, and only in a female (postero-anterior evolution). A similar progress of modification clearly takes place in many animals in the direction from the lower to the upper side of the body (infero-superior evolution).

In this way we can recognise that a definite evolution passes over the individual forms in a definite succession of stages by a natural necessity.

This conclusion is in a certain sense opposed to Darwin's, since it recognises a perfectly definite direction in the evolution and continuous modification of organisms, which even down to the smallest detail is prescribed by the material composition (constitution) of the body. According to this conclusion, the real Darwinian principle, that of selection depending on utility, is only effective within the limits which are prescribed by the material composition of the body, that is, by the fixed directions of evolution.

Accordingly there is nothing fortuitous, but everything in evolution to the smallest detail is governed by laws.

Chance only selects to a certain degree.

This view thus removes an objection which has so often been raised against the Darwinian attempt to explain the evolution of the organic world, namely, the predominance of chance, while on the other hand it does not undervalue the importance of Darwinism. It sees the essential factor in the origin of species in the progress towards, or the temporary continuance in, definite stages of evolution (genepistatic evolution).

And so we can imagine all forms which externally show their connection with one another—apart from the fact that they are also connected by their internal organisation, only in a manner which is not so clear to every one,—we can imagine the whole series of all the individuals of one line of descent at a given moment united together into a whole, if we conceive that the whole course of evolution which has taken place in organic nature in the course of endless periods is condensed into this moment. Then we should again have a whole, the original larva from which the manifold variety of organic nature has evolved itself. But as things now exist in nature we have in species only separate fragments of the whole, which have separated themselves more and more from the original—fragments, moreover, which are regarded at a particular stage in a continuous evolution the end of which cannot be seen. For if we accurately examine into the meaning of the propositions which I have established, we shall find that it follows from them that in a whole complete evolutionary series the succeeding must always be the more highly evolved, the preceding always the more lowly evolved —as it were, the larva of the succeeding.

And when we take the whole series, we find that the most highly evolved must pass through in its life, though in a brief and condensed fashion, the evolution of the whole series; the changes, to extend more generally the special case, passing for the most part in the direction from behind forwards over the whole body. We recognise therefore the most intimate relations one with another among all the single individuals. Thus the preceding is always as it were the larval form of the next following; and thus of him who considers himself the culmination of all things it can only be said that in all probability he likewise will be nothing more than the larva of a succeeding form which will be evolved from him.

When we thus place ourselves among other organic beings, and disdain to arrogate to ourselves an exceptional position in comparison with them, it is self-evident that we must of necessity apply to ourselves the laws which here reveal themselves.

Thus for the future the idea of a whole in organic nature completely disappears, whether we attempt to regard as such a whole single beings or groups of beings, let the latter be described as species, orders, or classes.

Thus the single being, as the German term for individual (*einzelwesen*) rightly implies, is but a fragment, not merely of its own species, but also of the totality of the animal kingdom.

In the light of this conception, the latter in connection with the rest of nature is seen as a harmonious whole consisting of many members, in which no part has any right to an absolute pre-eminence over another.

When we consider the animal world as such a whole, we reach the conception of our great philosopher Oken, who regarded individuals as the organs of the whole.

The manifold variety of individuals then appears as the result of division of labour as much as the formation of organs within the individual.

This conception implies at the same time the recognition of a life which is at least relatively eternal, immortal, a life in which the good continues to live through the effort of the individual, through the improvement of the offspring.

This conception disdains the idea of doing good for the sake of reward; it desires to do it for its own sake, for the benefit of all. Good, however, is not something prescribed from without; we call only that the good which is serviceable to the common wellbeing, and because it has become serviceable to the common wellbeing.

Only so far as is not inconsistent with the good of all will the individual put forth his might in the struggle for existence.

He must ever keep this in view if he will avoid injuring himself.

For even the utility principle does not demand brute strength throughout, but leaves room for universal love, the action of which, as reason tells the individual, will always reflect back upon himself.

Thus our conception leads not only to the complete recognition of the rights of our neighbour, but also to the most complete subordination in family and state.

It is the most uncompromising opponent of that confused idea of freedom so injurious to the common good which claims unlimited independence for the individual.

It takes in some sense the social life of the bees for its model, in which the work of the individual for the community has become automatic action.

And that in our civilised life we act to some extent in this way may be proved by the fact that we have come in ordinary morals to regard the good as something absolute, universally admitted, to make the conscience responsible for its violation, and to strive towards the ideal.

If the thinking man whose mental vision sees the infinite multitude of worlds, or if the student of nature is questioned concerning our position and significance in the universe beyond the limits of that nature which is accessible to him, he will point to the Infusorian whose horizon is limited to the drop of water which contains it.

Our duty is work; our right is free investigation; our satisfaction the establishment of a grain of truth for the benefit of mankind; our hope—knowledge.

THE END

Printed by R. & R. CLARK, *Edinburgh*

www.ingramcontent.com/pod-product-compliance
Lightning Source LLC
Chambersburg PA
CBHW030323020526
44117CB00030B/760